AF754094

Digital Signal Processing and Statistical Classification

For a listing of recent titles in the *Artech House Signal Processing Library*, turn to the back of this book.

Digital Signal Processing and Statistical Classification

George J. Miao
Mark A. Clements

Artech House
Boston • London
www.artechhouse.com

Library of Congress Cataloging-in-Publication Data
Miao, George J.
Digital signal processing and statistical classification / George J. Miao, Mark A. Clements.
p. cm. — (Artech House signal processing library)
Includes bibliographical references and index.
ISBN 1-58053-135-0 (alk. paper)
1. Signal processing—Digital techniques—Mathematics. 2. Signal processing—Statistical methods. 3. Stochastic processes. I. Clements, Mark A. II. Title. III. Series.

TK5102.9 .M52 2002
621.382'2—dc21

2002016429

British Library Cataloguing in Publication Data
Miao, George J.
Digital signal processing and statistical classification. —
(Artech House signal processing library)
1. Signal processing—Digital techniques 2. Discrete-time systems 3. Signal processing—Digital techniques—Statistics
I. Title II. Clements, Mark A.
621.3'822

ISBN 1-58053-135-0

Cover design by Igor Valdman

© 2002 ARTECH HOUSE, INC.
685 Canton Street
Norwood, MA 02062

All rights reserved. Printed and bound in the United States of America. No part of this book may be reproduced or utilized in any form or by any means, electronic or mechanical, including photocopying, recording, or by any information storage and retrieval system, without permission in writing from the publisher.

All terms mentioned in this book that are known to be trademarks or service marks have been appropriately capitalized. Artech House cannot attest to the accuracy of this information. Use of a term in this book should not be regarded as affecting the validity of any trademark or service mark.

International Standard Book Number: 1-58053-135-0
Library of Congress Catalog Card Number: 2002016429

10 9 8 7 6 5 4 3 2 1

To Lisa, Kathleen, and Julia
G. J. M.

To Melinda, Isaac, Chrissy, and Andi
M. A. C.

Contents

Preface

This is the first book to integrate the topics of digital signal processing and statistical classification, and the first to introduce some currently state-of-the-art transforms. Those transforms covered in this book include the discrete Fourier transform, the fast Fourier transform, the discrete cosine transform, the discrete sine transform, the discrete Hartley transform, the discrete Hilbert transform, the discrete fractional Hilbert transform, the discrete-time wavelet transform, the discrete Walsh transform, and the discrete Hadamard transform, as well as the principal components transform, canonical correlation transform, canonical discriminant transform, optimal discriminant transform, and optimal declustering transform. The relationships among the transforms are explored with an emphasis on minimizing computational complexity. The reader gets step-by-step guidance in discrete-time random processes, discrete-time domain signal processing and frequency domain signal analysis, discrete-time transforms, digital filter design and adaptive filtering, multirate digital signal processing, and statistical signal classification. The text also helps the reader overcome problems associated with multirate A/D and D/A converters.

Extensively referenced with over 1,065 equations, 142 illustrations, and numerous examples, the book furnishes an up-to-date, comprehensive, and coherent treatment of the fundamentals of this cutting-edge technology, based on the academic and industry experience of the authors. An excellent technical reference and research tool for both practicing engineers and graduate students in electrical, computer, and other engineering disciplines, this book offers assistance in applying digital signal processing knowledge and statistical signal classification to real-world situations. This authoritative volume includes critical concepts never before covered in such detail.

This book is intended for a reader who has some background in advanced calculus, linear system theory, probability and random variables, and linear algebra, as well as a basic knowledge of data conversion, as taught in many undergraduate programs. It is also assumed that the reader is familiar with the basic theory of digital signal processing, including discrete-time signals and systems, the z-transform, and frequency representations. The above material is sufficient for exploring the content of the entire book and is reviewed in detail in Chapters 2 and 3 as well as in the Appendix.

In Chapter 4, we introduce the fundamentals of discrete-time transforms, which are at the heart of discrete-time signal processing. This chapter provides a design and analysis tool for frequency representation of discrete-time signals and systems, as well as the implementation of discrete-time signal processing algorithms. Efficient computation methods are explored in terms of the relationship between the fast Fourier transform and many other transforms.

In Chapter 5, we carry out a detailed discussion of digital filter design. The design techniques for digital filters separate into those used for finite impulse response (FIR) and infinite impulse response (IIR) filters. The implementation structures of digital filters are also covered.

In Chapter 6, we present adaptive filtering. We introduce Wiener filter theory, the discrete-time Kalman filter, the least-mean square (LMS) algorithm, and the recursive least squares (RLS) algorithm, as well as their applications.

The focus of Chapter 7 is on multirate digital signal processing. We present decimation and interpolation systems for multiple sample rates and fractional sample rates. The efficient implementation structures based on noble identities and polyphase decomposition are discussed, along with proofs with respect to their theories. Exploring the designs of multiband filter and multistage signal systems allows for efficient implementation in real-world applications. In addition, we discuss the QMF bank and the uniform DFT filter bank. The multirate adaptive filter bank is also addressed.

In Chapter 8 we turn our attention to the design of multirate analog-to-digital (A/D) and digital-to-analog (D/A) converters. This chapter begins with conventional A/D and D/A converters and shows how multirate A/D and multirate D/A converters can increase the resolution by reducing the quantization noise. We then look at design problems of oversampling A/D converters and sigma-delta A/D modulators. Those methods provide effective ways of increasing the resolution for A/D and D/A converters. Finally, we introduce hybrid multirate QMF bank-based A/D and D/A converters. This technique enables us to develop a very-high-speed A/D and D/A converter by using a set of A/D and D/A converters with lower sampling rates that operate in parallel.

Multidimensional and transform-based statistical signal classifications are discussed in Chapters 9 and 10, respectively. In Chapter 9 we first introduce digital measurement features, feature extraction theory, cluster analysis, digital classification schemes, and estimation methods for probability of misclassification. Then, in Chapter 10, we contribute a detailed discussion of transform-based statistical signal classification architectures by using the principal components transform, canonical correlation transform, canonical discriminant transform, optimal discriminant transform, and optimal declustering transform. This method is used to reduce a multidimensional feature space into a lower space while not increasing probability of classification error.

We have been fortunate in being able to benefit from exceptional supervisors, peers, and colleagues who have made significant technical contributions to the field of digital signal processing at the Georgia Institute of Technology: Thomas Barnwell, Paul Benkeser, Monson Hayes, Vijay Madisetti, Russell Mersereau, James McClellan, Ronald Schafer, Mark Smith, Douglas Williams, and Tong Zhou. We would like to thank them for stimulating our own thinking and research in this field, as well as for sharing with us their wisdom and experience.

We would like to thank the following reviewers for their suggestions and comments on enhancing the technical presentation of this book: Dr. Wendem Beyene, Hewlett-Packard Company, for the chapter on statistical signal classification; Dr. Yassin Hasan,

Intel Corporation, for the chapters on digital filtering and adaptive filters; and Dr. Fu-Huei Lin, QuickSilver Technology Incorporation, for the chapter on discrete-time signal processing. We are particularly grateful to Dr. José Carlos M. Bermudez, who is a professor in the Department of Electrical Engineering of the Federal University of Santa Catarina, Brazil, and an associate editor for the *IEEE Transactions on Signal Processing*, for thoroughly reading the entire manuscript, and providing many detailed and valuable suggestions for the book. We truly appreciate his contributions. We are also wish to thank the technical reviewers at Artech House, who read the final manuscript, and provided the valuable comments for the book.

Many thanks are due to Artech House staff Kevin Danahy, Barbara Lovenvirth, Jessica McBride, Judi Stone, Jill Hodgson Stoodley, Mark Walsh, and John Wyzalek, for providing guidance throughout the development and publication of this book.

George Miao's deepest gratitude goes, last but by no means least, to his wife, Lisa, who provided infinite support, patience, and understanding from the book's initial stage to final manuscript. Countless were the evenings, weekends, holidays, and vacations during which she offered him the encouragement, seclusion, and peace of mind needed to pursue this ambitious project. Without her love and support, this book would not have become a reality.

1

Introduction

The ability to communicate with people on the move is key to our society. Communication involves the exchange of information including voice, image, video, and data. This exchange may be over short distances, such as a conversation in a meeting room, or it may be over a long distance through a telephone line, cable, fiber, or wireless link. The key means of carrying this information through the communication channel is a signal, which may assume a variety of different forms and may contain many different kinds of information.

The concept of signals occurs in an extremely wide variety of fields. The ideas and techniques associated with these concepts play an important role in many areas including telecommunications, aeronautics and astronautics, circuit design, consumer electronics, acoustics, seismology, biomedical engineering, sonar, radar, pattern recognition, speech, image, and video as well as data processing. Although the physical character of the signals in these various disciplines may be significantly different, they all have two fundamental features in common. The signals are functions of one or more independent variables, and they illustrate the behavior phenomenon in which the system responds to input signals by generating other signals.

Signal processing is about the representation, manipulation, and transformation of signals and the information that they

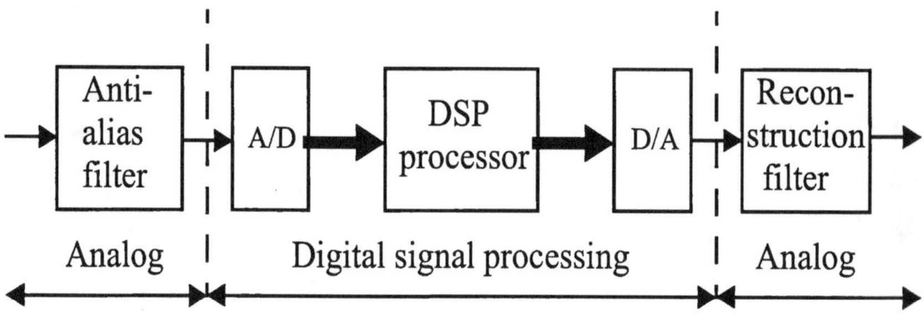

Figure 1.1 A generic digital signal processing system with a single input and a single output.

carry. The development of technology for signal processing was almost continuous time analog technology in the 1950s. The analog systems were implemented with electronic circuits or even with mechanical devices. The rapid evolution of digital computers and microprocessors, which caused a major shift from analog technologies to digital technologies and mixed analog-digital signal processing technologies, made possible the field of digital signal processing.

Digital signal processing is concerned with the digital representation of signals and the use of digital signal processors to analyze, filter, synthesize, or modify such signals. Digital signal processing also involves the conversion of analog signals into digital sequences and the conversion of digital sequences to analog signals.

Figure 1.1 shows the block diagram of a generic system for digital signal processing of analog signals. The input signal is filtered by a lowpass "anti-aliasing" analog filter, sampled at sampling rate F_s with precisely timed intervals, and converted to digital format. A digital signal processor performs the implementation of many algorithms in the discrete-time domain or in the frequency domain for the digital signals. At the output, a digital-to-analog converter produces a set of analog samples. It is followed by a lowpass smoothing (or reconstruction) filter that filters out frequencies higher than one-half the sampling frequency.

The revolution of digital signal processing perhaps began in the early 1990s when computers became powerful enough to perform real-time processing of complex algorithms. Over the last 10 years, digital signal processing has emerged as an enabling technology for a wide range of applications including digital modem communications over a telephone line and cable line, digital cellular communications over the frequency bands of 800 MHz and 1.9 GHz, wireless local area network (WLAN) over the frequency bands of 2.5 and 5.0 GHz, network communications, digital audio broadcasting, multimedia, Internet, speech, high-fidelity audio, video compression, statistical classification for machine learning, and biomatrics identification.

Digital signal processing and statistical classification is now a dynamic, rapidly growing field, and the technology has a vast array of current applications. It is a technology that should be understood and appreciated by a wide and diverse audience of design and systems engineers, scientists, researchers, computer programmers, project managers, undergraduate and graduate engineering students, and others. Therefore, our goal in this book is to educate readers in the core theory and algorithm development of digital signal processing and statistical classification, and thereafter demonstrate efficient computation methods for real-time implementation of digital signal processing strategies.

There are a number of texts on digital signal processing and systems. We cite the books by Oppenheim and Schafer [1, 2]; Oppenheim, Willsky, and Young [3]; and Dudgeon and Mersereau [4]. For practical applications of digital signal processing (DSP), we mention as examples the books by Quackenbush, Barnwell, and Clements [5], and Churchill and Brown [6].

1.1 Data Acquisition System

A generic data acquisition system usually contains an analog-to-digital (A/D) converter, a digital signal processor, and a digital-to-analog (D/A) converter as shown in Figure 1.1. The A/D converter is a vital component in many DSP systems. The architecture used to implement the A/D converter is determined by the combination of sample rate and resolution for which the converter is specified. Therefore, one of the important things that a systems engineer will have to face with respect to a data acquisition system is designing or choosing the right A/D converter.

There are many kinds of A/D converters on the market [7, 8]. These A/D converters can be classified as follows: the successive approximation A/D converter, the flash A/D converter, the semiflash A/D converter, the N-stage pipelined A/D converter, the sigma-delta A/D converter, and the hybrid quadrature-mirror filter (QMF) based A/D converter.

Successive approximation A/D converters, which require N sequential comparisons, are common technology, and have been used extensively throughout the last two decades. This type of A/D converter can produce a sample rate from 10 kHz to over 1 MHz with a resolution up to 16 bits. Because of this, successive approximation A/D converters are still the most popular type of A/D converters.

Flash A/D converters can achieve a sample rate from 10 MHz to over 1 GHz. However, flash A/D converters require $2^n - 1$ comparators to implement an n-bit converter. For example, 255 comparators are needed for an 8-bit A/D converter. As a result, flash A/D converters require a lot of processing power and occupy a relatively large area of silicon. In addition, flash A/D converters have a high input capacitance and are relatively expensive. On the other hand, a semiflash A/D converter can obtain a sample rate from 5 MHz to over 50 MHz with a resolution form 8 bits to 12 bits. This type of A/D converter generally provides a good compromise between lower power and the high speed of the flash method.

The N-stage pipelined A/D converter has a higher sample rate than a successive approximation A/D converter that is cost effective. This A/D converter can achieve sample rates from 1 MHz to 100 MHz with a resolution from 8 bits to 14 bits. It contains a number of individual n-bit resolution A/D converter stages that are cascaded to form the complete A/D converter. Typically, n is from 2 to 4 bits. Each stage consists of an n-bit A/D converter, an n-bit D/A converter, a sample-hold, and a data register. The sample-hold is n-bit precision where n denotes the overall resolution of the A/D converter. Such a pipelined multistage A/D converter can obtain a high sample rate with lower power consumption. However, the problem is that such an A/D converter produces pipeline latency that is the number of clock cycles between the conversion beginning on an input sample and the corresponding output data.

The sigma-delta A/D converter based on oversampling technologies has grown in recent years and is particularly used for sample rates below 20 MHz. They are used extensively for the conversion of audio signals in the field of DSP, and wideband signals in telecommunications. This converter lends itself well to oversampling with noise shaping technologies followed by decimation filtering. The sigma-delta A/D converter can achieve the highest resolution of A/D converters currently available, with a resolution from 16 to 24 bits.

The hybrid QMF-based A/D converter has been recently developed based on the quadrature-mirror filter bank techniques. This A/D converter achieves a very high sample rate using a set of lower sample rate A/D converters that are operated in parallel. Such an A/D converter has great application value in the field of wireline and wireless communications.

So far, we have introduced five types of A/D converters that can be interfaced with a DSP system. However, the question is how a systems engineer chooses the right A/D converter that can be the best matched with a DSP processor for a special application. Basically, selecting the right A/D converter begins with the DSP system definition and specification. Determining the overall system performance and signal characteristics are the first steps for choosing the right A/D converter. Then the frequency bandwidth of the signal will determine the sample rates of the A/D converter. The information content of the signal will be useful to establish the resolution of the A/D converter. In addition, the DSP system interface will decide the A/D converter data interface. These necessary steps result in an initial system specification for an A/D converter. Moreover, the A/D converter requirements can be further defined by a more detailed assessment of the DSP system requirements in an application.

The sampling theorem, which is a foundation of the A/D and D/A converter, was introduced by Nyquist [9] and later contributed by Shannon [10]. The conventional A/D and D/A converter techniques are treated in a book by Rudy [11]. The sigma-delta A/D and D/A converters have been treated in the technical literature. Particularly, we mention the work of Gray, Chou, and Wong [12], and Candy and Temes [13]. For hybrid QMF-based A/D converters, we cite the reference by Petraglia, Maloberti, and Mitra [14].

1.2 DSP Algorithms

DSP algorithms are a dynamic and rapidly growing field that has well-formulated fundamentals. DSP algorithms play a central role in the field of digital signal processing and have grown explosively in terms of their theory and application. The evolution and development of DSP algorithms was accelerated in 1975 by Oppenheim and Schafer after they published their textbook entitled *Digital Signal Processing* [1]. Everyday, new DSP algorithms are born with sophisticated applications for DSP-based products.

The scope of DSP algorithms has grown so large that there are a number of recognized subdisciplines. The following paragraphs provide a brief overview of the importance of these topics to applications that are covered in this book.

1.2.1 Overview of Transforms

Discrete-time transforms that are central to DSP algorithms have grown rapidly in terms of the variety of transforms available for various applications. These discrete-time transforms include the discrete Fourier transform, fast Fourier transform, discrete cosine transform, discrete sine transform, discrete Hartley transform, discrete Hilbert transform, discrete fractional Hilbert transform, and discrete-time wavelet transform, as well as discrete Walsh transform and discrete Hadamard transform. These transforms are useful as a design and analysis tool as well as for implementation of DSP algorithms. This is because discrete-time transforms can provide a frequency representation for discrete-time signals and systems.

Cooley and Tukey [15] published an efficient algorithm for computation of Fourier transforms. This algorithm is known as the fast Fourier transform. The fast Fourier transform is at the heart of computing all of these discrete-time transforms. One reason is that the fast Fourier transform algorithm is inherently in the computation of the Fourier transform of a discrete-time sequence exactly in the discrete-time domain. Another reason is that the fast Fourier transform algorithm makes it possible to calculate the discrete-time transforms efficiently. Furthermore, the fast Fourier transform algorithm serves as the benchmark and foundation for computing all of the other discrete-time transforms. Therefore, other transforms can be efficiently implemented by using the fast Fourier transform algorithm for many real-time applications.

1.2.2 Digital Filter Problem

Digital filters are one of the main elements in DSP algorithms and are used in every DSP-based system product. A digital filter usually performs a filtering function on sampled data by attenuating or reducing the energy of uninteresting bands of frequencies. In other words, digital filters are traditionally used to pick out signals of interest from noise, by using their differing frequency characteristics.

Digital filters are often classified into one of five types: lowpass, highpass, bandpass, bandstop, and allpass. For example, a lowpass digital filter can remove high-frequency noise from a digital speech signal. If input sampled data contains a rapidly varying waveform, which is biased by a slowly varying background and swamped by a low-frequency oscillation, the signal can be passed through a highpass digital filter. A bandpass digital filter can be used to emphasize a particular frequency band of interest in a musical signal. A bandstop digital filter is able to remove a 50-Hz hum from a heartbeat signal.

There are two main classes of digital filters, according to the filter length of the impulse response. The first is finite impulse response (FIR) filters. An FIR filter is a response to an impulse input that dies away in a finite number of samples. These digital filters are also referred to as moving average (MA) filters, because the filter output is simply a weighted average of the input samples. The second class of digital filters is infinite impulse response (IIR) filters. These digital filters sometimes are referred to as autoregressive (AR) filters and, in general, as autoregressive moving average (ARMA) filters. In these filters, a single input can produce an output of infinite duration. Sometimes a finite input can produce a diverging output. If this happens, the IIR digital filter is unstable.

The FIR digital filter design method contributed by McClellan and Parks [16] produces the optimum filter design in the Chebyshev sense. It minimizes the maximum error between the frequency response of the ideal filter and the response of the digital filter. This design method is based on equiripple approximations to produce the minimum filter length given a filter specification. The Parks-McClellan design method is often considered a benchmark for all of the other FIR filter design methods.

1.2.3 Adaptive Filters

Adaptive filters, which can be considered as digital filters with changing coefficients over time, have a rich history in DSP algorithms. Adaptive filters can also be viewed as a subdiscipline of statistical signal processing. They are recognized as being within the core of DSP algorithms. Most importantly, adaptive filters do not always require knowledge of the underlying statistics of the input data.

Adaptive filters are used in situations with either stationary or nonstationary signals. The stationary case assumes that the statistical properties of the signals are not varying in time. Unlike the stationary case, the nonstationary case means that the statistical properties of the signals are varying in time.

A Wiener filter is optimal in the minimum mean-square sense and can be considered as a precursor to the adaptive filter. Wiener filters are designed based on a priori information about the statistics of the input sampled data to be processed. This filter is optimal only when the statistical characteristic of the input sampled data matches a priori information when the filter is designed. This requires that the signals be stationary. However, this will never be the case in the real world. Therefore, for real-time implementation, the Wiener filter has the disadvantage of requiring excessively elaborate and costly hardware.

Kalman filters are also optimum filters in some sense. These filters operate by using a recursive algorithm to find the best estimate in terms of minimizing mean-square error. This assumes that the signals are nonstationary with time varying statistical properties. Kalman filters are very general and may be used to solve a variety of different problems such as prediction, smoothing, echo cancellation, channel equalization, and stochastic control. However, the operation of Kalman filters requires high computational complexity.

Another type of adaptive filter, which uses the stochastic gradient algorithm known as least mean square (LMS), has been used successfully in many applications. LMS, which uses a simple gradient estimate, is efficient and well understood in terms of its convergence properties. This adaptive algorithm requires the least computational complexity among the existing adaptive algorithms.

The recursive least squares (RLS)-based adaptive filter is a special case of the Kalman filter. This RLS adaptive filter enjoys a significant performance advantage; that is, the convergence rate is typically an order of magnitude faster than the LMS algorithm. However, this RLS adaptive filter pays the expense of a large increase in computational complexity.

As we can see, a wide variety of adaptive filters exist. Choosing the right adaptive filter for a particular application is always challenging. In general, selecting one adaptive filter over another is determined by one or more factors as follows: (1) rate of convergence, (2) misadjustment in terms of the minimum mean-squared error, (3) tracking ability of statistical variations of signals, (4) robustness of overcoming signal disturbances, (5) quantization effects in the filter coefficients, in the input sampled data, and in the calculations of the adaptive algorithm, (6) requirements of computation complexity, and (7) implementation structure.

As such a rich selection of literature exists on the design of adaptive filters, it is impossible to cite all the important references. We have selected books by Kailath [17], Brown and Hwang [18], Haykin [19], and Hayes [20].

1.2.4 Polyphase Decomposition

Multirate signal processing has been developed for many years and has tremendous application value in many DSP-based system products. Basically, multirate signal processing deals with the use of multiple sampling rates and fractional sampling rates. There are two building blocks in multirate signal processing, namely the M-fold decimator and the L-fold interpolator. The decimator and interpolator are used to avoid aliasing and eliminate images during multirate signal processing operations.

Multirate signal processing has become practical and attractive. One of the key reasons for this is that Bellanger, Bonnerot, and Coudreuse [21] popularized the concepts of polyphase decomposition. Polyphase decomposition enables a design and systems engineer to implement all computations at the "lowest sampling rate," resulting in dramatic computational savings for many products. It therefore reduces the speed requirements of DSP processors.

The word "polyphase" has gained a certain mystique in the theory of multirate signal processing. The meaning of the term can be explained by using an example as follows: when input data is downsampled by 2, its even-numbered samples are kept, and its odd-numbered samples are lost. These are the two phases including even $x_e[n]$ and odd $x_o[n]$. As the even samples $x_e[n]$ and odd samples $x_o[n]$ go through the filter bank, they are acted on by using two phases of filter, $h_e[n]$ and $h_o[n]$. We then watch these two phases during downsampling. The word "phase" is applied because the even filter with filter coefficients $h_e[0]$ and $h_e[2]$ has a different delay (or phase shift) from the odd phase filter with filter coefficients $h_o[1]$ and $h_o[3]$.

1.2.5 Multirate QMF Bank

The quadrature mirror filter (QMF) bank, which was introduced in the 1970s, is also known as a subband filter bank used for subband coding. This scheme was initially developed in the field of speech processing and has been very useful for speech coding because it can exploit the perceptual properties of the human ear. Early contributions to the QMF bank belong to Croisier, Esteban, and Galand [22], Crochiere [23], Esteban and Galand [24], Barnwell [25], and Crochiere and Rabiner [26].

Figure 1.2 shows a two-channel QMF bank. A discrete-time signal $x[n]$ is passed through a parallel connection of digital filters, $H_0(z)$ and $H_1(z)$, which are called analysis filters. The filtered signals $x_0[n]$ and $x_1[n]$ are thus approximately bandlimited. They are then downsampled by two. The output decimated signals are transmitted through the channel. At the receiver, these signals are recombined by using upsampling of the two synthesis filters, $R_0(z)$ and $R_1(z)$.

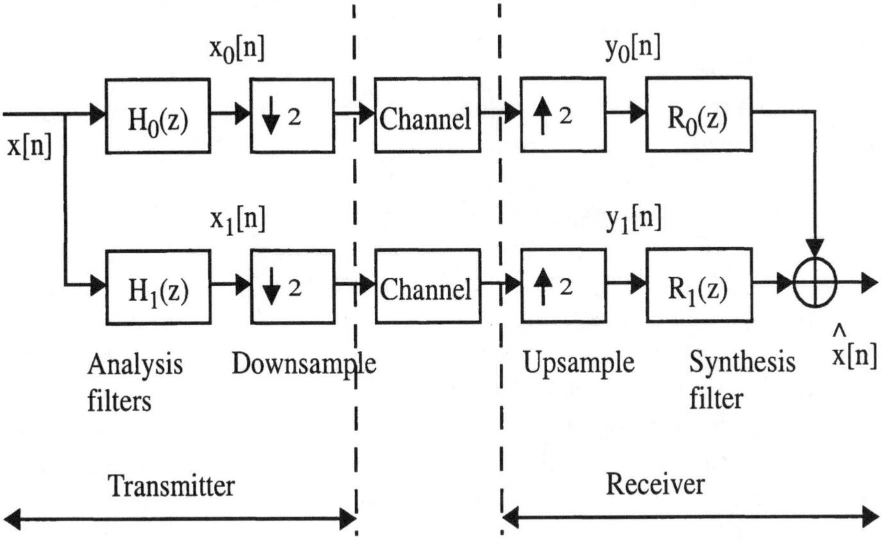

Figure 1.2 A two-channel quadrature mirror filter bank.

In this case, an approximation $\hat{x}[n]$ of the discrete-time signal $x[n]$ is produced. If the QMF bank has M subbands, such a system is called an M-channel maximally decimated filter bank, or simply an M-channel QMF bank.

The QMF bank has been so successful in many DSP-based systems due to the perfect reconstruction property. This property enables the QMF bank to be alias-free. Smith and Barnwell [27, 28] and Mintzer [29] proved a fundamental theory of perfect reconstruction for the two-channel QMF bank. Later, Vetterli [30] and Vaidyanathan [31, 32] introduced a polyphase decomposition approach for the QMF bank so that the theory of the QMF bank could be simplified for efficient implementation.

So far, the QMF bank has been applied to a wide range of problems. The theories and algorithms of QMF banks have been applied in many fields, such as speech coding, image and video coding, telecommunication transceivers, and multirate data converters.

Recently, the QMF bank with polyphase matrices and adaptive algorithms has enabled a reduction in computational complexity and an increase in the speed of convergence, while a conventionally adaptive filter suffers from its heavy computational complexity and slow convergence in some applications. This new area is now known as the multirate adaptive filter bank.

1.3 Statistical Signal Classification

Statistical signal classification is a core technology for a DSP-based machine intelligence system and is used for both sampled data processing and decision making. Statistical signal classification can be considered as the process whereby a received pattern or sampled signal is assigned to one of a prescribed number of classes. Statistical signal classification is also referred to as pattern recognition, because it is basically statistical in nature, with the patterns being represented by points in a multidimensional data space. The space is divided into regions, with each associated with a class. The training procedure determines decision boundaries between classes. The fundamental creation of these decision boundaries is made statistically by the inherent variability that exists within and between classes.

Typically, a statistical signal classification machine contains a feature extractor followed by a pattern classifier, as shown in Figure 1.3. A feature extractor is formed based on a multidimensional data space. If a space consists of a large number of dimensions for a feature extractor, computational complexity of the classification machine is extremely high. This leads to expensive products. One alternative solves this problem by using a linear transform, such as principal components, canonical correlation, canonical discriminant, optimal discriminant, and optimal declustering transforms published by Miao and Clements [33] and Miao [34]. These transforms can be used for mapping a higher multidimensional data space into a lower dimensionality, while the probability classification of error will not be increased. This approach is referred to as a transform-based statistical signal classification. This approach leads to DSP-based machine intelligence systems with less power consumption and more economic value.

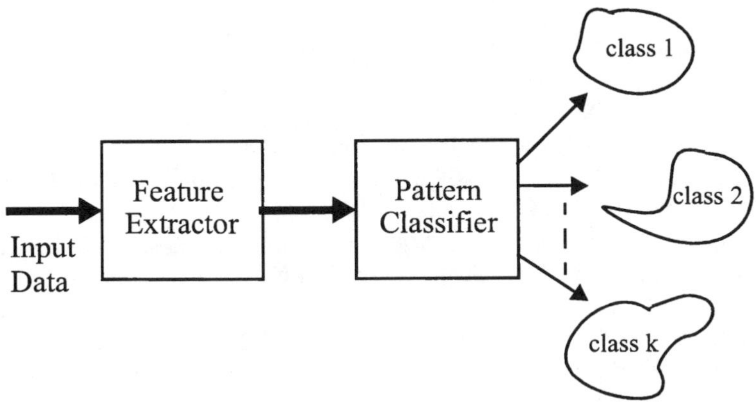

Figure 1.3 Block diagram of statistical signal classification machine.

As DSP processors rapidly develop, DSP-based system-on-chip (SOC) pattern intelligence systems make it possible for real-world applications. For instance, an SOC pattern intelligence system for biomatrics identification is useful for security purposes. This SOC can be built into a notebook or a wireless cellular phone to protect private data information.

1.4 Scope of This Book

The scope of this book is to provide a uniquely coherent treatment of the theory, algorithms, and implementation for a rapidly developing field of digital signal processing and statistical signal classification. The topics presented in this book include discrete-time signal processing, random processes, transforms, digital filters, adaptive filtering, multirate signal processing, multirate data converters, as well as multidimensional and transform-based signal classification. These topics should provide readers with sufficient knowledge for a wide range of applications, including multimedia products and digital communication transceivers, and the foundation for contributing to future designs in this challenging field.

References

[1] Oppenheim, A. V., and R. W. Schafer, *Digital Signal Processing*, Prentice Hall, Englewood Cliffs, New Jersey, 1975.

[2] Oppenheim, A. V., and R. W. Schafer, *Discrete-Time Signal Processing*, Prentice Hall, Englewood Cliffs, New Jersey, 1989.

[3] Oppenheim, A. V., A. S. Willsky, and I. T. Young, *Signal and Systems*, Prentice Hall, Englewood Cliffs, New Jersey, 1983.

[4] Dudgeon, D. E., and R. M. Mersereau, *Multidimensional Digital Signal Processing*, Prentice-Hall, Englewood Cliffs, New Jersey, 1984.

[5] Quackenbush, S. R., T. B. Barnwell, and M. A. Clements, *Objective Measures of Speech Quality*, Prentice Hall, Englewood Cliffs, New Jersey, 1988.

[6] Churchill, R. V., and J. W. Brown, *Introduction to Complex Variables and Application*, McGraw-Hill Book Company, New York, 1984.

[7] Texas Instruments (TI), "DSP/Analog Technologies," Seminar Series, 1998.

[8] Kenington, P. B., and L. Astier, "Power Consumption of A/D Converters for Software Radio Applications," *IEEE Trans. on Vehicular Technology*, Vol. 49, No. 2, pp. 643-650, March 2000.

[9] Nyquist, H., "Certain Topics in Telegraph Transmission Theory," *Trans. on AIEE*, Vol. 47, pp. 617-644, April 1928.

[10] Shannon, C. E., "Communication in the Presence of Noise," *Proc. of IRE*, pp. 10-21, January 1949.

[11] Rudy, V. D. P., *Integrated Analog-to-Digital and Digital-to-Analog Converters*, Kluwer Academic Publishers, The Netherlands, 1994.

[12] Gray, R. M., W. Chou, and P. W. Wong, "Quantization Noise in Single-Loop Sigma-Delta Modulation with Sinusoidal Inputs," *IEEE Trans. on Communications*, Vol. 37, No. 9, pp. 956-968, September 1989.

[13] Candy, J. C., and G. C. Temes (eds.), *Oversampling Delta-Sigma Data Converters: Theory, Design and Simulation*, IEEE Press, New York, 1992.

[14] Petraglia, A., F. Maloberti, and S. K. Mitra, "QMF-based A/D Converters: Overview and New Results," *Proc. Int. Conf. Analog to Digital and Digital to Analog Conversion*, pp. 112-117, Wales, U.K., September 1991.

[15] Cooley, J. W., and J. W. Tukey, "An Algorithm for the Machine Computation of Complex Fourier Series," *Mathematics of Computation*, Vol. 19, pp. 297-301, April 1965.

[16] McClellan, J. H., and T. R. Parks, "A Unified Approach to the Design of Optimum FIR Linear Phase Digital Filters," *IEEE Trans. on Circuit Theory*, Vol. CT-20, pp. 697-701, November 1973.

[17] Kailath, T., *Lectures on Wiener and Kalman Filtering*, 2d ed., Springer-Verlag, Wien, New York, 1981.

[18] Brown, R. G., and P. Y. C. Hwang, *Introduction to Random Signals and Applied Kalman Filtering*, 2d ed., John Wiley & Sons, Inc., New York, 1992.

[19] Haykin, S., *Adaptive Filter Theory*, 3d ed., Prentice Hall, Englewood Cliffs, New Jersey, 1996.

[20] Hayes, M. H., *Statistical Digital Signal Processing and Modeling*, John Wiley & Sons, Inc., New York, 1996.

[21] Bellanger, M., G. Bonnerot, and M. Coudreuse, "Digital Filtering by Polyphase Network: Application to Sample Rate Alteration and Filter Banks," *IEEE Trans. Acoustics, Speech and Signal Processing*, Vol. ASSP-24, pp. 109-114, April 1976.

[22] Croisier, A., D. Esteban, and C. Galand, "Perfect Channel Splitting by Use of Interpolation/Decimation/Tree Decomposition Techniques," *Int. Symp. Information Circuits and Systems*, Patras, Greece, 1976.

[23] Crochiere, R. E., "On the Design of Subband Coders for Low Bit Rate Speech Communication," *Bell System and Technology Journal*, Vol. 56, pp. 747-771, May-June 1977.

[24] Esteban, D., and C. Galand, "Application of Quadrature Mirror Filter to Split Band Voice Coding Schemes," *Proc. IEEE Int. Conf. Acoustics, Speech and Signal Processing*, pp. 191-195, May 1977.

[25] Barnwell, T. P., "Sub-band Coder Design Incorporating Recursive Quadrature Filters and Optimum ADPCM Coders," *IEEE Trans. on Acoustics, Speech and Signal Processing*, Vol. ASSP-30, pp. 751-765, October 1982.

[26] Crochiere, R. E., and L. R. Rabiner, *Multirate Digital Signal Processing*, Prentice Hall, Englewood Cliffs, New Jersey, 1983.

[27] Smith, M. J. T., and T. P. Barnwell, "A Procedure for Designing Exact Reconstruction Filter Banks for Tree Structure Subband Coders," *Proc. IEEE Int. Conf. Acoustics, Speech and Signal Processing*, pp. 27.1.1-27.1.4, March 1984.

[28] Smith, M. J. T., and T. P. Barnwell, "Exact Reconstruction Techniques for Tree-Structured Subband Coders," *IEEE Transactions on Acoustics, Speech and Signal Processing*, pp. 434-441, June 1986.

[29] Mintzer, F., "Filters for Distortion-Free Two-Band Multirate Filter Banks," *IEEE Trans. on Acoustics, Speech and Signal Processing*, Vol. ASSP-33, pp. 626-630, June 1985.

[30] Vetterli, M., "Filter Banks Allowing for Perfect Reconstruction," *Signal Processing*, Vol. 10, pp. 219-244, April 1986.

[31] Vaidyanathan, P. P., "Theory and Design of M-Channel Maximally Decimated Quadrature Mirror Filters with Arbitray M, Having Perfect Reconstruction Property," *IEEE Trans. on Acoustics, Speech and Signal Processing*, Vol. ASSP-35, pp. 476-492, April 1987.

[32] Vaidyanathan, P. P., *Multirate Systems and Filter Banks*, Prentice Hall, New Jersey, 1993.

[33] Miao, J., and M. A. Clements, "Optimal Discriminant Feature-Based Waveform Classification with Neural Networks," *Proc. IEEE Int. Conf. Acoustics, Speech and Signal Processing*, Vol. 4, pp. 3462-3465, Atlanta, Georgia, May 1996.

[34] Miao, J., "Component Feature-Based Digital Waveform Analysis and Classification," Ph.D. Dissertation, Georgia Institute of Technology, Atlanta, Georgia, April 1996.

2

Discrete-Time Signal Processing

2.1 Introduction

Signal processing has to do with the representation, manipulation, implementation, and transformation of signals as well as the information that they carry in the transformation media. Basically, signals can be classified into three categories. The first is made up of analog signals that are continuous both in time space and amplitude. The second, known as discrete-time signals, consists of signals discrete in time space and continuous in amplitude. The third involves digital signals in which the signals are discrete in both time space and amplitude. One common way to create digital or discrete-time signals is by sampling continuous-time signals. We refer to signal processing based on discrete-time signals as *discrete-time signal processing*.

The goal of this chapter is to provide the reader with a summary of, as well as a reference for, the fundamentals of discrete-time signal processing. The reader will then be able to explore advanced discrete-time signal processing as described in subsequent chapters. In the brief descriptions of discrete-time signal processing presented in this chapter, the primary emphasis will be on the specification and characterization of discrete-time signals and systems. This will require use of the z-transform and frequency-domain representations.

The discussions will be cursory, with most of the mathematical treatment left to the references (see Oppenheim and Schafer [1], Kailath [2], and Burrus et al. [3]). In this chapter, we only review topics introducing discrete-time signal notation and terminology.

The organization of this chapter is as follows. In this section we described three kinds of signal processing and pointed out the difference between discrete-time and other signals. In Sections 2.2 and 2.3, we introduce the basic concept of discrete-time signals and their basic operations. In Section 2.4, we discuss discrete-time systems. Linear time-invariant discrete-time systems have many applications due to their properties, as shown in Section 2.5. Subsequently, the z-transform and inverse z-transform are discussed in Sections 2.6 and 2.7. Then we present frequency-domain descriptions of discrete-time signals and systems in Section 2.8. The relationship between allphase and minimum-phase systems is presented in Section 2.9. Finally, a brief summary is given in Section 2.10.

2.2 Discrete-Time Signals

Discrete-time signals are represented as an indexed sequence of numbers, mathematically denoted by $x[n]$, where n is an integer from $-\infty$ to ∞. The signal $x[n]$ is generally referred to as a function of discrete-time. The graphical illustration of $x[n]$ is shown in the form of a sequence of lollipops in Figure 2.1, because discrete-time signals are only defined for integer values of n.

Discrete-time signals can often be considered as a result of sampling a continuous-time signal $x_a(t)$ by using an analog-to-digital (A/D) converter. The relationship between continuous-time frequency f and the sampling frequency F_s is the key point of the *Nyquist-Shannon* sampling theorem (see Nyquist [4] and Shannon [5]), which requires that the sampling frequency F_s must be at least twice the highest frequency in the signal, for perfect reconstruction.

Assume that a continuous-time signal $x_a(t)$ is sampled at a rate of $F_s = 1/T_s$ samples per second. The sampled discrete-time signal $x[n]$ is then equal to the value of the analog signal $x_a(t)$ at time nT_s as follows:

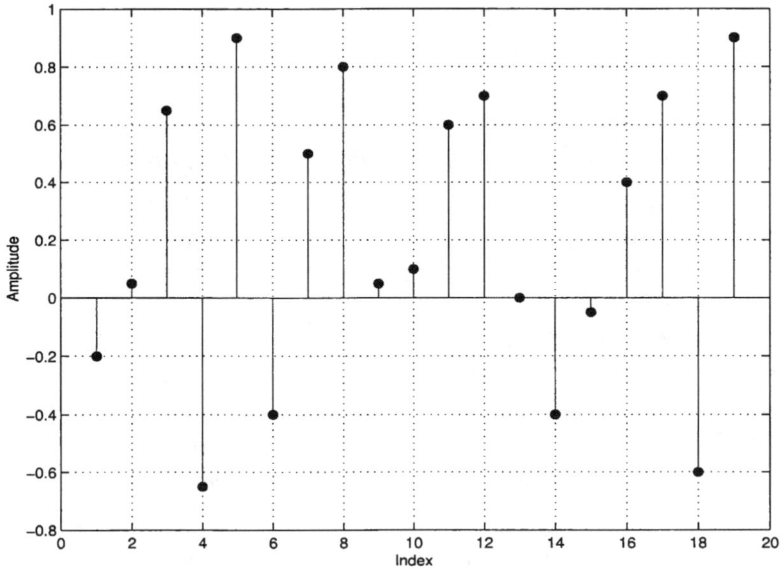

Figure 2.1 The graphical illustration of a discrete-time signal $x[n]$.

$$x[n] = x_a(nT_s), \qquad -\infty < n < \infty, \tag{2.1}$$

where the quantity T_s is known as the sampling period.

For example, a continuous-time sinusoid $s(t)$ is given by the mathematical formula

$$s(t) = A[sin(2\pi f_0 t + \phi)], \tag{2.2}$$

where A is the amplitude, f_0 is the frequency in hertz, and ϕ is the initial phase. If the discrete-time signal $s[n]$ is obtained by using an A/D converter, sampling the continuous-time signal $s(t)$ at a sampling rate of $F_s = 1/T_s$, then we have the discrete-time signal

$$\begin{aligned} s[n] &= s(t)\,|_{t=nT_s} \\ &= A[sin(2\pi f_0 T_s n + \phi)] \\ &= A\left[sin\left(2\pi \frac{f_0}{F_s} n + \phi\right)\right] \\ &= A[sin\,(\omega_0 n + \phi)], \end{aligned} \tag{2.3}$$

where $\omega_0 = 2\pi f_0 T_s$ is the normalized frequency in radians.

Figure 2.2(a) shows a 25-millisecond plot for the continuous-time signal $s(t)$ with amplitude $A = 2$, signal frequency $f_0 = 1{,}000$ Hz, and initial phase $\phi = 45$ degrees. Figure 2.2(b) shows a plot of $s[n]$ for 200 samples with sampling frequency at $F_s = 8{,}000$ Hz. From the Nyquist-Shannon sampling theorem, the original continuous-time signal can be reconstructed as accurately as desired from a corresponding discrete-time signal, since the sampling frequency F_s is greater than twice the highest frequency f_0 in the signal, $F_s > 2f_0$.

2.3 Basic Operations of Discrete-Time Signals

In discrete-time signal processing, sample sequences are manipulated in several basic ways. There are five simple discrete-time signals that are often used as building blocks for more complicated signals. These functions are the delayed or shifted version of a sequence, the unit sample, the unit step, the real exponential, the complex exponential, and the sinusoid.

If discrete-time signal $d[n]$ has the following form,

$$d[n] = x[n - n_0], \tag{2.4}$$

where n_0 is an integer, then sequence $d[n]$ is called a delayed or shifted version of a sequence $x[n]$.

The *unit sample*, which is denoted by $\delta[n]$, is defined by

$$\delta[n] = \begin{cases} 1, & n = 0 \\ 0, & \text{otherwise.} \end{cases} \tag{2.5}$$

The unit sample plays an important role for discrete-time signal processing. Sometimes, the unit sample is referred to as a *discrete-time impulse.* One of the important roles of the discrete-time impulse is to form an arbitrary sequence that can be represented as a sum of weighted and shifted unit samples as follows:

$$x[n] = \sum_{k=-\infty}^{\infty} x[k]\delta[n - k]. \tag{2.6}$$

For example, the sample sequence $y[n]$ as shown in Figure 2.3 can be expressed as

$$y[n] = a_3\delta[n - 3] + a_5\delta[n - 5] - a_6\delta[n - 6] + a_9\delta[n - 9]. \tag{2.7}$$

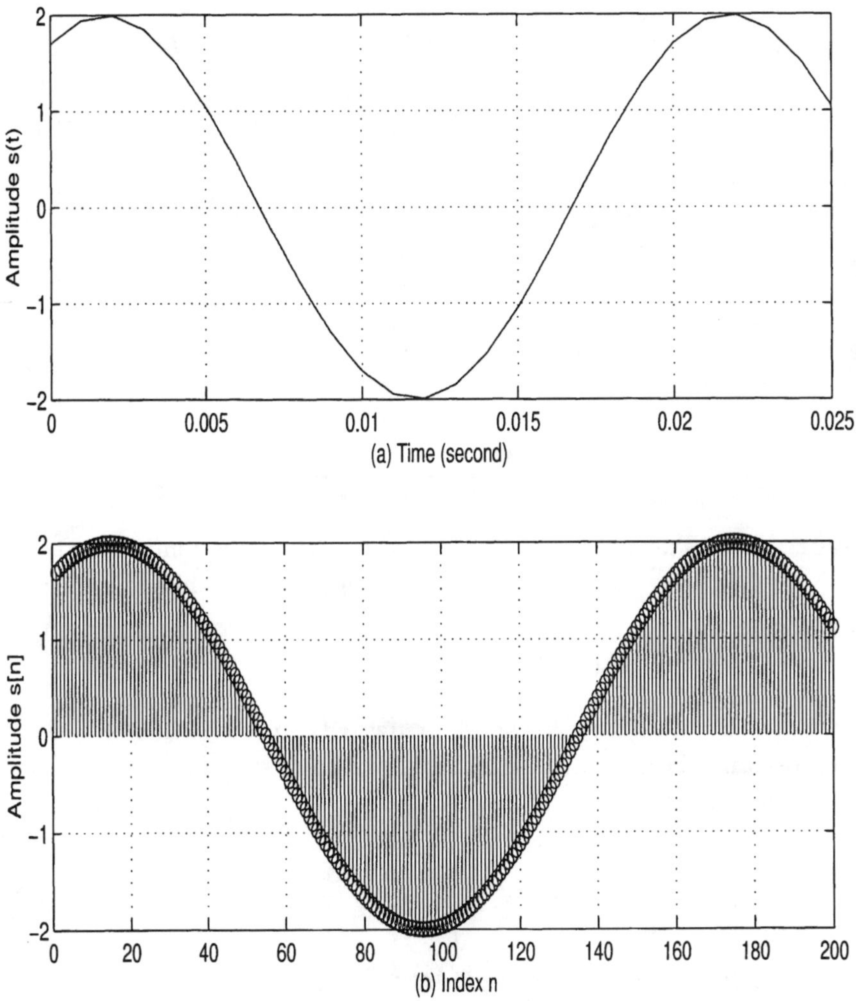

Figure 2.2 The relationship between the continuous-time signal $s(t)$ and the discrete-time signal $s[n]$: (a) The 25-millisecond segment of a continuous-time signal. (b) The 200 samples of discrete-time signal obtained from part (a) with the sampling frequency at F_s = 8,000 Hz.

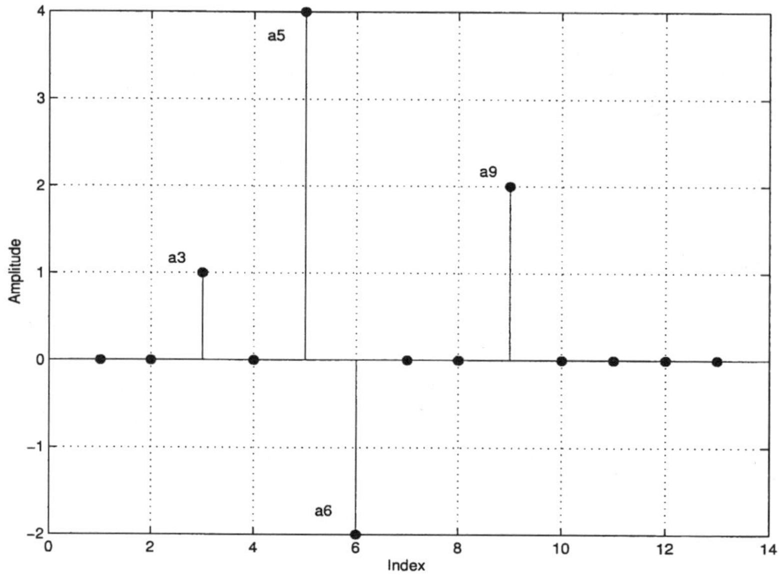

Figure 2.3 The illustration of a discrete-time sequence $y[n]$ to be represented as a sum of weighted and shifted impluses.

We will see that (2.6) is useful in describing such illustrations of discrete-time signals and linear time-invariant systems.

The *unit step*, $u[n]$, is defined by

$$u[n] = \begin{cases} 1, & n \geq 0 \\ 0, & n < 0, \end{cases} \tag{2.8}$$

and is related to the unit sample by

$$u[n] = \sum_{k=0}^{\infty} \delta[n-k]. \tag{2.9}$$

Also, the discrete-time unit sample can be expressed as the first backward difference of the unit step as follows:

$$\delta[n] = u[n] - u[n-1]. \tag{2.10}$$

The general form of an *exponential* is given by

$$x[n] = Ca^n, \tag{2.11}$$

where C and a are arbitrary constants. The right-sided exponential is expressed as

$$x[n] = Ca^n u[n], \tag{2.12}$$

and the left-sided exponential is defined by

$$x[n] = Ca^n u[-n]. \tag{2.13}$$

The exponential is useful in representation and analysis of a linear time-invariant discrete-time system. A special case of this is the *complex exponential* in which $a = e^{j\omega}$. This signal is periodical and can be expressed as

$$\begin{aligned} x[n] &= Ca^n \\ &= Ce^{j\omega_0 n} \\ &= C[cos(\omega_0 n) + jsin(\omega_0 n)], \end{aligned} \tag{2.14}$$

where ω_0 is real. This complex exponential is useful in the Fourier decomposition of signals.

A *sinusoidal* sequence has the form

$$\begin{aligned} x[n] &= A[cos(\omega_0 n + \phi)] \\ &= \frac{A}{2}\left[e^{j(\omega_0 n+\phi)} + e^{-j(\omega_0 n+\phi)}\right], \text{ for all } n, \end{aligned} \tag{2.15}$$

where A is real. Equation (2.15) contains two frequencies, ω_0 and $-\omega_0$. Hence, a sinusoid sequence is not a single frequency signal.

The formulas of complex exponential sequences are expressed with *sines* and *cosines* as follows:

$$e^{j\omega n} = cos(\omega n) + jsin(\omega n) \tag{2.16}$$

$$cos(\omega n) = \frac{1}{2}(e^{j\omega n} + e^{-j\omega n}) \tag{2.17}$$

$$sin(\omega n) = \frac{1}{2j}(e^{j\omega n} - e^{-j\omega n}). \tag{2.18}$$

These are useful in the calculation of the discrete-time Fourier transform for a closed-form solution.

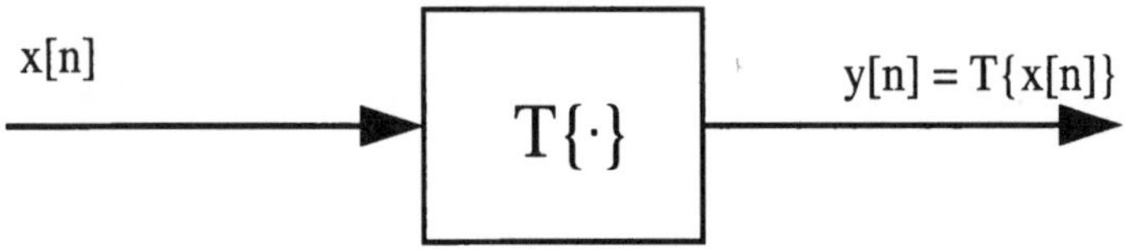

Figure 2.4 The illustration of a discrete-time system as a mapping function that transforms an input discrete-time sequence $x[n]$ into an output discrete-time sequence $y[n]$.

2.4 Discrete-Time Systems

Discrete-time systems can be considered mathematical functions or transformations, which map an input discrete-time sequence $x[n]$ into an output discrete-time sequence $y[n]$. This operation is defined by

$$y[n] = T\{x[n]\}, \tag{2.19}$$

and is illustrated in Figure 2.4. The mapping operation, denoted by $T\{\cdot\}$, can be either a *linear* or *nonlinear* function.

2.4.1 Difference Equations

Difference equations are computationally much easier to work with, and play an important role in the analysis and synthesis of discrete-time systems. For example, the relationship between the discrete-time input sequence $x[n]$ and discrete-time output sequence $y[n]$ may be expressed as the difference equation

$$y[n] = x^2[n], \tag{2.20}$$

or

$$y[n] = 0.8y[n-1] + 0.5x[n] + 0.2x[n-1]. \tag{2.21}$$

Equation (2.20) is referred to as memoryless, since the output $y[n]$ at every index of n depends only on the input $x[n]$ with the same index n.

It is also possible for discrete-time systems to have a set of difference equations that provide a sequence of instructions for operation between input sequence $x[n]$ and output sequence $y[n]$. For example, these discrete-time systems have the difference equations as follows:

$$\begin{aligned} y_1[n] &= 0.9y_1[n-1] + 0.5x_1[n] + 0.2x_2[n] - 0.8x_3[n] \\ y_2[n] &= 0.3y_2[n-1] - 0.5x_1[n] + 0.3x_2[n] + 0.7x_3[n] \\ y_3[n] &= 0.2y_3[n-1] + 0.2x_1[n] - 0.9x_2[n] - 0.5x_3[n]. \end{aligned} \quad (2.22)$$

Also let the final output $y[n]$ have a form

$$y[n] = 0.5\left(y_1[n] + y_2[n] + y_3[n]\right). \quad (2.23)$$

Equation (2.22) is sometimes called a multi-input and multi-output (MIMO) discrete-time system.

2.4.2 Block Diagrams

Block diagrams are useful in the design and analysis of discrete-time systems. The block diagrams for constructing discrete-time linear systems consist of three basic building blocks: the *N-delay*, the *amplifier*, and the *summer*.

N-delay

An N-delay element produces a discrete-time output sequence $y[n]$ that is equal to its discrete-time input sequence $x[n]$ delayed by N sample intervals as follows:

$$y[n] = x[n-N]. \quad (2.24)$$

The N-delay element is shown in Figure 2.5. The term z^{-N} is used to denote an N sample delay. When N is equal to one, the N-delay element is referred to as a *unit-delay* element.

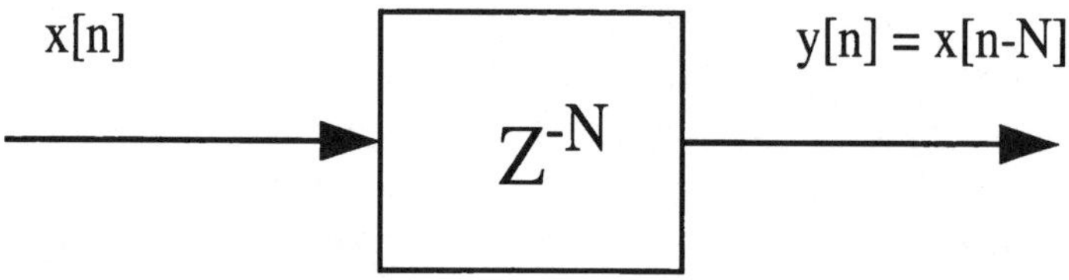

Figure 2.5 Block diagram representation of an N-delay element.

Amplifier

An amplifier provides a discrete-time output, which is the product of a constant C and the input $x[n]$,

$$y[n] = Cx[n], \tag{2.25}$$

and is drawn in Figure 2.6. Note that an amplifier can be represented by all three common forms.

Summer

A summer is used to add two or more input signals to generate the output signal,

$$\begin{aligned} y[n] &= x_1[n] + x_2[n] + \cdots + x_N[n], \\ &= \sum_{k=1}^{N} x_k[n], \end{aligned} \tag{2.26}$$

and is illustrated in Figure 2.7. If a negative sign is placed next to an input path in the summer, this indicates that the signal is to be subtracted rather than added.

Example 2.1 Block Diagram Representation of a Discrete-Time System

Assume that a difference equation describing a discrete-time system has the form

$$y[n] = x[n] + 2x[n-1] - 5x[n-2] + e[n], \tag{2.27}$$

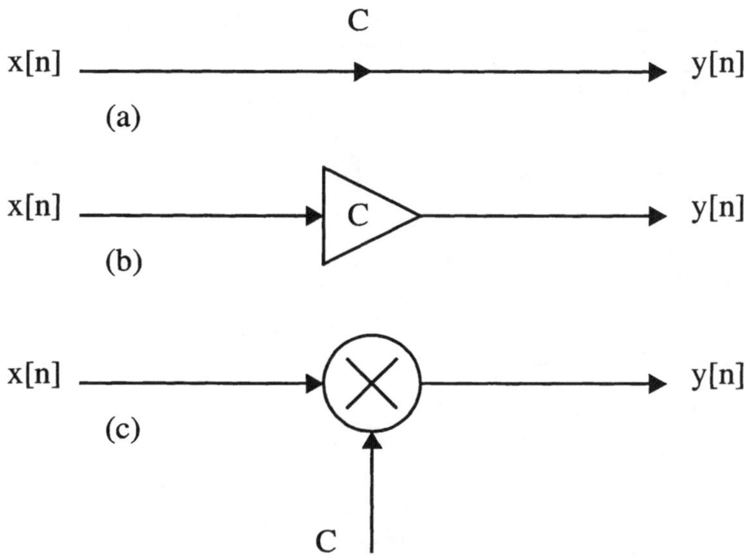

Figure 2.6 Block diagram representation of a multiplier element. There are three common representations as shown in (a), (b), and (c); but all are equivalent.

where $x[n]$ represents an input signal, $e[n]$ is noise, and $y[n]$ is the discrete-time output signal.

We can draw a block diagram to represent the above difference equation. Using the basic building blocks of unit-delay, amplifier, and summer, the block diagram representation of this discrete-time system can be constructed as shown in Figure 2.8.

2.5 Linear Time-Invariant Discrete-Time Systems

Two important properties that systems can have are *linearity* and *time-invariance*. As described in the section below, we find that if a system possesses both properties, many useful applications are possible. The origin of this great utility is the ability to describe such systems by linear, constant, and coefficient difference equations.

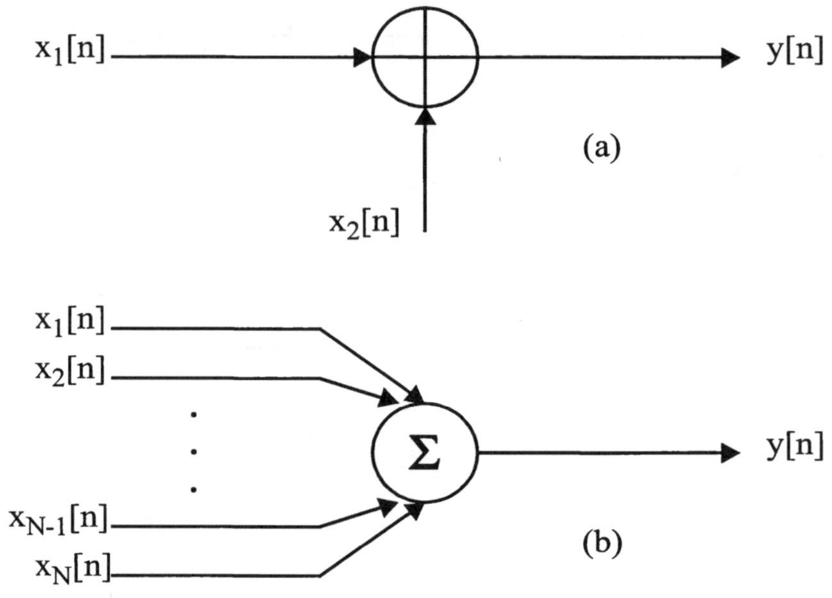

Figure 2.7 Block diagram representation of a summer: (a) A summer adds two discrete-time input signals, $x_1[n]$ and $x_2[n]$. (b) A summer adds more discrete-time input signals, $x_1[n], ..., x_N[n]$.

2.5.1 Linear Systems

A discrete-time system is said to be linear if and only if it satisfies two conditions: *additivity* property and *homogeneity* (or *scaling*). Assume that $y_1[n]$ and $y_2[n]$ are the outputs of a discrete-time system when $x_1[n]$ and $x_2[n]$ are input signals. Then the system is called linear if and only if

$$\begin{aligned} y_1[n] + y_2[n] &= T\{x_1[n]\} + T\{x_2[n]\} \\ &= T\{x_1[n] + x_2[n]\}, \end{aligned} \tag{2.28}$$

and

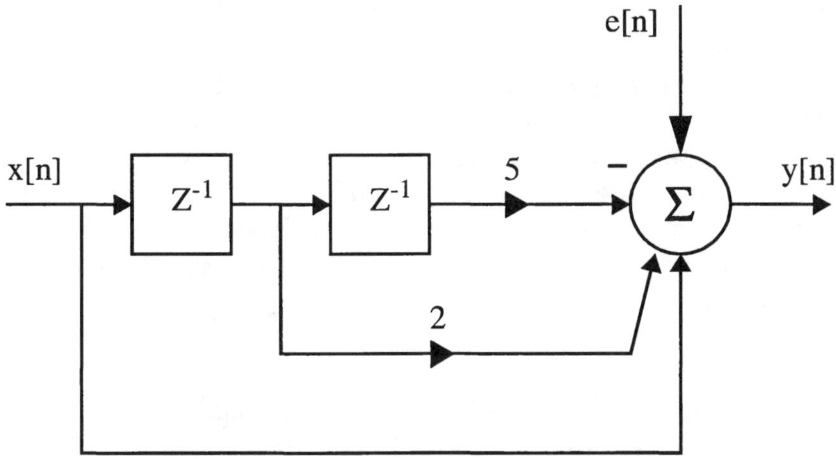

Figure 2.8 Block diagram for the difference equation in Example 2.1.

$$ay[n] = aT\{x[n]\} = T\{ax[n]\}, \tag{2.29}$$

where a is an arbitrary constant. Equations (2.28) and (2.29) obey the *principle of superposition* as follows:

$$\begin{aligned} ay_1[n] + by_2[n] &= T\{ax_1[n] + bx_2[n]\} \\ &= aT\{x_1[n]\} + bT\{x_2[n]\}, \end{aligned} \tag{2.30}$$

where a and b are arbitrary constants.

2.5.2 Time-Invariant Systems

A discrete-time system is said to be time-invariant if a shift in the discrete-time input by m results in shifting the output by m as well for all integers m. In other words, given the condition, $y[n] = T\{x[n]\}$, the discrete-time system $T\{\cdot\}$ is time-invariant if and only if $T\{x[n-m]\} = y[n-m]$ for all discrete-time input sequences $x[n]$ and all integers shift by m.

2.5.3 Causality

A discrete-time system is said to be *causal* if, for any given m, the output sequence at index $n = m$ depends only on the input sequence for $n \leq m$. For example, the system $y[n] = x[n] - 2x[n-1]$ is causal because the value of the output at any time $n = m$ depends only on the input $x[n]$ at time m and at time $m - 1$. However, the backward difference system, defined by $y[n] = x[n] + 2x[n+1] + x[n+2]$, is noncausal, since the output at time $n = m$ depends on the value of the input at time $m + 1$. Causal systems are important for real-time DSP implementation.

2.5.4 Stability

A discrete-time system in which the output $y[n]$ is bounded in amplitude whenever the input $x[n]$ is bounded is said to be *stable.* We call this bounded-input bounded-output (BIBO) stability. More specifically, a discrete-time system is stable if and only if, for any bounded input $|x[n]| \leq B < \infty$, the output is bounded $|y[n]| \leq C < \infty$.

2.5.5 Convolution Sum

A discrete-time system that is both linear and time-invariant is referred to as a linear and time-invariant (LTI) system. The importance of this property is evidenced in the observation that if $h[n] = T\{\delta[n]\}$ and if the input $x[n]$ is decomposed into a superposition of weighted and shifted unit samples,

$$x[n] = \sum_{k=-\infty}^{\infty} x[k]\delta[n-k], \tag{2.31}$$

then the output $y[n]$ becomes

$$\begin{aligned} y[n] &= \sum_{k=-\infty}^{\infty} x[k]T\{\delta[n-k]\} \\ &= \sum_{k=-\infty}^{\infty} x[k]h[n-k]. \end{aligned} \tag{2.32}$$

Equation (2.32) is called the *convolution sum.* In order to simplify mathematical notation, the convolution sum is often expressed as

$$y[n] = x[n] * h[n]. \tag{2.33}$$

The operation of convolution takes two sequences $x[n]$ and $h[n]$, and produces a third sequence $y[n]$.

An LTI system is said to be causal if and only if the impulse response $h[n]$ satisfies the condition

$$h[n] = 0, \qquad n < 0. \tag{2.34}$$

For an LTI system, BIBO stability is equivalent to the condition

$$\sum_{n=-\infty}^{\infty} |h[n]| < \infty. \tag{2.35}$$

In other words, the discrete-time impulse response $h[n]$ must be absolutely summable.

2.5.6 Properties of Convolution

The convolution operation is *commutative*, *associative*, and *distributive* over addition. These lead to identities including *cascade* and *parallel connections*, which are useful for implementation of a discrete-time LTI system.

Commutative Property

The commutative property of convolution operation implies that

$$x[n] * h[n] = h[n] * x[n]. \tag{2.36}$$

The roles of signal $x[n]$ and impulse response $h[n]$ in the summation of (2.33) are interchanged. In other words, the order of the discrete-time sequences in a convolution is unimportant, and hence the output is the same even if the functions of input $x[n]$ and impulse response $h[n]$ are reversed. In addition, the convolution operation distributes over addition, such that

$$x[n] * (h_1[n] + h_2[n]) = x[n] * h_1[n] + x[n] * h_2[n]. \tag{2.37}$$

Cascade Property

In a cascade connection system, the impulse response of a cascade combination of $h_1[n]$ and $h_2[n]$ is independent of the order in which they are cascaded. For example, the output of two cascaded systems $h_1[n]$ and $h_2[n]$ is

$$\begin{aligned} h[n] &= h_1[n] * h_2[n] \\ &= h_2[n] * h_1[n]. \end{aligned} \tag{2.38}$$

Equation (2.38) indicates that the output impulse response $h[n]$ is the same even after interchanging the order of $h_1[n]$ and $h_2[n]$.

Parallel Property

In a parallel connection, two impulse responses of $h_1[n]$ and $h_2[n]$ in parallel are equivalent to a single system whose impulse response $h[n]$ is the sum of the individual impulse responses. For example,

$$h[n] = h_1[n] + h_2[n]; \tag{2.39}$$

this is illustrated in Figure 2.9. As we can see, Figure 2.9 (a) and (b) are equivalent.

2.5.7 The Closed Form for Convolution

The convolution sum can be computed analytically when discrete-time input signal $x[n]$ and impulse response $h[n]$ are given by simple formulas. In such cases, the convolution sum may have a compact form that may be derived by using closed-form formulas. The following closed-form identities will often be useful in the evaluation of convolution operations:

$$\sum_{k=0}^{N} k = \frac{1}{2}N(N+1) \tag{2.40}$$

$$\sum_{k=0}^{N} k^2 = \frac{1}{6}N(N+1)(2N+1) \tag{2.41}$$

$$\sum_{k=0}^{N} \alpha^k = \frac{1-\alpha^{N+1}}{1-\alpha} \tag{2.42}$$

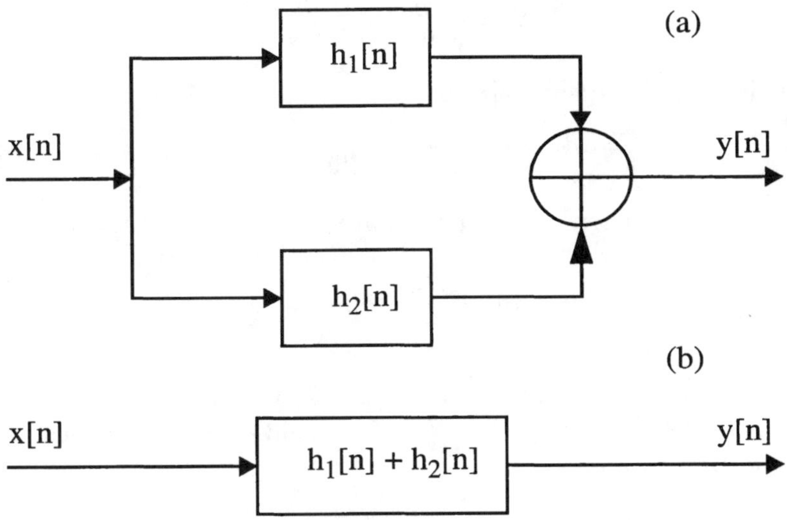

Figure 2.9 Block diagram representation for implementation of two impulse responses $h_1[n]$ and $h_2[n]$.

$$\sum_{k=N_1}^{N_2} \alpha^k = \frac{\alpha^{N_1} - \alpha^{N_2+1}}{1-\alpha} \quad (2.43)$$

$$\sum_{k=0}^{N} k\alpha^k = \frac{\alpha}{(1-\alpha)^2}\left[(1-\alpha^N) - (1-\alpha)N\alpha^N\right] \quad (2.44)$$

$$\sum_{k=0}^{N} k^2\alpha^k = \frac{\alpha}{(1-\alpha)^3}\left[(1+\alpha)(1-\alpha^N)\right] \quad (2.45)$$

$$- \frac{\alpha}{(1-\alpha)^2}\left[2N\alpha^N + (1-\alpha)N^2\alpha^N\right], \quad (2.46)$$

where $|\alpha| < 1$ and $|\alpha| \neq 0$, and $N_2 \geq N_1$.

Example 2.2 Calculation of the Convolution Sum

Assume that a discrete-time system with impulse response $h[n]$

is

$$h[n] = \begin{cases} \frac{1}{N}, & 0 \le n \le N, \\ 0, & \text{otherwise}, \end{cases} \tag{2.47}$$

and the input sequence $x[n]$ is

$$x[n] = b^n u[n], \tag{2.48}$$

where $|b| < 1$. Then, the convolution sum $y[n]$ is expressed as

$$y[n] = \sum_{k=0}^{\infty} x[k]h[n-k]. \tag{2.49}$$

For $n < 0$, the convolution sum is $y[n] = 0$, since $x[k]$ and $h[n-k]$ do not overlap. For $0 \le n \le N$, using (2.42), the closed form of the convolution sum is

$$\begin{aligned} y[n] &= \sum_{k=0}^{n} \frac{1}{N} b^n \\ &= \frac{1}{N}\left(\frac{1-b^{n+1}}{1-b}\right). \end{aligned} \tag{2.50}$$

For $n > N$, the convolution sum $y[n]$ is

$$\begin{aligned} y[n] &= \sum_{k=n-N}^{n} \frac{1}{N} b^n \\ &= \frac{1}{N}\left(\frac{b^{n-N}-b^{n+1}}{1-b}\right). \end{aligned} \tag{2.51}$$

The discrete-time output sequence of convolution sum $y[n]$ is plotted in Figure 2.10 for $b = 0.85$ and $N = 5$.

2.6 The z-Transform

The z-transform is useful for analysis of discrete-time signals and systems, and is often a convenient notation in analytical problems. The z-transform allows the designer to bring the power of complex variable theory to bear on problems of discrete-time signals and systems.

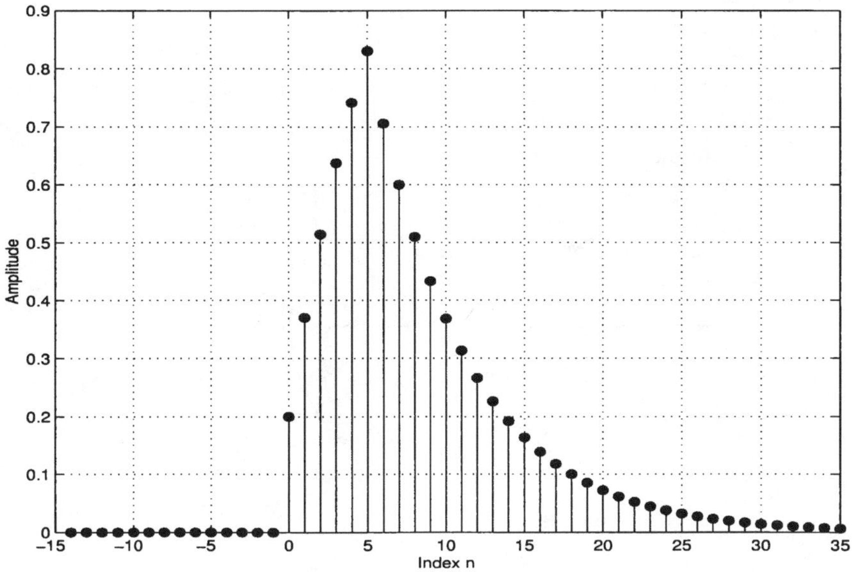

Figure 2.10 The discrete-time output sequence of convolution sum $y[n]$ as a function of n for $b = 0.85$ and $N = 5$.

The z-transform $X(z)$ of a sequence $x[n]$ is defined by

$$X(z) = \sum_{n=-\infty}^{\infty} x[n]z^{-n}, \tag{2.52}$$

where z is a complex variable. It is convenient for a designer to describe and interpret it by using the complex z-plane, because the z-transform is a function of a complex variable.

2.6.1 Region of Convergence

In the complex z-plane, the unit circle plot is illustrated in Figure 2.11 when the contour corresponding to $|z| = 1$ is a circle of unit radius and ω is the angle between the vector to a point z on the unit circle and the real axis of the complex z-plane. The unit circle plot is useful for determining the region of convergence for the z-transform, because the z-transform does not converge for all

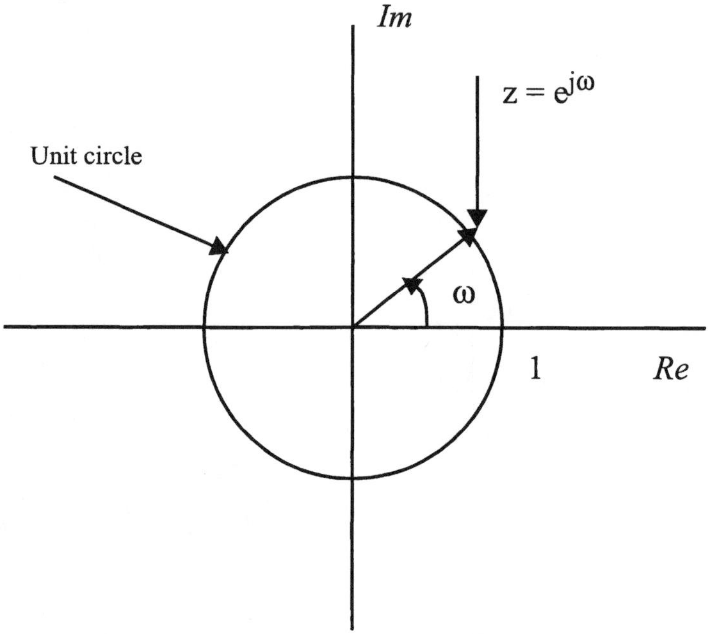

Figure 2.11 In the complex z-plane of the unit circle, the contour corresponding to $z = 1$ is a circle of unit radius and ω is the angle between the vector to a point and the real axis of the complex z-plane.

sequences. For any given discrete-time sequence $x[n]$, the collection of values of z for which the z-transform converges is referred to as the *region of convergence* (ROC). The z-transform of sequence $x[n]$ converges absolutely only for values of z in its ROC.

The convergence in (2.52) depends on the magnitude of z, because the z-transform $|X(z)| < \infty$ if and only if the following equation holds:

$$\sum_{n=-\infty}^{\infty} |x[n]||z^{-n}| < \infty. \tag{2.53}$$

In this case, the ROC will always be bounded by circles centered at the origin of the complex z-plane. For example, if $z = z_1$ is within the ROC, then all values of z on the circle defined by $|z| = |z_1|$ will also be in the ROC. Basically, the ROC can be classified into four configuration regions for the z-transform, including the interior of a circle, the exterior of a circle, the annulus, and the entire z-plane as shown in Figure 2.12.

The z-transform is most useful when the infinite sum can be expressed in a closed form of a rational function inside the ROC, such as

$$X(z) = \frac{P(z)}{Q(z)}, \tag{2.54}$$

where $P(z)$ and $Q(z)$ are polynomials of z. When the z-transform $X(z) = 0$, the values of z are called the *zeros* of $X(z)$. When the z-transform $X(z)$ is infinite, the values of z are called the *poles* of $X(z)$. In addition, the poles may take place at $z = 0$ or $z = \infty$. Therefore, by definition, the ROC should not contain any poles, since the z-transform $X(z)$ becomes infinite at the poles.

Example 2.3 The ROC of the z-Transform

Consider a discrete-time signal $x[n] = \left(\frac{1}{2}\right)^n u[n]$. Using (2.52), the z-transform of $x[n]$ is

$$\begin{aligned} X(z) &= \sum_{n=-\infty}^{\infty} \left(\frac{1}{2}\right)^n u[n] z^{-n} \\ &= \sum_{n=0}^{\infty} \left(\frac{1}{2} z^{-1}\right)^n . \end{aligned} \tag{2.55}$$

For convergence of $X(z)$, we need to have

$$\sum_{n=0}^{\infty} \left|\frac{1}{2} z^{-1}\right|^n < \infty. \tag{2.56}$$

Thus, the ROC is the range of values of z for $|z| > \frac{1}{2}$. In the ROC, the infinite series converges to

$$X(z) = \frac{1}{1 - \frac{1}{2} z^{-1}}, \quad \text{for } |z| > \tfrac{1}{2}. \tag{2.57}$$

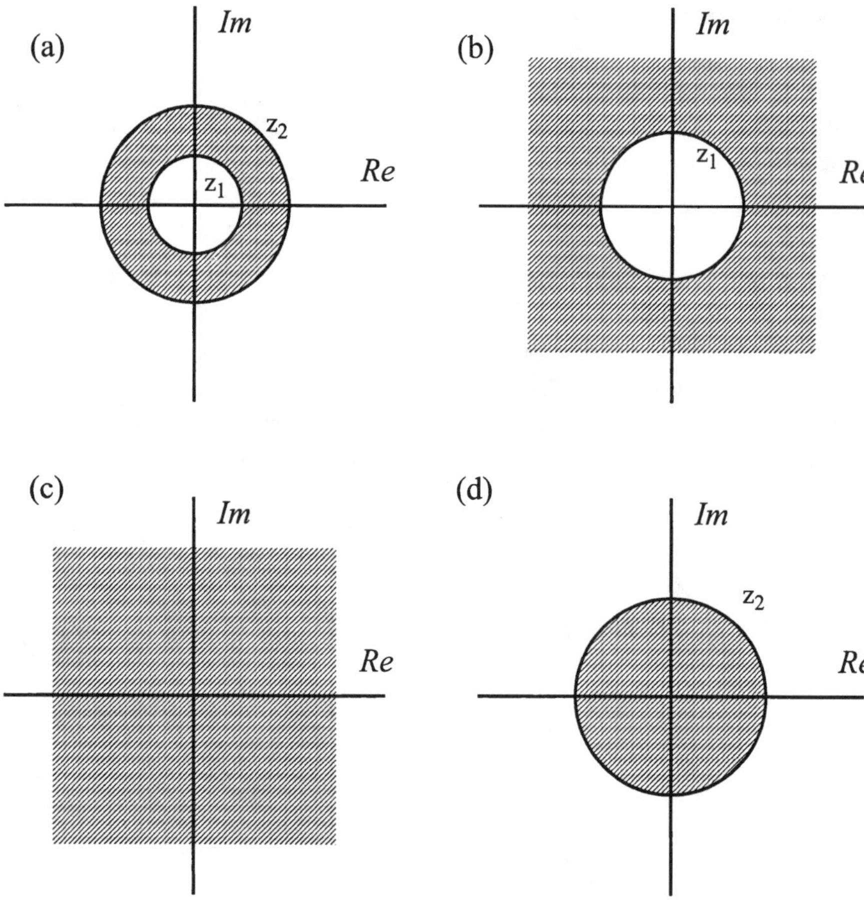

Figure 2.12 The z-transform configuration of the region of convergence: (a) an annulus, $|z_1| < |z| < |z_2|$; (b) exterior of a circle, $|z| > |z_1|$; (c) the entire plane; and (d) interior of a circle, $|z| < |z_2|$.

Example 2.4 The ROC of the z-Transform for Two Exponentials

Let a discrete-time signal be the sum of two real exponentials as follows:

$$x[n] = a^n u[n] + b^n u[n], \tag{2.58}$$

where $a < 1$, $b < 1$, and $a \geq b$. Then, the z-transform is obtained by

$$\begin{aligned} X(z) &= \sum_{n=-\infty}^{\infty} \{a^n u[n] + b^n u[n]\} z^{-n} \\ &= \sum_{n=-\infty}^{\infty} a^n u[n] z^{-n} + \sum_{n=-\infty}^{\infty} b^n u[n] z^{-n} \qquad (2.59) \\ &= \sum_{n=0}^{\infty} (az^{-1})^n + \sum_{n=0}^{\infty} (bz^{-1})^n \\ &= \frac{1}{1-az^{-1}} + \frac{1}{1-bz^{-1}} \\ &= \frac{2-(a+b)z^{-1}}{(1-az^{-1})(1-bz^{-1})}. \qquad (2.60) \end{aligned}$$

In order to converge, both sums in (2.59) must converge; that is, we require that both $|az^{-1}| < 1$ and $|bz^{-1}| < 1$. Hence, this is equivalent to $|z| > a$ and $|z| > b$. Therefore, the ROC is $|z| > a$, since $a \geq b$.

2.6.2 z-Transform Properties

The z-transform has a number of properties that make it a valuable tool in the design and analysis of discrete-time signals and systems. These properties are often useful for obtaining the inverse z-transform of more complicated expressions. A number of properties of the z-transform are summarized without giving the verification in detail. The theory and treatment of z-transform properties can be found in many references (see Oppenheim and Willsky [1], Dudgeon and Mersereau [6], and Oppenheim and Schafer [7]).

In order to simplify the notation, we denote $X(z)$ as the z-transform of $x[n]$, and R_x as the ROC of $X(z)$,

$$x[n] \stackrel{Z}{\Longleftrightarrow} X(z), \tag{2.61}$$

where the ROC is R_x, representing a set of values of z with the condition $r_R < |z| < r_L$.

Linearity

The linearity property states that if

$$x_1[n] \overset{Z}{\Longleftrightarrow} X_1(z), \qquad \text{ROC} = R_{x_1}, \tag{2.62}$$

and

$$x_2[n] \overset{Z}{\Longleftrightarrow} X_2(z), \qquad \text{ROC} = R_{x_2}, \tag{2.63}$$

then

$$ax_1[n] + bx_2[n] \overset{Z}{\Longleftrightarrow} aX_1(z) + bX_2(z), \tag{2.64}$$

where the ROC includes $R_{x_1} \bigcap R_{x_2}$.

Time Shifting

Time shifting results in: if

$$x[n] \overset{Z}{\Longleftrightarrow} X(z), \qquad \text{ROC} = R_x, \tag{2.65}$$

then

$$x[n-d] \overset{Z}{\Longleftrightarrow} z^{-d}X(z) \tag{2.66}$$

where d is an integer. If d is positive, the discrete-time sequence $x[n]$ is shifted right, and if d is negative, then the sequence $x[n]$ is shifted left. The ROC in (2.66) is R_x except for the possible addition or deletion of $z = 0$ or $z = \infty$.

Frequency Shifting or Modulation

Frequency shifting results in: if

$$x[n] \overset{Z}{\Longleftrightarrow} X(z), \qquad \text{ROC} = R_x, \tag{2.67}$$

then

$$e^{j\omega n}x[n] \overset{Z}{\Longleftrightarrow} X(e^{-j\omega}z). \tag{2.68}$$

Note that if $x[n]$ is real, then $e^{j\omega n}x[n]$ will not be real unless ω is an integer multiple of π. In addition, if the poles and zeros of $X(z)$ are in complex conjugate pairs, they may not have this symmetry after frequency shifting.

Multiplication of Exponential Sequence

When a signal is multiplied by an exponential sequence, the following results in: if

$$x[n] \overset{Z}{\Longleftrightarrow} X(z), \qquad \text{ROC} = R_x, \tag{2.69}$$

then

$$z_0^n x[n] \overset{Z}{\Longleftrightarrow} X(z_0^{-1} z), \tag{2.70}$$

where the ROC is $|z_0|R_x$. If R_x has a set of values of z with a range of $r_R < |z| < r_L$, then $|z_0|R_0$ is the set of values of z such that $|z_0|r_R < |z| < |z_0|r_L$.

Time Reversal

When a signal displays time reversal, if

$$x[n] \overset{Z}{\Longleftrightarrow} X(z), \qquad \text{ROC} = R_x, \tag{2.71}$$

then

$$x[-n] \overset{Z}{\Longleftrightarrow} X(z^{-1}), \tag{2.72}$$

where the ROC is $\frac{1}{R_x}$. The ROC indicates that $R_x(z)$ is inverted. In other words, if z_0 is in the ROC for $x[n]$, then $1/z_0$ is in the ROC for $x[-n]$.

Convolution of Sequences

The convolution of sequences results in: if

$$x_1[n] \overset{Z}{\Longleftrightarrow} X_1(z), \qquad \text{ROC} = R_{x_1}, \tag{2.73}$$

and

$$x_2[n] \overset{Z}{\Longleftrightarrow} X_2(z), \qquad \text{ROC} = R_{x_2}, \tag{2.74}$$

then

$$x_1[n] * x_2[n] \overset{Z}{\Longleftrightarrow} X_1(z)X_2(z), \tag{2.75}$$

where the ROC includes $R_{x_1} \bigcap R_{x_2}$. That is, the ROC contains the intersection of the regions of convergence of $X_1(z)$ and $X_2(z)$. In some cases, the ROC may be larger if a pole that borders on the

ROC of one of the z-transforms is canceled by a zero of the other ROC of the z-transform.

Conjugation of a Complex Sequence

The conjugation of a complex sequence results in: if

$$x[n] \stackrel{Z}{\Longleftrightarrow} X(z), \qquad \text{ROC} = R_x, \tag{2.76}$$

then

$$x^*[n] \stackrel{Z}{\Longleftrightarrow} X^*(z^*) \tag{2.77}$$

$$Re\{x[n]\} \stackrel{Z}{\Longleftrightarrow} \frac{1}{2}[X(z) + X^*(z^*)] \tag{2.78}$$

$$Im\{x[n]\} \stackrel{Z}{\Longleftrightarrow} \frac{1}{2j}[X(z) - X^*(z^*)] \tag{2.79}$$

The ROC is equal to R_x. The above properties of the z-transform can be proved by using (2.52). We leave these proofs to interested readers as a straightforward exercise.

Differentiation of X(z)

It is differentiated in z, and if

$$x[n] \stackrel{Z}{\Longleftrightarrow} X(z), \qquad \text{ROC} = R_x, \tag{2.80}$$

then

$$nx[n] \stackrel{Z}{\Longleftrightarrow} -z\frac{dX(z)}{dz}, \tag{2.81}$$

where the ROC is R_x except for the possible addition or deletion of $z = 0$ or $z = \infty$.

2.6.3 Common z-Transform Pairs

A number of the basic z-transform pairs that involve values of $x[n]$ only for $n \geq 0$ are listed in Table 2.1. These z-transform pairs are useful in finding z-transforms $X(z)$ given a discrete-time sequence $x[n]$. Also, these z-transform pairs are useful in finding the discrete-time sequence $x[n]$ corresponding to a given z-transform $X(z)$.

Table 2.1 The Common z-Transform Pairs

Sequence	z-Transform	ROC
$\delta[n]$	1	All z
$\delta[n-m]$	z^{-m}	If $m > 0$, then all z except 0; If $m < 0$, then all z except ∞.
$u[n]$	$\frac{1}{1-z^{-1}}$	$\|z\| > 1$
$-u[-n-1]$	$\frac{1}{1-z^{-1}}$	$\|z\| < 1$
$nu[n]$	$\frac{z}{(z-1)^2}$	$\|z\| > 1$
$n^2u[n]$	$\frac{z(z+1)}{(z-1)^3}$	$\|z\| > 1$
$n^3u[n]$	$\frac{z(z^2+4z+1)}{(z-1)^4}$	$\|z\| > 1$
$a^nu[n]$	$\frac{1}{1-az^{-1}}$	$\|z\| > \|a\|$
$na^nu[n]$	$\frac{az^{-1}}{(1-az^{-1})^2}$	$\|z\| > \|a\|$
$n^2a^nu[n]$	$\frac{az(z+a)}{(z-a)^3}$	$\|z\| > \|a\|$
$n^3a^nu[n]$	$\frac{az(z^2+4az+a^2)}{(z-a)^4}$	$\|z\| > \|a\|$
$(n+1)a^nu[n]$	$\frac{z^2}{(z-a)^2}$	$\|z\| > \|a\|$
$\frac{(n+1)(n+2)}{2}a^nu[n]$	$\frac{z^3}{(z-a)^3}$	$\|z\| > \|a\|$
$\frac{(n+1)(n+2)(n+3)}{3}a^nu[n]$	$\frac{z^4}{(z-a)^4}$	$\|z\| > \|a\|$
$\frac{a^n}{n}u[n]$	$e^{a/z}$	All z
$-a^nu[-n-1]$	$\frac{1}{1-az^{-1}}$	$\|z\| < \|a\|$
$-na^nu[-n-1]$	$\frac{az^{-1}}{(1-az^{-1})^2}$	$\|z\| < \|a\|$
$a^n(u[n]-u[n-N])$	$\frac{1-a^Nz^{-N}}{1-az^{-1}}$	$\|z\| > 0$
$[cos(\omega_0 n)]u[n]$	$\frac{1-(cos\omega_0)z^{-1}}{1-(2cos\omega_0)z^{-1}+z^{-2}}$	$\|z\| > 1$
$[sin(\omega_0 n)]u[n]$	$\frac{(sin\omega_0)z^{-1}}{1-(2cos\omega_0)z^{-1}+z^{-2}}$	$\|z\| > 1$
$[r^n cos(\omega_0 n)]u[n]$	$\frac{1-(rcos\omega_0)z^{-1}}{1-(2rcos\omega_0)z^{-1}+r^2z^{-2}}$	$\|z\| > r$
$[r^n sin(\omega_0 n)]u[n]$	$\frac{(rsin\omega_0)z^{-1}}{1-(2rcos\omega_0)z^{-1}+r^2z^{-2}}$	$\|z\| > r$
$\alpha^{\|n\|}$	$\frac{1-\alpha^2}{(1-\alpha z^{-1})(1-\alpha z)}$	$\alpha < \|z\| < \frac{1}{\alpha}$

2.6.4 The z-Domain Discrete-Time Transfer Function

The z-transform has an important application in the analysis and representation of LTI discrete-time systems, since

$$Y(z) = H(z)X(z), \tag{2.82}$$

where $H(z)$, $X(z)$, and $Y(z)$ are the z-transforms of discrete-time system impulse response, input and output, respectively. $H(z)$ is referred to as the discrete-time *system function* or discrete-time *transfer function.*

The transfer function can also be derived from the linear, constant coefficient difference equation of the following form,

$$y[n] + \sum_{k=1}^{N-1} a_k y[n-k] = \sum_{k=0}^{M-1} b_k x[n-k]. \tag{2.83}$$

By taking the z-transform of each term in the above equation and by using the linearity and the time-shifting properties, we obtain:

$$Y(z) + Y(z)\sum_{k=1}^{N-1} a_k z^{-k} = X(z)\sum_{k=0}^{M-1} b_k z^{-k}. \tag{2.84}$$

Solving (2.84) for $H(z) = \frac{Y(z)}{X(z)}$ yields

$$\begin{aligned} H(z) &= \frac{Y(z)}{X(z)} \\ &= \frac{\sum_{k=0}^{M-1} b_k z^{-k}}{1 + \sum_{k=1}^{N-1} a_k z^{-k}}. \end{aligned} \tag{2.85}$$

It is often convenient to express (2.85) in factored form as

$$H(z) = b_0 \left[\frac{\prod_{k=1}^{M-1}(1 - c_k z^{-1})}{\prod_{k=1}^{N-1}(1 - d_k z^{-1})} \right]. \tag{2.86}$$

In the numerator, each of the factors $(1 - c_k z^{-1})$ contributes a zero at $z = c_k$ and a pole at $z = 0$. Similarly, in the denominator, each of the factors $(1 - d_k z^{-1})$ contributes a pole at $z = d_k$ and a zero at $z = 0$.

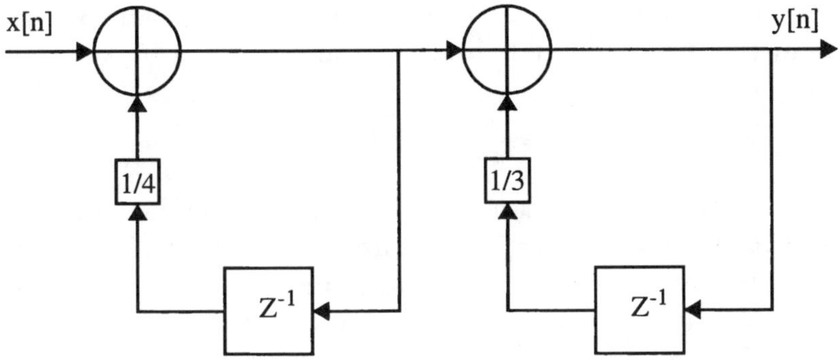

Figure 2.13 Block diagram of LTI discrete-time system.

Example 2.5 The Discrete-Time Transfer Function

Consider the LTI discrete-time system as shown in Figure 2.13. The input and output relationship can be expressed in difference equation form as follows:

$$y[n] - \frac{1}{3}y[n-1] = x[n] + \frac{1}{4}x[n-1]. \tag{2.87}$$

Applying the z-transform to both sides of (2.87) and using the linearity and time-shifting properties, we obtain

$$Y(z) - \frac{1}{3}z^{-1}Y(z) = X(z) + \frac{1}{4}z^{-1}X(z). \tag{2.88}$$

Then, the z-domain discrete-time transfer function is

$$H(z) = \frac{Y(z)}{X(z)} = \frac{1 + \frac{1}{4}z^{-1}}{1 - \frac{1}{3}z^{-1}}. \tag{2.89}$$

In this example, the ROC has two different possibilities: $|z| > 1/3$ is associated with the assumption that impulse response $h[n]$ is right-sided, and $|z| < 1/3$ is associated with the assumption that the impulse response $h[n]$ is left-sided.

2.7 The Inverse z-Transform

The inverse z-transform can be derived using the Cauchy integral theorem (see Churchill and Brown [8]) and is given by the contour integral

$$x[n] = \frac{1}{2\pi j} \oint_C X(z) z^{n-1} dz, \tag{2.90}$$

where the integral notation with C is a counterclockwise closed contour in the ROC of $X(z)$ and encircling the origin of the complex z-plane. In the next section, we introduce the direct calculation of (2.90) by using contour integration. Furthermore, to simplify, we also consider using well-known procedures to solve the inverse z-transforms. These methods include the established common z-transform pairs (also called the examination method) and partial fractions expansion.

2.7.1 Contour Integration

In this section, we illustrate inversing the z-transform directly by using contour integration based on the Cauchy residue theorem.

Let $X(z)$ be a function of the complex variable z; C is a counterclockwise closed contour in the ROC of $X(z)$. If the $(k+1)$-order derivative $\frac{dX(z)}{dz}$ exists on and inside the contour C, and if $X(z)$ does not have any poles at $z = z_0$, then the inverse z-transform is obtained by

$$\begin{aligned} x[n] &= \frac{1}{2\pi j} \oint_C \frac{X(z)}{(z - z_0)^k} dz \\ &= \frac{1}{(k-1)!} \left. \frac{d^{k-1} X(z)}{dz^{k-1}} \right|_{z=z_0}, \end{aligned} \tag{2.91}$$

if z_0 is inside C; otherwise, (2.91) is equal to zero. Thus, the result in (2.91) is the form of the Cauchy residue theorem.

2.7.2 Examination Method

The examination method uses the established common z-transform pairs, as shown in Table 2.1, to discover the relationship between the z-transform $X(z)$ and the discrete-time sequence $x[n]$. We illustrate this method by example as follows.

Consider the z-transform $X(z)$ given by

$$X(z) = \frac{1}{1+\frac{1}{4}z^{-1}} + \frac{1}{1-\frac{1}{3}z^{-1}}, \tag{2.92}$$

where the ROC is the annular region $\frac{1}{4} < |z| < \frac{1}{3}$. Using the common z-transform pairs, we obtain the discrete-time sequences

$$\left(-\frac{1}{4}\right)^n u[n] \overset{Z}{\Longleftrightarrow} \frac{1}{1+\frac{1}{4}z^{-1}}, \quad |z| > \tfrac{1}{4} \tag{2.93}$$

and

$$-\left(\frac{1}{3}\right)^n u[-n-1] \overset{Z}{\Longleftrightarrow} \frac{1}{1-\frac{1}{3}z^{-1}}, \quad |z| < \tfrac{1}{3}. \tag{2.94}$$

Thus, by using the linearity property, we obtain the discrete-time sequence,

$$x[n] = \left(-\frac{1}{4}\right)^n u[n] - \left(\frac{1}{3}\right)^n u[-n-1]. \tag{2.95}$$

As we can see, the discrete-time sequence $x[n]$ grows exponentially when n goes to $-\infty$.

2.7.3 Partial Fractions Expansion

The method of partial fractions expansion is used to obtain an alternative expression for the z-transform $X(z)$ as a sum of simpler terms. Each of the terms is tabulated to identify the discrete-time sequence $x[n]$ corresponding to the individual terms by using the common z-transform pairs in Table 2.1.

The method of partial fractions expansion differs slightly depending on whether the transfer function has all distinct poles or some multiple poles. The multiple-pole procedure of partial fractions expansion is more complicated. However, Oppenheim and Schafer [1] give a good mathematical treatment of this special case. We only describe the procedure for transfer functions with single poles, since most filter designs in practice have this property.

The procedure for partial fractions expansion of function with distinct poles first requires factoring the denominator of $H(z)$ to produce

$$H(z) = \frac{\sum_{k=0}^{M} b_k z^{M-k}}{\prod_{k=1}^{N}(z-d_k)}. \tag{2.96}$$

Then, we form the discrete-time sequence $h[n]$ as given by

$$h[n] = c_0\delta[n] + \sum_{k=1}^{N} c_k(d_k)^n, \tag{2.97}$$

where $h[n]$ is the inverse z-transform of transfer function $H(z)$.

In order to have the discrete-time sequence $h[n]$, we need to determine the coefficients c_k, for $k = 0, 1, ..., N$. First, the coefficient c_0 is given by

$$\begin{aligned} c_0 &= H(z)\,|_{z=0} \\ &= \frac{b_M}{\prod_{k=1}^{N}(-d_k)}. \end{aligned} \tag{2.98}$$

Next, the coefficients c_k for $k = 1, 2, ..., N$ can be obtained by

$$c_k = \left(\frac{z - d_k}{z}\right) H(z)\,|_{z=d_k}\,. \tag{2.99}$$

Example 2.6 Inverse z-Transform Using Partial Fractions Expansion

Consider the z-transform given by

$$H(z) = \frac{z^2}{z^2 + z - 6}, \tag{2.100}$$

where the ROC is the exterior of a circle region $|z| > 3$. Factoring the denominator of $H(z)$ produces

$$H(z) = \frac{z^2}{(z-2)(z+3)}. \tag{2.101}$$

The coefficient c_0 can be determined by

$$\begin{aligned} c_0 &= H(z)|_{z=0} \\ &= 0. \end{aligned} \tag{2.102}$$

The values of coefficients c_1 and c_2 are obtained by

$$\begin{aligned} c_1 &= \left[\frac{z-2}{z}\frac{z^2}{(z-2)(z+3)}\right]\Bigg|_{z=2} \\ &= \frac{2}{5}, \end{aligned} \tag{2.103}$$

and

$$
\begin{aligned}
c_2 &= \left[\frac{z+3}{z}\frac{z^2}{(z-2)(z+3)}\right]\Bigg|_{z=-3} \\
&= \frac{3}{5}. \qquad (2.104)
\end{aligned}
$$

Therefore, the inverse z-transform, $h[n]$, is given by

$$h[n] = \frac{2}{5}2^n u[n] + \frac{3}{5}(-3)^n u[n]. \qquad (2.105)$$

2.8 Frequency Domain of Discrete-Time Signals and Systems

The frequency-domain representation of discrete-time signals and systems provides an important analysis and design tool, and contributes insights into the solution of problems that would not otherwise be obvious. The most important frequency-domain analysis for discrete-time signals and systems is the *discrete-time Fourier transform* (DTFT). The DTFT representation of a sequence $x[n]$ is given by this equation:

$$X(e^{j\omega}) = \sum_{n=-\infty}^{\infty} x[n]e^{-j\omega n}, \qquad (2.106)$$

where $X(e^{j\omega})$ is periodic in ω with period of 2π. The DTFT is an invertible transformation. Given the DTFT representation $X(e^{j\omega})$ of $x[n]$, the discrete-time signal $x[n]$ can be recovered using the inverse DTFT formula,

$$x[n] = \frac{1}{2\pi}\int_{-\pi}^{\pi} X(e^{j\omega})e^{j\omega n}d\omega. \qquad (2.107)$$

Equations (2.106) and (2.107) are a special case of (2.52) and (2.90). Note that $z = re^{j\omega}$ is a complex variable. When $r = 1$ and $z = e^{j\omega}$, the z-transform becomes the DTFT representation

$$X(e^{j\omega}) = X(z)|_{z=e^{j\omega}}. \qquad (2.108)$$

Therefore, a sufficient condition for the convergence of DTFT in (2.106) is that the discrete-time signal $x[n]$ must be absolutely summable:

$$\sum_{n=-\infty}^{\infty} |x[n]| < \infty. \tag{2.109}$$

In general, the DTFT is a complex-valued function of ω. Frequency response is often represented in rectangular form in terms of its real and imaginary parts

$$X(e^{j\omega}) = X_R(e^{j\omega}) + jX_I(e^{j\omega}), \tag{2.110}$$

or in polar form in terms of its magnitude and phase

$$X(e^{j\omega}) = |X(e^{j\omega})|e^{j\phi_x(\omega)}. \tag{2.111}$$

In (2.111), the quantities $|X(e^{j\omega})|$ and $\phi_x(\omega)$ are called the *magnitude* and *phase*, respectively. For a real discrete-time signal $x[n]$, the DTFT representation $X(e^{j\omega})$ is conjugate symmetric. This implies that the magnitude is an even function, $|X(e^{j\omega})| = |X(e^{-j\omega})|$, and that the phase is an odd function, $\phi_x(\omega) = -\phi_x(-\omega)$.

Example 2.7 The Discrete-Time Fourier Transform

Consider the discrete-time signal

$$x[n] = b^n u[n], \tag{2.112}$$

with $b = 0.8$. The DTFT of $x[n]$ is

$$\begin{aligned} X(e^{j\omega}) &= \sum_{n=-\infty}^{\infty} b^n u[n] e^{-j\omega n} \\ &= \sum_{n=0}^{\infty} (be^{-j\omega})^n \\ &= \frac{1}{1 - be^{-j\omega}}. \end{aligned} \tag{2.113}$$

Equation (2.113) can be expressed in rectangular form in terms of its real and imaginary parts

$$X(e^{j\omega}) = \frac{1}{1 - be^{-j\omega}}$$

$$= \frac{1}{1-be^{-j\omega}}\left(\frac{1-be^{j\omega}}{1-be^{j\omega}}\right)$$
$$= \frac{1-be^{j\omega}}{1-b(e^{j\omega}+e^{-j\omega})+b^2}$$
$$= \frac{1-b(cos\omega+jsin\omega)}{1+b^2-2bcos\omega}$$
$$= \frac{1-bcos\omega}{1+b^2-2bcos\omega}+j\left(\frac{-bsin\omega}{1+b^2-2bcos\omega}\right). \quad (2.114)$$

Also, the magnitude is

$$|X(e^{j\omega})| = \frac{1}{(1+b^2-2bcos\omega)^{1/2}}, \quad (2.115)$$

and the phase is obtained by

$$\phi_x(\omega) = tan^{-1}\left(\frac{-bsin\omega}{1-bcos\omega}\right). \quad (2.116)$$

These functions, including $x[n]$, its magnitude response $X(e^{j\omega})$, and its phase $\phi_x(\omega)$ are plotted in Figure 2.14 for $b = 0.8$.

2.8.1 The DTFT in the Analysis of Linear Time-Invariant Systems

A special application of the DTFT is for linear time-invariant (LTI) discrete-time systems,

$$H(e^{j\omega}) = \sum_{n=-\infty}^{\infty} h[n]e^{-j\omega n}. \quad (2.117)$$

Equation (2.117) is referred to as the *frequency response* of the LTI system. Absolute summability is a sufficient condition for existence of a DTFT. It follows that the DTFT of an impulse response $h[n]$ exists for BIBO stable systems.

$H(e^{j\omega})$ is periodic with a period of 2π since $e^{-j\omega n}$ is also periodic in ω with a period of 2π. If a discrete-time impulse response $h[n]$ is real-value, then $H(e^{j\omega})$ is conjugate symmetric

$$H(e^{j\omega}) = H^*(e^{-j\omega}). \quad (2.118)$$

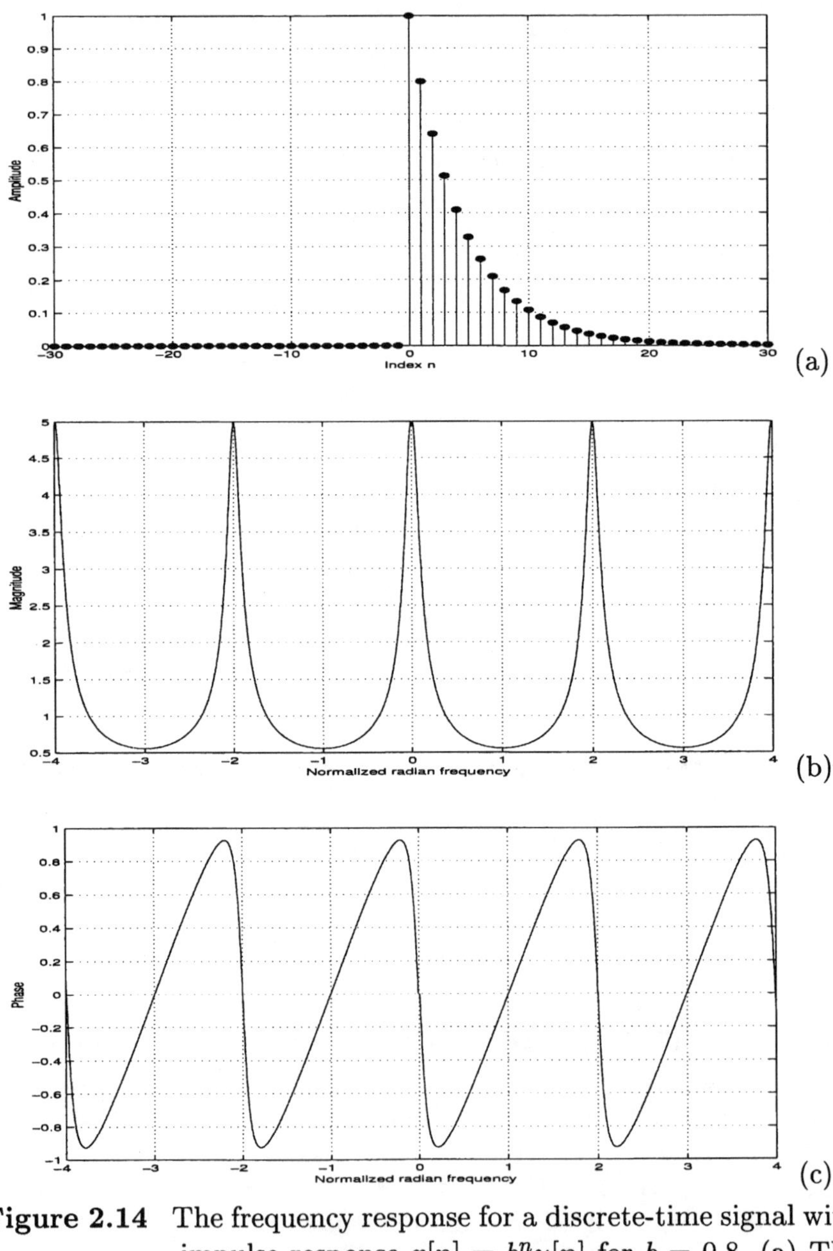

Figure 2.14 The frequency response for a discrete-time signal with impulse response $x[n] = b^n u[n]$ for $b = 0.8$. (a) The impulse response $x[n]$. (b) Magnitude $|X(e^{j\omega})|$. (c) Phase $\phi_x(\omega)$.

$H(e^{j\omega})$ also follows the convolution theorem. If there is a convolution of discrete-time input signal $x[n]$ and impulse response $h[n]$

$$y[n] = x[n] * h[n], \tag{2.119}$$

then the DTFT is equal to the product of the transforms

$$Y(e^{j\omega}) = X(e^{j\omega})H(e^{j\omega}). \tag{2.120}$$

Equation (2.120) is very useful in the design and analysis of LTI discrete-time systems.

Example 2.8 The DTFT for LTI Discrete-Time Systems

Consider the discrete-time impulse response

$$h[n] = \begin{cases} 1, & -M \leq n \leq N \\ 0, & \text{otherwise.} \end{cases} \tag{2.121}$$

and the discrete-time input signal $x[n]$

$$x[n] = \delta[n - n_d]. \tag{2.122}$$

Then the DTFT of the impulse response, $h[n]$, is

$$H(e^{j\omega}) = \sum_{n=-M}^{N} e^{-j\omega n}. \tag{2.123}$$

Equation (2.123) can be expressed in the closedform by using (2.43), resulting in

$$\begin{aligned} H(e^{j\omega}) &= \frac{e^{j\omega M} - e^{-j\omega(N+1)}}{1 - e^{-j\omega}} \\ &= e^{-j\omega(N-M+1)/2}\left[\frac{e^{j\omega(N+M+1)/2} - e^{-j\omega(N+M+1)/2}}{1 - e^{-j\omega}}\right] \\ &= e^{-j\omega(N-M)/2}\left[\frac{e^{j\omega(N+M+1)/2} - e^{-j\omega(N+M+1)/2}}{e^{j\omega/2} - e^{-j\omega/2}}\right] \\ &= e^{-j\omega(N-M)/2}\left[\frac{sin[\omega(M+N+1)/2]}{sin(\omega/2)}\right]. \end{aligned} \tag{2.124}$$

The DTFT of the input signal $x[n]$ is

$$X(e^{j\omega}) = e^{-j\omega n_d}. \tag{2.125}$$

Therefore, the DTFT of the output signal $y[n]$ is

$$\begin{aligned} Y(e^{j\omega}) &= X(e^{j\omega})H(e^{j\omega}) \\ &= e^{-j\omega(N-M+2n_d)/2} \\ &\quad \cdot \left[\frac{sin[\omega(M+N+1)/2]}{sin(\omega/2)}\right]. \end{aligned} \tag{2.126}$$

As we can see, the DTFT $Y(e^{j\omega})$ of the output signal is a frequency-shifted version of $H(e^{j\omega})$. Note that (2.124) is a rough approximation of a lowpass filter so that the LTI system $h[n]$ will smooth out rapid variations in the input sequence $x[n]$.

2.8.2 Properties of the DTFT

The DTFT operation has many properties that are useful in design and analysis of LTI systems. These are summarized below.

Linearity

The linearity property states that if

$$x_1[n] \overset{DTFT}{\Longleftrightarrow} X_1(e^{j\omega}) \tag{2.127}$$

and

$$x_2[n] \overset{DTFT}{\Longleftrightarrow} X_2(e^{j\omega}), \tag{2.128}$$

then

$$ax_1[n] + bx_2[n] \overset{DTFT}{\Longleftrightarrow} aX_1(e^{j\omega}) + bX_2(e^{j\omega}), \tag{2.129}$$

for any complex number a and b.

Time Shifting

The time shifting property states that if

$$x[n] \overset{DTFT}{\Longleftrightarrow} X(e^{j\omega}), \tag{2.130}$$

then

$$x[n-d] \overset{DTFT}{\Longleftrightarrow} e^{-j\omega d}X(e^{j\omega}), \tag{2.131}$$

where d is an integer. If d is positive, the discrete-time sequence $x[n]$ is shifted right, and if d is negative, then the discrete-time sequence $x[n]$ is shifted left.

Frequency Shifting or Modulation

The frequency shifting property states that if

$$x[n] \overset{DTFT}{\Longleftrightarrow} X(e^{j\omega}), \tag{2.132}$$

then

$$e^{j\omega_0 n}x[n] \overset{DTFT}{\Longleftrightarrow} X(e^{j(\omega-\omega_0)}). \tag{2.133}$$

Multiplying a discrete-time sequence by a complex sinusoidal sequence corresponds to shifting its DTFT transform.

Multiplication

The multiplication property states that if

$$x[n] \overset{DTFT}{\Longleftrightarrow} X(e^{j\omega}) \tag{2.134}$$

and

$$w[n] \overset{DTFT}{\Longleftrightarrow} W(e^{j\omega}), \tag{2.135}$$

then

$$x[n]w[n] \overset{DTFT}{\Longleftrightarrow} \frac{1}{2\pi}\int_{-\pi}^{\pi} X(e^{j\phi})W(e^{j(\omega-\phi)})d\phi. \tag{2.136}$$

The multiplication of two discrete-time sequences results in a periodic convolution of their DTFT transforms.

Time Reversal

The time reversal property states that if

$$x[n] \overset{DTFT}{\Longleftrightarrow} X(e^{j\omega}), \tag{2.137}$$

then

$$x[-n] \overset{DTFT}{\Longleftrightarrow} X(e^{-j\omega}). \tag{2.138}$$

Convolution of Sequences

The convolution property of sequences states that if

$$x[n] \overset{DTFT}{\Longleftrightarrow} X(e^{j\omega}) \tag{2.139}$$

and

$$h[n] \overset{DTFT}{\Longleftrightarrow} H(e^{j\omega}), \tag{2.140}$$

then

$$x[n] * h[n] \overset{DTFT}{\Longleftrightarrow} X(e^{j\omega})H(e^{j\omega}). \tag{2.141}$$

The convolution of two discrete-time sequences results in the multiplication of the corresponding DTFT.

Conjugation of a Complex Sequence

The conjugation property of a complex sequence states that if

$$x[n] \overset{DTFT}{\Longleftrightarrow} X(e^{j\omega}), \tag{2.142}$$

then

$$x^*[n] \overset{DTFT}{\Longleftrightarrow} X^*(e^{-j\omega}) \tag{2.143}$$

$$Re\{x[n]\} \overset{DTFT}{\Longleftrightarrow} \frac{1}{2}[X(e^{j\omega}) + X^*(e^{-j\omega})] \tag{2.144}$$

$$Im\{x[n]\} \overset{DTFT}{\Longleftrightarrow} \frac{1}{2j}[X(e^{j\omega}) - X^*(e^{-j\omega})] \tag{2.145}$$

$$Re\{X(e^{j\omega})\} \overset{DTFT}{\Longleftrightarrow} \frac{1}{2}\{x[n] + x^*[n]\} \tag{2.146}$$

$$Im\{X(e^{j\omega})\} \overset{DTFT}{\Longleftrightarrow} \frac{1}{2}\{x[n] - x^*[n]\}. \tag{2.147}$$

In a special case, if $x[n]$ is a real-valued sequence, then these relationships imply that

$$X(e^{j\omega}) = X^*(e^{-j\omega}) \tag{2.148}$$

$$Re\{X(e^{j\omega})\} = Re\{X(e^{-j\omega})\} \tag{2.149}$$

$$Im\{X(e^{j\omega})\} = -Im\{X(e^{-j\omega})\}. \tag{2.150}$$

Differentiation of $X(e^{j\omega})$

The differentiation property of $X(e^{j\omega})$ states that if

$$x[n] \overset{DTFT}{\Longleftrightarrow} X(e^{j\omega}), \tag{2.151}$$

then

$$nx[n] \overset{DTFT}{\Longleftrightarrow} -j\frac{dX(e^{j\omega})}{d\omega}. \tag{2.152}$$

Parseval's Theorem

Parseval's theorem states that if

$$x[n] \overset{DTFT}{\Longleftrightarrow} X(e^{j\omega}), \tag{2.153}$$

then

$$\sum_{n=-\infty}^{\infty} |x[n]|^2 = \frac{1}{2\pi}\int_{-\pi}^{\pi} |X(e^{j\omega})|^2 d\omega. \tag{2.154}$$

Equation (2.154) can be interpreted as the total energy in the discrete-time sequence $x[n]$.

2.9 The Relationship of Allpass and Minimum-Phase Systems

In this section, we introduce the allpass and minimum-phase systems, which are useful in inverse filtering, compensation for phase distortion, deconvolution, and system identification. Our discussion assumes that the system functions are stable and causal.

2.9.1 Allpass Systems

A stable allpass system, denoted by $H_{ap}(z)$, has a function form

$$H_{ap}(z) = \frac{z^{-1} - a^*}{1 - az^{-1}}. \tag{2.155}$$

Note that (2.155) has a frequency-response magnitude that is independent of ω. This can be shown by writing (2.155) into the frequency-domain in the form

$$H_{ap}(e^{j\omega}) = \frac{e^{-j\omega} - a^*}{1 - ae^{-j\omega}}$$

$$= e^{-j\omega}\frac{1 - a^* e^{j\omega}}{1 - ae^{-j\omega}}. \tag{2.156}$$

As can be seen, in the expression of (2.156), the term $e^{-j\omega}$ has unity magnitude. The remaining term's numerator and denominator are complex conjugates of each other so that this term has the same magnitude. As a result, $|H_{ap}(e^{j\omega})| = 1$. In other words, the system $H_{ap}(z)$ in (2.155) passes all of the frequency components of its input without changing magnitude gain. Therefore, (2.155) is referred to as an allpass or allphase system.

The properties of the allphase system are as follows: (1) the group delay of a rational allphase system is always positive, and (2) the phase of an allphase system is negative in the frequency range of $0 < \omega < \pi$.

2.9.2 Minimum-Phase Systems

For a finite impulse response (FIR) system $H(z)$ of arbitrary length, its frequency response can be expressed as

$$H(e^{j\omega}) = C_0(1 - z_1 e^{-j\omega})(1 - z_2 e^{-j\omega}) \cdots (1 - z_M e^{-j\omega}), \tag{2.157}$$

where $\{z_i\}$ denotes the zeros and C_0 is an arbitrary constant. The FIR system is called a *minimum-phase system* if all the zeros are inside the unit circle. On the other hand, the FIR system is called a *maximum-phase system* if all the zeros are outside the unit circle. In addition, if the FIR system has some of its zeros inside the unit circle and its remaining zeros outside the unit circle, then the FIR system is referred to as a *mixed-phase system.*

For an infinite impulse response (IIR) system $H(z) = \frac{B(z)}{A(z)}$, the IIR system is called a minimum-phase system if all the zeros and poles are inside the unit circle. The IIR system is called a maximum-phase system if all the zeros are outside the unit circle and all the poles are inside the unit circle. Moreover, the IIR system is referred to as a mixed-phase system if some, but not all, of the zeros are outside the unit circle, and all the poles are inside the unit circle.

Note that a minimum-phase system indicates a minimum delay function, while a maximum-phase system implies a maximum delay characteristic. In addition, if we are given a magnitude-square function $C(z) = H(z)H^*(1/z)$ and we know that the system $H(z)$ is a minimum-phase system, then we can determine the system function $H(z)$ uniquely.

2.9.3 Decomposition of Nonminimum-Phase Systems

Any rational system function $H(z)$ in the z-transform domain can be decomposited as

$$H(z) = H_{min}(z)H_{ap}(z), \tag{2.158}$$

where $H_{min}(z)$ is referred to as a minimum-phase system and $H_{ap}(z)$ is an allphase system as given by Oppenheim and Schafer [1] and Proakis and Manolakis [9].

A nonminimum-phase system can be formed into a minimum-phase system $H_{min}(z)$ by reflecting one or more zeros within the unit circle into their conjugate reciprocal locations outside the unit circle by using (2.158). On the other hand, a minimum-phase system can be created from a nonminimum-phase system by reflecting all the zeros lying outside the unit circle into their conjugate reciprocal locations inside the unit circle. However, the frequency response system will have the same magnitudes in the case of both the minimum-phase and nonminimum-phase systems.

2.10 Summary

In this chapter, we have introduced the basic theory of discrete-time signal processing, including discrete-time signals and systems and corresponding building blocks for operations. We presented the definition of linear time-invariant discrete-time systems in terms of linear and time-invariant systems. We also described causality, stability, convolution sums, and closedform expressions for their calculation. Linear time-invariant discrete-time systems have many signal processing applications, and provide solutions for many practical problems. The simplicity of these systems enables efficient analysis and design of discrete-time systems.

We have also defined the z-transform for discrete-time sequences and their corresponding inverse z-transforms. The region of convergence for the z-transform is explored in detail. Much of the discussion focused on the properties of the z-transform that are particularly useful for obtaining z-transforms of commonly used sequences. In addition, we discussed techniques for finding the inverse z-transform, such as the contour integration, examination method, and partial fractions expansion.

The discrete-time Fourier transform is introduced, leading to a definition of the frequency response that enables one to interpret the relationship between the impulse response and the frequency-domain representation. We also presented many properties of the discrete-time Fourier transform that are useful in finding the common Fourier transform pairs. In addition, we showed that the z-transform is a generalization of the discrete-time Fourier transform. The relationship of allphase, minimum-phase, maximum-phase, and mixed-phase systems was discussed as well.

Important applications of the z-transform and the discrete-time Fourier transform involve analyzing signals. They can also be used to represent linear time-invariant systems and allow frequency-domain interpretations. These properties are also useful in discrete-time transforms, digital filtering, and discrete-time multirate signal processing, which we will introduce in the following chapters.

References

[1] Oppenheim, A. V., and R. W. Schafer, *Discrete-Time Signal Processing*, Prentice Hall, New Jersey, 1989.

[2] Kailath, T., *Linear Systems*, Prentice Hall, Englewood Cliffs, New Jersey, 1980.

[3] Burrus, C. S., J. H. McClellan, A. V. Oppenheim, T. W. Parks, R. W. Schafer, and H. W. Schuessler, *Computer-Based Exercises for Signal Processing*, Prentice Hall, Englewood Cliffs, New Jersey, 1994.

[4] Nyquist, H., "Certain Topics in Telegraph Transmission Theory," *Trans. on AIEE*, Vol. 47, pp. 617-644, April 1928.

[5] Shannon, C. E., "Communication in the Presence of Noise," *Proc. of IRE*, Vol. 37, pp. 10-21, January 1949.

[6] Oppenheim, A. V., A. S. Willsky, and I. T. Young, *Signal and Systems*, Prentice Hall, Englewood Cliffs, New Jersey, 1983.

[7] Dudgeon, D. E., and R. M. Mersereau, *Multidimensional Digital Signal Processing*, Prentice Hall, Englewood Cliffs, New Jersey, 1984.

[8] Churchill, R. V., and J. W. Brown, *Introduction to Complex Variables and Application*, McGraw-Hill Book Company, New York, 1984.

[9] Proakis, J. G., and D. G. Manolakis, *Digital Signal Processing: Principles, Algorithms, and Applications*, 3d ed., Prentice Hall, Upper Saddle River, New Jersey, 1996.

3

Discrete-Time Random Processes

3.1 Introduction

In Chapter 2, we reviewed the mathematical representation of discrete-time signal processing. Discrete-time signals and systems have both time-domain and frequency-domain representations that play an important role in the theory and design of discrete-time signal processing. These developments assume that the discrete-time signals are deterministic. A discrete-time *deterministic* signal is one in which each value of an indexed sequence is uniquely determined by a mathematical expression.

In many cases, discrete-time signals that are generated from complex processes cannot be described precisely by such mathematical representation. We refer to such signals as discrete-time *random* signals or *stochastic* signals. A discrete-time random signal, or random process, is a signal that is not in general repeatable in a predictable manner. For example, quantization noise that is created by an analog-to-digital converter or fixed-point digital filter can be modeled as a random process. In another example, the discrete-time Kalman filter is developed based on discrete-time random process assumption.

A discrete-time random or stochastic signal can be considered as a member of an ensemble of signals that is characterized by a set of probabilities. A fundamental mathematical representation of such signals often uses their description in terms of averages. Therefore, in this chapter, we provide the stochastic background that is necessary to understand how a discrete-time random process can be represented and how its statistical properties are affected by linear shift-invariant systems. We anticipate that readers have had some prior exposure to the basic theory of probability and random processes. Our treatment serves primarily as review and notation definition. Detailed mathematical treatments in this subject can be found in the references by Davenport and Root [1], Davenport [2], and Papoulis [3].

This chapter is organized as follows. The introduction is given here in Section 3.1. In Section 3.2, we begin with a brief review of probability and random processes, focusing on concepts and definitions such as intuitive, axiomatic, conditional probabilities, independence, and random variable notions. Distribution and density functions are presented in Section 3.3, along with discussion of distribution and density functions, joint distribution and density functions, statistical averages and joint moments, as well as Gaussian and uniform random variables, chi-square and $\mathcal{F}$ distributions, sums of random variables, and central limit theorems. In Section 3.4, stochastic processes of discrete-time random signals are discussed, addressing random variables, stationary processes, estimated ensemble averages, autocorrelations, cross-correlations, power spectra, filtering of random processes, and mean square error and estimation. Finally, a short summary is given in Section 3.5.

3.2 Probability and Random Variables

In this section, we will introduce some results of probability theory that are important for random variables. Our treatment of probability theory is brief, taking into account the specific needs of subsequent chapters, with emphasis on random variables.

3.2.1 Intuitive Probability

Probability theory deals with averages of many phenomena occurring sequentially or simultaneously. One purpose of this theory is to describe and predict averages in terms of probabilities of the events.

For example, in a coin-tossing experiment, given a fair coin equally likely to result in heads or tails if the coin is flipped, the outcome would be heads approximately half of the time and tails the other half. Therefore, in the intuitive approach, if we assume that all possible outcomes of a chance experiment are equally likely, the probability of a particular event, $\mathcal{A}$, is defined as

$$p(\mathcal{A}) = \frac{\mathcal{N}_{\mathcal{A}}}{\mathcal{N}_{\mathcal{T}}}, \tag{3.1}$$

provided that $N_{\mathcal{T}}$ is *sufficiently large*, where $N_{\mathcal{A}}$ and $N_{\mathcal{T}}$ are possible outcomes favoring event $\mathcal{A}$ and total possible outcomes, respectively.

Note that the term "sufficiently large" in (3.1) has no clear meaning. However, this lack of precision statement cannot be avoided. Although intuitive probabilities have limitations, they still play an important role in probability theory. The concept of ratio for possible events is a useful problem-solving tool in many cases.

For example, if a coin is flipped 2,000 times and heads show 1,089 times, then the probability of heads equals 0.5445. If the coin is fair, then the probability of heads equals $\frac{1}{2}$.

If the outcome of probability process is numerical, the outcome is referred to as a random variable. The value of the outcome will vary randomly from one trial to the next. Thus, when we discussed the number of heads obtained in $N_{\mathcal{T}}$ tosses of a coin, we were talking about the random variable number of heads. In the same way, when we discussed the number of tails obtained in $N_{\mathcal{T}}$ tosses of a coin, we were talking about the random variable number of tails.

3.2.2 Axiomatic Probability

Axiomatic probability begins with the concept of a *sample space*. The sample space is the set of all possible outcomes of the experiment. We denote the sample space as $\mathcal{S}$ and its individual outcomes or elements in the sample space as $\{s_1, s_2, s_3, ...\}$. The elements in the sample space are *mutually exclusive* or *disjoint*. In other words, there is no overlap of elements in a sample space. Also, the elements are collectively exhaustive. Every possible outcome is included.

An *event* is a special subset of the sample space $\mathcal{S}$. We denote the events as $\mathcal{A}$, $\mathcal{B}$, $\mathcal{C}$, ..., and so on. Axiomatic probability theory now is stated as follows:

1. The probability of an event $\mathcal{A}$ is a positive number

$$p(\mathcal{A}) \geq 0 \tag{3.2}$$

2. The probability of the certain event equals 1,

$$p(S) = 1 \tag{3.3}$$

3. Assume that $\mathcal{A}$, $\mathcal{B}$, $\mathcal{C}$, ..., are mutually exclusive events, then

$$p(\mathcal{A} + \mathcal{B} + \mathcal{C} + \cdots) = p(\mathcal{A}) + p(\mathcal{B}) + p(\mathcal{C}) + \cdots \tag{3.4}$$

The above three definitions together with the traditional axioms of set theory are fundamental to axiomatic probability theory and provide the theoretical basis for the formal solution of a wide variety of probability problems. A simple consequence is that the probability of the impossible event is always 0: $p(\emptyset) = 0$.

In addition to the above definitions in axiomatic probability theory, the operation of intersection is also useful. The intersection of two events $\mathcal{A}$ and $\mathcal{B}$ is the event consisting of elements that are common to both $\mathcal{A}$ and $\mathcal{B}$. We denote the intersection of two events $\mathcal{A}$ and $\mathcal{B}$ as $\mathcal{AB}$. Therefore, for any $\mathcal{A}$ and $\mathcal{B}$, the probability can be written by

$$p(\mathcal{A} + \mathcal{B}) = p(\mathcal{A}) + p(\mathcal{B}) - p(\mathcal{AB}). \tag{3.5}$$

The subtractive term $p(\mathcal{AB})$ in (3.5) is required because the probabilities in the intersection region have been included twice in the summation of $p(\mathcal{A})$ and $p(\mathcal{B})$. The probability $p(\mathcal{AB})$ is referred to as the *joint probability* of the events $\mathcal{A}$ and $\mathcal{B}$, and can be expressed by the probability that both $\mathcal{A}$ and $\mathcal{B}$ occurred.

3.2.3 Conditional Probability

Consider an experiment in which a joint event occurs with probability $p(\mathcal{AB})$. Given that the event $\mathcal{B}$ has occurred, we want to determine the probability of occurrence of the event $\mathcal{A}$. This probability can be shown to be

$$p(\mathcal{A}|\mathcal{B}) = \frac{p(\mathcal{AB})}{p(\mathcal{B})}, \tag{3.6}$$

where $p(\mathcal{B}) > 0$.

Note that if $\mathcal{B} \subset \mathcal{A}$ then $p(\mathcal{A}|\mathcal{B}) = 1$. The notation $\subset$ denotes the meaning of the subset. This is because

$$\begin{aligned} p(\mathcal{A}|\mathcal{B}) &= \frac{p(\mathcal{AB})}{p(\mathcal{B})} \\ &= \frac{p(\mathcal{B})}{p(\mathcal{B})} \\ &= 1. \end{aligned} \tag{3.7}$$

Similarly, if $\mathcal{A} \subset \mathcal{B}$ then

$$\begin{aligned} p(\mathcal{A}|\mathcal{B}) &= \frac{p(\mathcal{A})}{p(\mathcal{B})} \\ &\geq p(\mathcal{A}). \end{aligned} \tag{3.8}$$

If events $\mathcal{A}_i$, $i = 1, 2, ..., n$, are mutually exclusive events in such a way that

$$\bigcup_{i=1}^{n} \mathcal{A}_i = S, \tag{3.9}$$

and $\mathcal{B}$ is an arbitrary event with $p(\mathcal{B}) > 0$, then

$$p(\mathcal{BA}_i) = p(\mathcal{B}|\mathcal{A}_i)p(\mathcal{A}_i), \tag{3.10}$$

and

$$p(\mathcal{B}) = p(\mathcal{BA}_1) + p(\mathcal{BA}_2) + \cdots + p(\mathcal{BA}_n). \tag{3.11}$$

Equation (3.11) is known as the *total probability theorem*.

Since $p(\mathcal{BA}_i) = p(\mathcal{A}_i|\mathcal{B})p(\mathcal{B})$, (3.10) can be rewritten as follows:

$$p(\mathcal{A}_i|\mathcal{B}) = p(\mathcal{B}|\mathcal{A}_i)\frac{p(\mathcal{A}_i)}{p(\mathcal{B})}. \tag{3.12}$$

Substituting (3.11) into (3.12), we obtain

$$p(\mathcal{A}_i|\mathcal{B}) = \frac{p(\mathcal{B}|\mathcal{A}_i)p(\mathcal{A}_i)}{\sum_{j=1}^{n} p(\mathcal{B}|\mathcal{A}_j)p(\mathcal{A}_j)}, \tag{3.13}$$

where the probabilities $p(\mathcal{A}_i)$ and $p(\mathcal{A}_i|\mathcal{B})$ are often referred to as *a priori* and *a posteriori*, respectively. Equation (3.13) is referred to as

Bayes' theorem. Note that the numerator is one of the terms in the denominator.

Example 3.1 Probability Calculation

Suppose that there are eight balls in a dish and you want to pick up two of them, choosing the balls randomly. Four of the balls are red and four are yellow. You prefer the red ones. The question is, What is the probability that you will get two red ones?

Let $\mathcal{R}_i$, and $\mathcal{Y}_i$, $i = 1, 2, 3, 4$, denote the red and yellow balls, respectively. The probability of getting two red ones is

$$p(\mathcal{R}_1\mathcal{R}_2) = p(\mathcal{R}_1)p(\mathcal{R}_2|\mathcal{R}_1). \tag{3.14}$$

since there are four red balls and eight balls together, then

$$p(\mathcal{R}_1) = \frac{1}{2}. \tag{3.15}$$

If you obtain a red one on the first selection, there will be three red ones left out of seven balls in the dish. Then you obtain

$$p(\mathcal{R}_2|\mathcal{R}_1) = \frac{3}{7}, \tag{3.16}$$

and

$$\begin{aligned} p(\mathcal{R}_1\mathcal{R}_2) &= p(\mathcal{R}_1)p(\mathcal{R}_2|\mathcal{R}_1) \\ &= \left(\frac{1}{2}\right)\left(\frac{3}{7}\right) = \frac{3}{14}. \end{aligned}$$

Moreover, if you get two red balls, what is the probability that you will get two more red ones? Since two red balls have been removed, there are two red left out of six balls. Therefore, the probability is

$$\begin{aligned} p(\mathcal{R}_3\mathcal{R}_4|\mathcal{R}_1\mathcal{R}_2) &= p(\mathcal{R}_3|\mathcal{R}_1\mathcal{R}_2)p(\mathcal{R}_4|\mathcal{R}_1\mathcal{R}_2\mathcal{R}_3) \\ &= \left(\frac{2}{6}\right)\left(\frac{1}{5}\right) = \frac{2}{30}. \end{aligned} \tag{3.17}$$

3.2.4 Independence

Two events $\mathcal{A}$ and $\mathcal{B}$ are said to be *independent* if

$$p(\mathcal{A}\mathcal{B}) = p(\mathcal{A})p(\mathcal{B}). \tag{3.18}$$

The independence of n events can be defined inductively. The events $\mathcal{A}_1$, $\mathcal{A}_2$,...,$\mathcal{A}_n$ are called independent if

$$p(\mathcal{A}_1 \cdots \mathcal{A}_n) = p(\mathcal{A}_1) \cdots p(\mathcal{A}_n). \tag{3.19}$$

This completes the generalization definition of independence for n events.

3.2.5 Random Variable

A *random variable* is a function that maps every point in the sample space $\mathcal{S}$ onto a real number. We denote a *boldface letter* $\mathbf{x}$ as a random variable. The process function must satisfy two conditions, as follows:

1. The set $\{\mathbf{x} \leq x\}$ is an event for every real value x.
2. The probabilities of the events $\{\mathbf{x} = \infty\}$ and $\{\mathbf{x} = -\infty\}$ are $p\{\mathbf{x} = \infty\} = 0$ and $p\{\mathbf{x} = -\infty\} = 0$, respectively.

For example, given x_1 and x_2 real numbers, the notational meaning of $\{x_1 \leq \mathbf{x} \leq x_2\}$ represents a subset of space consisting of all outcomes ξ such that $x_1 \leq \mathbf{x} \leq x_2$. The notation $\{\mathbf{x} = x\}$ is a subset of space consisting of all outcomes ξ such that $\mathbf{x} = x$, where x is a given number.

A *complex random variable* $\mathbf{z}$ is defined by

$$\mathbf{z} = \mathbf{x} + j\mathbf{y}, \tag{3.20}$$

where $\mathbf{x}$ and $\mathbf{y}$ are real random variables.

A *random vector* is a vector

$$\mathbf{X} = [\mathbf{x}_1, \mathbf{x}_2, ..., \mathbf{x}_n], \tag{3.21}$$

where components $\mathbf{x}_i$, $i = 1, 2, ..., n$, are random variables.

3.3 Distribution and Density Functions

3.3.1 Distribution Function

The distribution function of the random variable $\mathbf{x}$ is defined by

$$F_x(x) = p\{\mathbf{x} \leq x\}, \tag{3.22}$$

for every x with range from $-\infty$ to ∞. Equation (3.22) is known as the *probability distribution function* of the random variable $\mathbf{x}$. It is also called the *cumulative distribution function* (cdf) of random variable $\mathbf{x}$.

The cumulative distribution function has the following properties:

1. Since $F_x(x)$ is a probability, its interval is limited to the range
$$0 \leq F_x(x) \leq 1. \tag{3.23}$$
In fact, $F_x(\infty) = 1$ and $F_x(-\infty) = 0$.
2. It is a nondecreasing function of x. If $x_1 < x_2$, then
$$F_x(x_1) \leq F_x(x_2). \tag{3.24}$$
3. The probability $p\{\mathbf{x} > x\}$ of the event $\{\mathbf{x} > x\}$ is given by
$$p\{\mathbf{x} > x\} = 1 - F_x(x). \tag{3.25}$$
4. The probability $p\{x_1 < \mathbf{x} \leq x_2\}$ of the event $\{x_1 < \mathbf{x} \leq x_2\}$ is obtained by
$$p\{x_1 < \mathbf{x} \leq x_2\} = F_x(x_2) - F_x(x_1). \tag{3.26}$$

3.3.2 Density Function

The derivative of the cumulative distribution function is called the *density function* of the random variable $\mathbf{x}$:

$$f_x(x) = \frac{dF_x(x)}{dx}, \quad -\infty < x < \infty. \tag{3.27}$$

This is equivalent to

$$F_x(x) = \int_{-\infty}^{x} f_x(u)du, \quad -\infty < x < \infty. \tag{3.28}$$

The density function has the following properties:

1. Since $F_x(x)$ is monotonic, it follows that
$$f_x(x) \geq 0, \tag{3.29}$$

2. Since $F_x(\infty) = 1$, (3.28) yields

$$\int_{-\infty}^{\infty} f_x(x)dx = 1. \tag{3.30}$$

3. From (3.28), we obtain

$$F_x(x_1) - F_x(x_2) = \int_{x_1}^{x_2} f_x(x)dx. \tag{3.31}$$

3.3.3 Joint Distributions and Densities

The joint cumulative distribution function of two random variables $\mathbf{x}_1$ and $\mathbf{x}_2$ is defined as

$$\begin{aligned} F_{x_1x_2}(x_1, x_2) &= p\{\mathbf{x}_1 \leq x_1, \mathbf{x}_2 \leq x_2\} \\ &= \int_{-\infty}^{x_2}\int_{-\infty}^{x_1} f_{x_1x_2}(u_1, u_2)du_1du_2, \end{aligned} \tag{3.32}$$

where $f_{x_1x_2}(x_1, x_2)$ is the joint density function. The joint density function of random variables $\mathbf{x}_1$ and $\mathbf{x}_2$ is obtained by

$$f_{x_1x_2}(x_1, x_2) = \frac{\partial^2 F_{x_1x_2}(x_1, x_2)}{\partial x_1 \partial x_2}. \tag{3.33}$$

In the study of two random variables $\mathbf{x}_1$ and $\mathbf{x}_2$, the marginal distribution and the marginal density are defined by

$$F_{x_1}(x_1) = F_{x_1x_2}(x_1, \infty), \qquad F_{x_2}(x_2) = F_{x_1x_2}(\infty, x_2), \tag{3.34}$$

and

$$f_{x_1}(x_1) = \int_{-\infty}^{\infty} f_{x_1x_2}(x_1, x_2)dx_2, \qquad f_{x_2}(x_2) = \int_{-\infty}^{\infty} f_{x_1x_2}(x_1, x_2)dx_1. \tag{3.35}$$

If joint density function $f_{x_1x_2}(x_1, x_2)$ is integrated over both variables x_1 and x_2, then we obtain

$$\int_{-\infty}^{\infty}\int_{-\infty}^{\infty} f_{x_1x_2}(x_1, x_2)dx_1dx_2 = F_{x_1x_2}(\infty, \infty) = 1. \tag{3.36}$$

In addition, we also know that $F_{x_1x_2}(-\infty, -\infty) = F_{x_1x_2}(-\infty, x_2)$ $= F_{x_1x_2}(x_1, -\infty) = 0$.

The generalization of multidimensional random variables is straightforward. For n random variables, $\mathbf{x}_1, \mathbf{x}_2, ..., \mathbf{x}_n$, the joint cumulative distribution function is given by

$$\begin{aligned} F_{x_1...x_n}(x_1, ..., x_n) &= p(\mathbf{x}_1 \le x_1, ..., \mathbf{x}_n \le x_n) \\ &= \int_{-\infty}^{x_1} \cdots \int_{-\infty}^{x_n} f_{x_1...x_n}(u_1, ..., u_n) \\ &\quad \cdot du_1 \cdots du_n, \end{aligned} \tag{3.37}$$

where $f_{x_1...x_n}(x_1, ..., x_n)$ is the joint density function, and the joint density function is obtained as follows:

$$f_{x_1...x_n}(x_1, ..., x_n) = \frac{\partial^n F_{x_1...x_n}(x_1, ..., x_n)}{\partial x_1 \cdots \partial x_n}. \tag{3.38}$$

The n random variables, $\mathbf{x}_1, \mathbf{x}_2, ..., \mathbf{x}_n$, are statistically independent if and only if

$$F_{x_1...x_n}(x_1, ..., x_n) = F_{x_1}(x_1) \cdots F_{x_n}(x_n), \tag{3.39}$$

or

$$f_{x_1...x_n}(x_1, ..., x_n) = f_{x_1}(x_1) \cdots f_{x_n}(x_n). \tag{3.40}$$

3.3.4 Statistical Averages and Joint Moments

In this section, we define statistical averages and joint moments, which play an important role in the characterization of the outcomes of experiments. Of great importance are the first and the second moments of a single random variable. We also define joint moments, such as the correlation and covariance, between pairs of random variables within a set of n random variables.

The *mean* or *expected value* of random variable $\mathbf{x}$ with density function $f_x(x)$ is defined by

$$m_x = E\{\mathbf{x}\} = \int_{-\infty}^{\infty} x f_x(x) dx, \tag{3.41}$$

where $E\{\}$ denotes expectation or statistical average.

The *variance* of random variable $\mathbf{x}$ is defined by

$$\sigma_x^2 = \int_{-\infty}^{\infty} (x - m_x)^2 f_x(x) dx, \tag{3.42}$$

where $m_x = E\{\mathbf{x}\}$. The value σ_x is called the *standard deviation* of random variable $\mathbf{x}$. From the definition in (3.42), the variance σ_x^2 can be considered to be the mean of the random variable $(\mathbf{x} - m_x)^2$. The relationship between the variance and mean is

$$\begin{aligned} \sigma_x^2 &= E\{(\mathbf{x} - m_x)^2\} \\ &= E\{\mathbf{x}^2 - 2\mathbf{x}m_x + m_x^2\} \\ &= E\{\mathbf{x}^2\} - 2m_x E\{\mathbf{x}\} + m_x^2 \\ &= E\{\mathbf{x}^2\} - E^2\{\mathbf{x}\}. \end{aligned} \tag{3.43}$$

Consider a new random variable $\mathbf{y} = (\mathbf{x} - m_x)^n$, where m_x is the mean value of $\mathbf{x}$. The mean of the random variable $\mathbf{y}$ is

$$\begin{aligned} E\{\mathbf{y}\} &= E\{(\mathbf{x} - m_x)^n\} \\ &= \int_{-\infty}^{\infty} (\mathbf{x} - m_x)^n f_x(x) dx. \end{aligned} \tag{3.44}$$

Equation (3.44) defines the nth *central moment* of random variable $\mathbf{x}$.

Let $f_{x_i x_j}(x_i, x_j)$ be the joint density function of the random variables $\mathbf{x}_i$ and $\mathbf{x}_j$. The *correlation*, denoted by $R_{\mathbf{x}_i\mathbf{x}_j}$, between random variables $\mathbf{x}_i$ and $\mathbf{x}_j$ is defined by the joint moment as follows:

$$\begin{aligned} R_{\mathbf{x}_i\mathbf{x}_j} &= E\{\mathbf{x}_i\mathbf{x}_j\} \\ &= \int_{-\infty}^{\infty}\int_{-\infty}^{\infty} x_i x_j f_{x_i x_j}(x_i, x_j) dx_i dx_j. \end{aligned} \tag{3.45}$$

The *covariance* of the random variables $\mathbf{x}_i$ and $\mathbf{x}_j$ is given by

$$\begin{aligned} C_{\mathbf{x}_i\mathbf{x}_j} &= E\{(\mathbf{x}_i - m_i)(\mathbf{x}_j - m_j)\} \\ &= \int_{-\infty}^{\infty}\int_{-\infty}^{\infty} (x_i - m_i)(x_j - m_j) f_{x_i x_j}(x_i, x_j) dx_i dx_j \\ &= \int_{-\infty}^{\infty}\int_{-\infty}^{\infty} x_i x_j f_{x_i x_j}(x_i, x_j) dx_i dx_j - m_i m_j \\ &= E\{\mathbf{x}_i\mathbf{x}_j\} - m_i m_j. \end{aligned} \tag{3.46}$$

Clearly, the covariance function is equal to the correlation function, $C_{\mathbf{x}_i\mathbf{x}_j} = R_{\mathbf{x}_i\mathbf{x}_j}$, when $m_i = 0$ and/or $m_j = 0$.

Related to the covariance and correlation is the *correlation coefficient.* The correlation coefficient is defined to be the covariance divided by the standard deviations of random variables $\mathbf{x}_i$ and $\mathbf{x}_j$ as follows,

$$\rho_{x_i x_j} = \frac{C_{\mathbf{x}_i\mathbf{x}_j}}{\sigma_{x_i}\sigma_{x_j}}. \tag{3.47}$$

When two random variables, $\mathbf{x}_i$ and $\mathbf{x}_j$, are uncorrelated, their covariance and correlation coefficient are equal to zero.

Note that if $E\{\mathbf{x}_i\mathbf{x}_j\} = E\{\mathbf{x}_i\}E\{\mathbf{x}_j\} = m_i m_j$, then random variables $\mathbf{x}_i$ and $\mathbf{x}_j$ are uncorrelated. This implies that the covariance $C_{\mathbf{x}_i\mathbf{x}_j} = 0$. If random variables $\mathbf{x}_i$ and $\mathbf{x}_j$ are statistically independent, then they are uncorrelated. If random variables $\mathbf{x}_i$ and $\mathbf{x}_j$ are uncorrelated, then they may not be statistically independent. If random variables $\mathbf{x}_i$ and $\mathbf{x}_j$ have $E\{\mathbf{x}_i\mathbf{x}_j\} = 0$, then they are referred to as orthogonal. When random variables are uncorrelated, they are sometimes referred to as linearly independent.

3.3.5 Gaussian Density Function

A random variable $\mathbf{x}$ is called *Gaussian* or *normal* if its density function is

$$f_x(x) = \frac{1}{\sigma_x\sqrt{2\pi}} exp\left\{-\frac{(x-m_x)^2}{2\sigma_x^2}\right\}, \tag{3.48}$$

where m_x is known as its mean defined by (3.41) and σ_x^2 is its variance defined by (3.42). The symbol σ_x is also known as the standard deviation. The function as defined by (3.48) is a symmetric around m_x. Since $f_x(x)$ is a density function, then

$$\int_{-\infty}^{\infty} f_x(x)dx = 1. \tag{3.49}$$

The corresponding distribution function can be obtained by taking the integral of the Gaussian density function,

$$\begin{aligned} F_x(x) &= \int_{-\infty}^{x} f_x(u)du \\ &= \int_{-\infty}^{x} \frac{1}{\sigma_x\sqrt{2\pi}} exp\left\{-\frac{(u-m_x)^2}{2\sigma_x^2}\right\} du. \end{aligned} \tag{3.50}$$

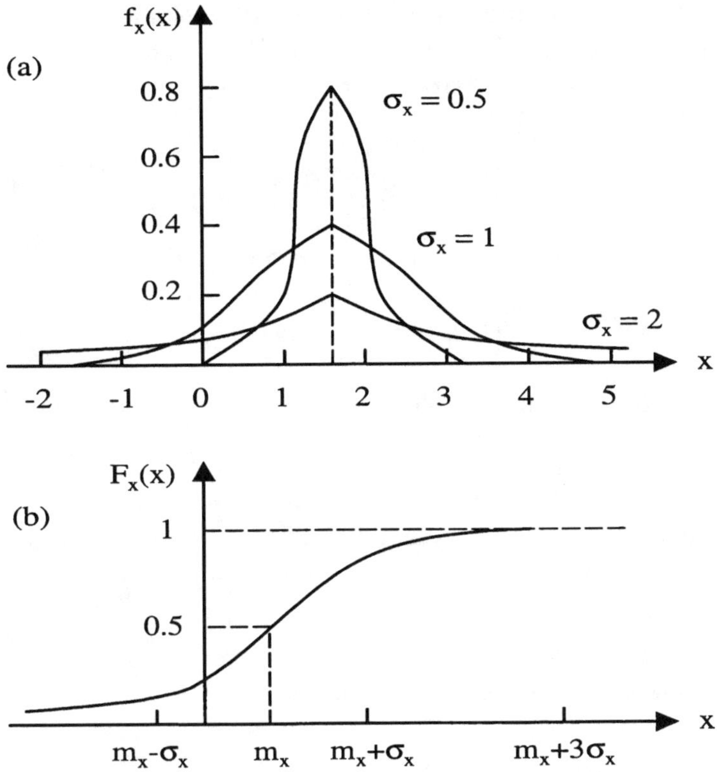

Figure 3.1 The Gaussian random variable $\mathbf{x}$. (a) Gaussian density function $f_x(x)$. (b) Gaussian distribution function $F_x(x)$.

Figure 3.1(a) and (b) show the Gaussian density and distribution functions, respectively. One can readily check that the mode of a Gaussian density occurs at $x = m_x$ and inflection points happen at $m_x - \sigma$ and $m_x + \sigma$.

We often use a shorthand notation

$$\mathbf{x} \sim N(m_x, \sigma_x^2) \tag{3.51}$$

to indicate that a random variable $\mathbf{x}$ has a Gaussian density function. We can also use the notation $\phi_{m_x,\sigma_x^2}(x)$ for the Gaussian density function and the notation $\Phi_{m_x,\sigma_x^2}(x)$ for the Gaussian distribution function.

If the Gaussian random variable has zero-mean and unit-variance, it is called a standard or normalized Gaussian random variable. The normalized Gaussian density function is

$$\phi_{0,1}(x) = \frac{1}{\sqrt{2\pi}} exp\left\{-\frac{x^2}{2}\right\}, \tag{3.52}$$

and the normalized Gaussian distribution function is

$$\Phi_{0,1}(x) = \int_{-\infty}^{x} \phi_{0,1}(u)du. \tag{3.53}$$

Gaussian random variables have many important properties. We list two of these below.

Property 1. If the random variable $\mathbf{x} \sim N(m_x, \sigma_x^2)$, then the probability of the random variable falling in the interval $(a, b]$ is

$$p\{a < \mathbf{x} \leq b\} = \Phi_{0,1}\left(\frac{b - m_x}{\sigma_x}\right) - \Phi_{0,1}\left(\frac{a - m_x}{\sigma_x}\right). \tag{3.54}$$

This property can be proved as follows:

$$\begin{aligned} p\{a < \mathbf{x} \leq b\} &= \int_a^b \frac{1}{\sigma_x\sqrt{2\pi}} exp\left\{-\frac{(u - m_x)^2}{2\sigma_x^2}\right\} du \\ &= \int_{(a-m_x)/\sigma_x}^{(b-m_x)/\sigma_x} \frac{1}{\sqrt{2\pi}} exp\left\{-\frac{z^2}{2}\right\} dz \\ &= \Phi_{0,1}\left(\frac{b - m_x}{\sigma_x}\right) - \Phi_{0,1}\left(\frac{a - m_x}{\sigma_x}\right). \end{aligned} \tag{3.55}$$

Property 2. For the normalized Gaussian distribution function, we have $\Phi_{0,1}(x) = 1 - \Phi_{0,1}(-x)$. This property follows from the Gaussian symmetry.

The normalized Gaussian distribution appears to be a reasonable model of the behavior of certain random phenomena. A number of the processes in applications are Gaussian or approximately Gaussian as shown in central limit theorems, discussed in Section 3.3.10.

3.3.6 Uniform Density Function

A random variable $\mathbf{x}$ is called *uniform* if its density function is a constant in the range from a to b and 0 elsewhere:

$$f_x(x) = \begin{cases} \frac{1}{b-a} & a < x \leq b, \text{ and } b > a \\ 0 & \text{otherwise} \end{cases} \tag{3.56}$$

The corresponding distribution function is obtained by taking the integral of the density function,

$$\begin{aligned} F_x(x) &= \int_{-\infty}^{x} f_x(u)du \\ &= \begin{cases} 0, & -\infty < x < a \\ \frac{x-a}{b-a}, & a < x \leq b \\ 1, & \text{otherwise} \end{cases} \end{aligned} \tag{3.57}$$

Figure 3.2(a) and (b) show the uniform density and distribution functions, respectively.

The mean of the uniform random variable $\mathbf{x}$ is

$$\begin{aligned} m_x &= E\{\mathbf{x}\} \\ &= \int_a^b \frac{x}{b-a} dx \\ &= \frac{a+b}{2}, \end{aligned} \tag{3.58}$$

and the variance is

$$\begin{aligned} \sigma_x^2 &= E\{\mathbf{x}^2\} - E^2\{\mathbf{x}\} \\ &= \int_a^b \frac{x^2}{b-a} dx - \frac{(a+b)^2}{4} \\ &= \frac{b^3 - a^3}{3(b-a)} - \frac{(a+b)^2}{4} \\ &= \frac{(b-a)^2}{12}. \end{aligned} \tag{3.59}$$

The uniform distribution function gets its name from the fact that its density function is constant over the interval range $(a, b]$. Sometimes, the uniform distribution function is also called rectangular because of the shape of its density function.

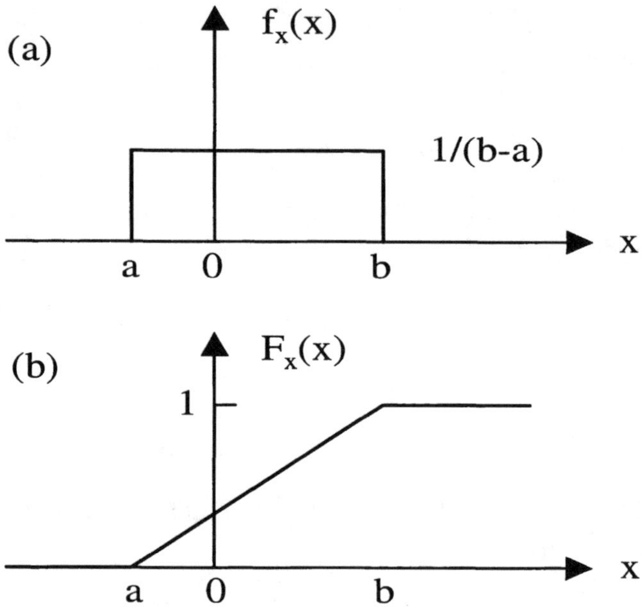

Figure 3.2 The uniform random variable **x**. (a) Uniform density function $f_x(x)$. (b) Uniform distribution function $F_x(x)$.

3.3.7 Chi-Square Distribution

A random variable **x** is called *chi-square density* if its density function is

$$f_x(x) = \frac{1}{\Gamma(k/2)} \left(\frac{1}{2}\right)^{k/2} x^{k/2-1} e^{-\frac{1}{2}x}, \tag{3.60}$$

where k is called the *degrees of freedom* and is a positive integer, and $\Gamma(\cdot)$ is called the *gamma function*, defined by

$$\Gamma(t) = \int_0^\infty x^{t-1} e^{-x} dx, \quad t > 0. \tag{3.61}$$

The gamma function has properties as follows:

$$\Gamma(t+1) = t\Gamma(t), \tag{3.62}$$

and if $t = n$ and n is an integer, then

$$\Gamma(n+1) = n!, \tag{3.63}$$

$$\Gamma\left(n+\frac{1}{2}\right) = \frac{1\cdot 3\cdot 5\cdots(2n-1)}{2^n}\sqrt{\pi}, \tag{3.64}$$

and, in particular,

$$\Gamma\left(\frac{1}{2}\right) = 2\Gamma\left(\frac{3}{2}\right) = \sqrt{\pi}. \tag{3.65}$$

It can be shown that the mean of chi-square random variable $\mathbf{x}$ is

$$m_x = E\{\mathbf{x}\} = k, \tag{3.66}$$

and the variance is

$$\sigma_x^2 = 2k. \tag{3.67}$$

The corresponding chi-square distribution function is obtained by taking the integral of the density function,

$$F_x(x) = \int_0^x \frac{1}{\Gamma(k/2)}\left(\frac{1}{2}\right)^{k/2} u^{k/2-1}e^{-\frac{1}{2}u}du, \tag{3.68}$$

where k is equal to 1, 2, ..., 30. For larger values of k, an approximation of normal distribution with zero mean and unit variance can be used and is quite accurate. Thus, x_α, the αth quantity point of the distribution, may be calculated by

$$x_\alpha = \frac{1}{2}(z_\alpha + \sqrt{2k-1})^2, \tag{3.69}$$

where z_α is the αth quantity point of the standard normal distribution. Equation (3.69) was shown in error by less than 1 percent by Mood, Graybill, and Boes [4].

3.3.8 $\mathcal{F}$ Distribution

Let U be a chi-square random variable with m degrees of freedom and V be a chi-square random variable with n degrees of freedom. In addition, let U and V be statistically independent. Then, the new random variable,

$$X = \frac{U/m}{V/n}, \tag{3.70}$$

is distributed as an $\mathcal{F}$ distribution with m and n degrees of freedom. Equation (3.70) is sometimes referred to as the variance ratio. The density function of the random variable X is given by

$$f_X(x) = \frac{\Gamma[(m+n)/2]}{\Gamma(m/2)\Gamma(n/2)}\left(\frac{m}{n}\right)^{m/2}\left\{\frac{x^{(m-2)/2}}{[1+(m/n)x]^{(m+n)/2}}\right\}. \quad (3.71)$$

The mean of the $\mathcal{F}$ density function of the random variable X is obtained by

$$\begin{aligned} E\{X\} &= E\left\{\frac{U/m}{V/n}\right\} \\ &= \left(\frac{n}{m}\right)E\{U\}E\left\{\frac{1}{V}\right\}. \end{aligned} \quad (3.72)$$

By using (3.66), we obtain $E\{U\} = m$, and

$$\begin{aligned} E\left\{\frac{1}{V}\right\} &= \int_0^\infty \frac{1}{v}\frac{1}{\Gamma(n/2)}\left(\frac{1}{2}\right)^{n/2} v^{n/2-1}e^{-\frac{1}{2}v}dv \\ &= \frac{1}{\Gamma(n/2)}\left(\frac{1}{2}\right)^{n/2}\int_0^\infty v^{(n-4)/2}e^{-\frac{1}{2}v}dv \\ &= \frac{\Gamma[(n-2)/2]}{\Gamma(n/2)}\left(\frac{1}{2}\right)^{n/2}\left(\frac{1}{2}\right)^{-(n-2)/2} \\ &= \frac{1}{n-2}. \end{aligned} \quad (3.73)$$

Therefore, the mean is

$$\begin{aligned} E\{X\} &= \left(\frac{n}{m}\right)E\{U\}E\left\{\frac{1}{V}\right\} \\ &= \frac{n}{m}\frac{m}{n-2} \\ &= \frac{n}{n-2}. \end{aligned} \quad (3.74)$$

In a similar way, the variance is derived by

$$\sigma_X^2 = \frac{2n^2(m+n-2)}{m(n-2)^2(n-4)}, \quad n > 4. \quad (3.75)$$

The corresponding $\mathcal{F}$ distribution function is obtained by taking the integral of the density function,

$$F_X(x) = \int_0^x \frac{\Gamma[(m+n)/2]}{\Gamma(m/2)\Gamma(n/2)} \left(\frac{m}{n}\right)^{m/2} \left\{ \frac{u^{(m-2)/2}}{[1+(m/n)u]^{(m+n)/2}} \right\} du. \tag{3.76}$$

The direct calculation of (3.76) may not be necessary. One should always be able to obtain the values of $\mathcal{F}$ distribution by using the table in [4].

Note that the mean of the $\mathcal{F}$ distribution depends only on the degrees of freedom of the denominator. This is because the density of the $\mathcal{F}$ distribution is not symmetrical in the degrees of freedom m and n. As the $\mathcal{F}$ distribution of the ratio of two independent chi-square random variables U and V are divided by their respective degrees of freedom m and n, we shall later find this to be of practical interest when we introduce the statistical signal classification.

3.3.9 Sums of Random Variables

In this section, we discuss the summations of n random variables, $\mathbf{x}_1, \mathbf{x}_2, ..., \mathbf{x}_n$. Some of the results of these summations are given with a proof.

Theorem 1. Given n random variables, $\mathbf{x}_1, \mathbf{x}_2, ..., \mathbf{x}_n$, the mean of the sum is the sum of the means:

$$E\left\{\sum_{i=1}^{n} \mathbf{x}_i\right\} = \sum_{i=1}^{n} E\{\mathbf{x}_i\}. \tag{3.77}$$

Also

$$var\left\{\sum_{i=1}^{n} \mathbf{x}_i\right\} = \sum_{i=1}^{n} var\{\mathbf{x}_i\} + 2\sum_{i}\sum_{j(i<j)} cov\{\mathbf{x}_i, \mathbf{x}_j\}, \tag{3.78}$$

where the notation *var* and *cov* denote variance and covariance of the appropriate random variables.

Proof: For (3.77), we note that

$$E\left\{\sum_{i=1}^{n} \mathbf{x}_i\right\} = E\{\mathbf{x}_1 + \mathbf{x}_2 + \cdots + \mathbf{x}_n\}$$

$$= E\{\mathbf{x}_1\} + E\{\mathbf{x}_2\} + \cdots + E\{\mathbf{x}_n\}$$
$$= \sum_{i=1}^{n} E\{\mathbf{x}_i\}. \qquad (3.79)$$

Equation (3.78) can be proven as follows:

$$\begin{aligned}
var\left\{\sum_{i=1}^{n}\mathbf{x}_i\right\} &= E\left\{\left[\sum_{i=1}^{n}\mathbf{x}_i - E\left\{\sum_{i=1}^{n}\mathbf{x}_i\right\}\right]^2\right\} \\
&= E\left\{\left[\sum_{i=1}^{n}(\mathbf{x}_i - E\{\mathbf{x}_i\})\right]^2\right\} \\
&= E\left\{\sum_{i=1}^{n}\sum_{j=1}^{n}(\mathbf{x}_i - E\{\mathbf{x}_i\})(\mathbf{x}_j - E\{\mathbf{x}_j\})\right\} \\
&= \sum_{i=1}^{n}\sum_{j=1}^{n}E\left\{(\mathbf{x}_i - E\{\mathbf{x}_i\})(\mathbf{x}_j - E\{\mathbf{x}_j\})\right\} \\
&= \sum_{i=1}^{n} var\{\mathbf{x}_i\} + 2\sum_{i}\sum_{j(i<j)} cov\{\mathbf{x}_i,\mathbf{x}_j\}. \qquad (3.80)
\end{aligned}$$

From theorem 1, it follows that

$$E\{\mathbf{x}_1 \pm \mathbf{x}_2\} = E\{\mathbf{x}_1\} \pm E\{\mathbf{x}_2\}, \qquad (3.81)$$

and

$$var\{\mathbf{x}_1 \pm \mathbf{x}_2\} = var\{\mathbf{x}_1\} + var\{\mathbf{x}_2\} \pm 2cov\{\mathbf{x}_1,\mathbf{x}_2\}. \qquad (3.82)$$

These give the mean and variance of the sum or the difference of two random variables.

Theorem 2. Consider n random variables, say, $\mathbf{x}_1, \mathbf{x}_2, ..., \mathbf{x}_n$. If $\mathbf{x}_1, \mathbf{x}_2, ..., \mathbf{x}_n$ are uncorrelated random variables, then

$$var\left\{\sum_{i=1}^{n}\mathbf{x}_i\right\} = \sum_{i=1}^{n} var\{\mathbf{x}_i\}. \qquad (3.83)$$

Proof: This theory can be easily proved by using theorem 1. Since $\mathbf{x}_1, \mathbf{x}_2, ..., \mathbf{x}_n$ are uncorrelated random variables, then $cov\{\mathbf{x}_i, \mathbf{x}_j\} =$

0, for all i, j, $i \neq j$. As a result, the second term in (3.78) equals zero. Therefore,

$$var\left\{\sum_{i=1}^{n} \mathbf{x}_i\right\} = \sum_{i=1}^{n} var\{\mathbf{x}_i\}. \tag{3.84}$$

Theorem 3. Given n random variables, $\mathbf{x}_1, \mathbf{x}_2, ..., \mathbf{x}_n$, and n constants, $c_1, c_2, ..., c_n$, then

$$\begin{aligned} var\left\{\sum_{i=1}^{n} c_i\mathbf{x}_i\right\} &= \sum_{i=1}^{n}\sum_{j=1}^{n} c_ic_jcov\{\mathbf{x}_i, \mathbf{x}_j\} \\ &= \sum_{i=1}^{n} c_i^2 var\{\mathbf{x}_i\} \\ &\quad + \sum_{i}\sum_{j(j\neq i)} c_ic_jcov\{\mathbf{x}_i, \mathbf{x}_j\}. \end{aligned} \tag{3.85}$$

This theorem states a result that is somewhat related to theorem 1 as well as its proof. Hence, the proof of this theorem is left as an exercise.

Theorem 4. Consider n random variables, $\mathbf{x}_1, \mathbf{x}_2, ..., \mathbf{x}_n$. If $\mathbf{x}_1, \mathbf{x}_2, ..., \mathbf{x}_n$ are independent and identically distributed random variables with mean m_x and variance σ_x^2, and if

$$\bar{\mathbf{x}}_n = \frac{1}{n}\sum_{i=1}^{n} \mathbf{x}_i, \tag{3.86}$$

then

$$E\{\bar{\mathbf{x}}_n\} = m_x, \tag{3.87}$$

and

$$var\{\bar{\mathbf{x}}_n\} = \frac{\sigma_x^2}{n}. \tag{3.88}$$

Proof: Equation (3.87) can be proven as follows:

$$E\{\bar{\mathbf{x}}_n\} = E\left\{\frac{1}{n}\sum_{i=1}^{n} \mathbf{x}_i\right\}$$

$$= \frac{1}{n}\sum_{i=1}^{n} E\{\mathbf{x}_i\}$$
$$= \frac{1}{n}\sum_{i=1}^{n} m_x$$
$$= m_x. \qquad (3.89)$$

In order to prove (3.88), we derive

$$var\{\bar{\mathbf{x}}_n\} = var\left\{\frac{1}{n}\sum_{i=1}^{n}\mathbf{x}_i\right\}$$
$$= \left(\frac{1}{n}\right)^2 var\left\{\sum_{i=1}^{n}\mathbf{x}_i\right\}. \qquad (3.90)$$

Since $\mathbf{x}_1, \mathbf{x}_2, ..., \mathbf{x}_n$ are independent, then $\mathbf{x}_1, \mathbf{x}_2, ..., \mathbf{x}_n$ are uncorrelated random variables. Therefore, by using theorem 2, (3.90) can be written as

$$var\{\bar{\mathbf{x}}_n\} = \left(\frac{1}{n}\right)^2 \sum_{i=1}^{n} var\{\mathbf{x}_i\}$$
$$= \left(\frac{1}{n}\right)^2 \sum_{i=1}^{n} \sigma_x^2$$
$$= \frac{\sigma_x^2}{n}. \qquad (3.91)$$

3.3.10 Central Limit Theorem

The most widely applied central limit theorem states that if $\mathbf{x}_1, ..., \mathbf{x}_n$ are independent and identically distributed random variables with mean m_x and variance σ_x^2, then, for each $\mathbf{z}_n$,

$$\mathbf{z}_n = \frac{\bar{\mathbf{x}}_n - E\{\bar{\mathbf{x}}_n\}}{\sqrt{var\{\bar{\mathbf{x}}_n\}}}$$
$$= \frac{\bar{\mathbf{x}}_n - m_x}{\sigma_x/\sqrt{n}}, \qquad (3.92)$$

the distribution function $F_{\mathbf{z}_n}(z)$ converges to the normalized Gaussian distribution function $\Phi_{0,1}(z)$, when n approaches ∞.

This central limit theorem tells us that the distribution function $F_{\mathbf{z}_n}(z)$ approximates a normalized Gaussian distribution function.

In other words, this theorem tells us that $\bar{\mathbf{x}}_n$ is approximately, or asymptotically, distributed as a normalized distribution with mean m_x and variance σ_x^2/n.

Note that this theorem assumes nothing about the form of the original density function except that it has finite variance. Whatever the distribution function with finite variance, the sample mean $\bar{\mathbf{x}}_n$ will have, approximately, the Gaussian distribution, given large enough samples. In practice, most density functions encountered have finite variance.

The nature of this central limit theorem gives a specific error bound if the random variables $\mathbf{x}_1, ..., \mathbf{x}_n$ are independent and identically distributed with n = 30 for many applications [3]. However, based on practical experience, we suggest using the value $n = 60$ for many applications. In some cases, we may need a large value of n. In general, this central limit theorem models densities extremely well, with a few random variables of the mean.

In this section, we did not attempt to prove this central limit theorem, because this theorem requires rather advanced statistical techniques. If the reader is interested in this topic, please refer to [3-5].

3.4 Stochastic Processes

In this section, we discuss stochastic processes for the characterization and analysis of discrete-time random processes. The extension of these concepts to such processes is straightforward, once an understanding of random variables as presented in the above sections is obtained. A discrete-time random process can be simply considered as an indexed sequence of random variables.

3.4.1 Discrete-Time Definitions

A discrete-time random process is a sequence of random variables, denoted by $\mathbf{x}[n]$, where the brackets indicate that n is an integer. Since a discrete-time random process is an indexed sequence of random variables, the calculation of the mean of each of these random variables is obtained by

$$m_x[n] = E\{\mathbf{x}[n]\}. \tag{3.93}$$

Similarly, the variance of each random variable in the sequence $\mathbf{x}[n]$ is given by

$$\begin{aligned} \sigma_x^2[n] &= E\{(\mathbf{x}[n] - m_x[n])(\mathbf{x}[n] - m_x[n])^*\} \\ &= E\{\mathbf{x}^2[n]\} - m_x^2[n]. \end{aligned} \tag{3.94}$$

Two important functions in the study of discrete-time random processes are the autocorrelation

$$R_x[k,l] = E\{\mathbf{x}[k]\mathbf{x}^*[l]\} \tag{3.95}$$

and the autocovariance

$$C_x[k,l] = R[k,l] - m_x[k]m_x^*[l]. \tag{3.96}$$

Therefore, the autocovariance and autocorrelation are equal, $C_x[k,l] = R_x[k,l]$, when $m_x[k] = 0$ and/or $m_x[l] = 0$. Also note that if $k = l$, then the autocovariance function in (3.96) is equal to the variance in (3.94), as follows:

$$C_x[k,k] = \sigma_x^2[k]. \tag{3.97}$$

As we can see, the mean in (3.93) defines the average value of the random process at index n, while the variance in (3.94) represents the average squared deviation of the random process away from the mean at index n. In addition, the autocorrelation function in (3.95) and autocovariance function in (3.96) provide the degree of linear dependence between two random variables, $\mathbf{x}[k]$ and $\mathbf{x}[l]$.

Example 3.2 The Harmonic Random Process

Consider the complex harmonic random process

$$\mathbf{x}[n] = Ae^{j(n\omega+\phi)}, \tag{3.98}$$

where A and ω are fixed constants and ϕ is a random variable that is uniform over the interval $(-\pi, \pi]$.

The density function of ϕ is

$$f(\phi) = \begin{cases} \frac{1}{2\pi}, & -\pi \le \phi \le \pi \\ 0, & \text{otherwise.} \end{cases} \tag{3.99}$$

So, the mean of the process is

$$\begin{aligned} m_x[n] &= E\{\mathbf{x}[n]\} \\ &= E\{Ae^{j(n\omega+\phi)}\} \\ &= \int_{-\infty}^{\infty} Ae^{j(n\omega+\phi)} f(\phi) d\phi \\ &= \int_{-\pi}^{\pi} \frac{A}{2\pi} e^{j(n\omega+\phi)} d\phi \\ &= 0. \end{aligned} \tag{3.100}$$

Thus, the random variable $\mathbf{x}[n]$ is a zero mean process for all n. The autocorrelation of the process is

$$\begin{aligned} R_x[k,l] &= E\{\mathbf{x}[k]\mathbf{x}^*[l]\} \\ &= E\{Ae^{j(k\omega+\phi)} A^* e^{-j(l\omega+\phi)}\} \\ &= |A|^2 E\{e^{j(k-l)\omega}\}. \end{aligned} \tag{3.101}$$

Note that the mean is zero and the autocorrelation depends only on the difference between k and l. This means that the mean and autocorrelation do not change even if the random process is shifted in time index. In fact, the complex harmonic process is a *wide-sense stationary* process, which is discussed in the following section.

3.4.2 Stationary Processes

A stochastic process $\mathbf{x}[n]$ is called *stationary* if its statistical properties are invariant to a shift of the origin. This indicates that the processes $\mathbf{x}[n]$ and $\mathbf{x}[n+k]$ have the same statistics for any k.

Two random processes $\mathbf{x}[n]$ and $\mathbf{y}[n]$ are called *jointly stationary process* if the joint statistics of $\mathbf{x}[n]$ and $\mathbf{y}[n]$ are the same as the statistics of $\mathbf{x}[n+k]$ and $\mathbf{y}[n+k]$ for any k.

In general, if the nth joint density of random variables is

$$f_{\mathbf{x}_1[n],\ldots,\mathbf{x}_n[n]}(l_1,\ldots,l_n) = f_{\mathbf{x}_1[n+k],\ldots,\mathbf{x}_n[n+k]}(l_1,\ldots,l_n) \tag{3.102}$$

for all k and all n, then the stochastic process is said to be *stationary in the strict sense* (SSS). If (3.102) holds for values of n up to and including N, the process is Nth-order stationary.

A stochastic process $\mathbf{x}[n]$ is called wide-sense stationary (WSS) if the following conditions are satisfied:

1. Its mean is a constant $E\{\mathbf{x}[n]\} = m_x$.
2. Its autocorrelation function depends only on the difference, $\tau = k - l$, so that

$$R_x[\tau] = R_x[k, l] = E\{\mathbf{x}[k]\mathbf{x}^*[l]\}, \tag{3.103}$$

where τ is known as the *lag*.

In general, if a stochastic process is stationary in the strict sense, then it is also wide-sense stationary. The converse is not true. WSS does not even imply first-order stationary. However, in the case of a Gaussian random process, wide-sense stationary is equal to stationary in the strict sense. This is an important property of Gaussian random processes (see [3] for proof).

3.4.3 Estimated Functions

Ensemble averages such as means, variances, autocorrelations, and autocovariances as well as individual density functions are not generally known in practice. As such, estimating these averages from a realization of a discrete-time random process becomes an important problem. Therefore, in this section we present estimation methods for the mean, variance, and autocorrelation of a discrete-time random process.

Consider the problem of estimating the ensemble averages of a discrete-time random process $\mathbf{x}[n]$. If a large number of samples N from the process are available, then an estimated mean is obtained by

$$\hat{m}_x = \frac{1}{N} \sum_{n=0}^{N-1} \mathbf{x}[n], \tag{3.104}$$

and estimated variance is given by

$$\hat{\sigma}_x^2 = \frac{1}{N} \sum_{n=0}^{N-1} |\mathbf{x}[n] - \hat{m}_x|^2. \tag{3.105}$$

The autocorrelation can be estimated as follows:

$$\hat{R}_x[k] = \frac{1}{N} \sum_{n=0}^{N-1} \mathbf{x}[n]\mathbf{x}^*[n-k]. \tag{3.106}$$

Equations (3.104) and (3.105) are also referred to as the *sample mean* and *sample variance*. For these estimates to be accurate, the processes must be stationary or, ideally, *ergodic*, which is a stronger condition than stationary. Discussion of ergodicity is out of the scope of this text. The interested reader can refer to Papoulis [3] for a precise mathematical definition.

3.4.4 Power Spectrum

The analysis tool of the discrete-time Fourier transform is important for deterministic discrete-time signals, as presented in Chapter 2. It also plays an important role in the description and analysis of discrete-time random processes. However, we may not calculate the discrete-time Fourier transform for random processes directly, because a random process is an ensemble of discrete-time random signals. Therefore, in this section we discuss the power spectrum or power spectral density for a discrete-time random process.

Given a WSS process $\mathbf{x}[n]$, we can determine the autocorrelation sequence $R_x[k]$ in the discrete-time domain. Since $R_x[k]$ is a deterministic sequence, we can compute its discrete-time Fourier transform as follows:

$$P_x(e^{j\omega}) = \sum_{k=-\infty}^{\infty} R_x[k]e^{-jk\omega}. \tag{3.107}$$

Equation (3.107) is referred to as the *power spectrum* or *power spectral density* of the discrete-time random process. In the reverse direction, given the power spectrum in (3.107), which is periodic with period 2π, the autocorrelation sequence $R_x[k]$ can be computed by taking the inverse discrete-time Fourier transform of $P_x(e^{j\omega})$,

$$R_x[k] = \frac{1}{2\pi}\int_{-\pi}^{\pi} P_x(e^{j\omega})e^{jk\omega}d\omega. \tag{3.108}$$

Hence, the power spectrum contributes a frequency domain description and analysis of the second-order moment of the random process.

It is also convenient to represent the power spectrum by using the z-transform instead of the discrete-time Fourier transform as follows:

$$P_x(z) = \sum_{k=-\infty}^{\infty} R_x[k]z^{-k}. \tag{3.109}$$

Equation (3.109) is also referred to as the power spectrum of the random variable $\mathbf{x}[n]$.

If the discrete-time random variable $\mathbf{x}[n]$ is a real process, then $R_x[-k] = R_x[k]$. This means that $R_x[k]$ is real and even. Thus, (3.107) yields:

$$P_x(e^{j\omega}) = R_x[0] + 2\sum_{k=0}^{\infty} R_x[k]cos(k\omega). \tag{3.110}$$

Equation (3.110) shows that the power spectrum of a real process is real and even, because $cos(k\omega)$ is a real and even function.

Example 3.3 Calculation of the Power Spectrum

Consider the autocorrelation function $R_x[k] = a^{|k|}$, $a < 1$ and real, for the discrete-time random process $\mathbf{x}[n]$. The power spectrum in (3.109) is obtained by

$$\begin{aligned} P_x(z) &= \sum_{k=-\infty}^{\infty} R_x[k]z^{-k} \\ &= \sum_{k=-\infty}^{-1} a^{-k}z^{-k} + \sum_{k=0}^{\infty} a^k z^{-k} \\ &= \frac{az}{1-az} + \frac{z}{z-a} \\ &= \frac{a^{-1}-a}{(a^{-1}+a)-(z^{-1}+z)}. \end{aligned} \tag{3.111}$$

Thus, substituting $z = e^{j\omega}$ into (3.111), the power spectrum in (3.107) is given by

$$P_x(e^{j\omega}) = \frac{1-a^2}{1-2acos\omega + a^2}. \tag{3.112}$$

As can be seen, the power spectrum is real, even, and positive.

3.4.5 Linear Filtering of Stochastic Processes

In this section, we derive the relationship between the second-order statistics (mean, autocorrelation, and power spectrum) of a random process before and after linear time-invariant filtering.

Let $\mathbf{x}[n]$ be a WSS random process, and let $h[n]$ be the impulse response of a stable linear time-invariant system to which $\mathbf{x}[n]$ is input. The output $\mathbf{y}[n]$ is a random process whose values are the discrete-time convolution of input $\mathbf{x}[n]$ with impulse response $h[n]$,

$$\begin{aligned} \mathbf{y}[n] &= \mathbf{x}[n] * h[n] \\ &= \sum_{k=-\infty}^{\infty} h[k]\mathbf{x}[n-k]. \end{aligned} \tag{3.113}$$

The mean of the output random process can be computed by taking the expected value of (3.113),

$$\begin{aligned} m_y[n] &= E\{\mathbf{y}[n]\} \\ &= \sum_{k=-\infty}^{\infty} h[k]E\{\mathbf{x}[n-k]\} \\ &= m_x[n] \sum_{k=-\infty}^{\infty} h[k]. \end{aligned} \tag{3.114}$$

Equation (3.114) indicates that the mean $m_y[n]$ of the output random process $\mathbf{y}[n]$ is also a constant and is related to $m_x[n]$ by a scale factor equal to the sum of the impulse response values.

An equivalent form of (3.114) in terms of frequency response is given by

$$m_y[n] = m_x[n]H(e^{j0}). \tag{3.115}$$

This means that the mean $m_y[n]$ is equal to the mean $m_x[n]$ scaled by the value of the frequency response of the filter at $\omega = 0$.

The autocorrelation function of the output random process is obtained by

$$\begin{aligned} R_y[n, n+l] &= E\{\mathbf{y}[n]\mathbf{y}^*[n+l]\} \\ &= \sum_{k=-\infty}^{\infty} h[k] \sum_{r=-\infty}^{\infty} h^*[r] \\ &\quad \cdot E\{\mathbf{x}[n-k]\mathbf{x}^*[n+l-r]\}. \end{aligned} \tag{3.116}$$

Since the input random process $\mathbf{x}[n]$ is WSS, the term of $E\{\mathbf{x}[n-k]\mathbf{x}^*[n+l-r]\}$ depends only on the index $l+k-r$. Thus, (3.116) can be rewritten as follows:

$$\begin{aligned} R_y[n, n+l] &= \sum_{k=-\infty}^{\infty} h[k] \sum_{r=-\infty}^{\infty} h^*[r] R_x[l+k-r] \\ &= R_y[l]. \end{aligned} \tag{3.117}$$

Note that the autocorrelation function $R_y[l]$ in (3.117) depends only on the index difference of l. For the variance of the output random process, $\mathbf{y}[n]$, we set $l = 0$ in (3.117),

$$\begin{aligned} \sigma_y^2[n] &= R_y[0] \\ &= \sum_{k=-\infty}^{\infty} \sum_{r=-\infty}^{\infty} h[k]h^*[r]R_x[k-r]. \end{aligned} \tag{3.118}$$

Hence, it follows from (3.115), (3.117), and (3.118) that if input random variable $\mathbf{x}[n]$ is WSS, then the output process $\mathbf{y}[n]$ will also be WSS with the condition $\sigma_y^2[n] < \infty$ for a stable linear time-invariant system.

For the power spectrum, we take the discrete-time Fourier transform on both sides of (3.117),

$$\begin{aligned} P_y(e^{j\omega}) &= H(e^{j\omega})H^*(e^{j\omega})P_x(e^{j\omega}) \\ &= |H(e^{j\omega})|^2 P_x(e^{j\omega}). \end{aligned} \tag{3.119}$$

Note that if the impulse response $h[n]$ is real, then we obtain $H^*(e^{j\omega}) = H(e^{-j\omega})$. In this case, the power spectrum of the output random process in terms of the z-transform is

$$P_y(z) = H(z)H(1/z)P_x(z). \tag{3.120}$$

Equation (3.120) provides a convenient way for analysis of the spectral factorization of the power spectrum $P_y(z)$.

One interesting point to note is that the total average power in the output can be computed by

$$\begin{aligned} E\{\mathbf{y}^2[n]\} &= R_y[0] \\ &= \frac{1}{2\pi}\int_{-\pi}^{\pi} P_y(e^{j\omega})d\omega \\ &= \frac{1}{2\pi}\int_{-\pi}^{\pi} |H(e^{j\omega})|^2 P_x(e^{j\omega})d\omega. \end{aligned} \tag{3.121}$$

This indicates that the area under the curve $|H(e^{j\omega})|^2 P_y(e^{j\omega})$ for $-\pi \leq \omega \leq \pi$ represents the mean-square value of the input random process.

The cross-correlation function between the input random process and output random process of a linear time-invariant system is also an important subject, with applications to estimating the impulse response or frequency response of a linear system. This cross-correlation function is given by

$$\begin{aligned} R_{xy}[l] &= E\{\mathbf{x}[n]\mathbf{y}^*[n+l]\} \\ &= \sum_{k=-\infty}^{\infty} h[k]E\{\mathbf{x}[n]\mathbf{x}^*[n+l-k]\} \\ &= \sum_{k=-\infty}^{\infty} h[k]R_x[l-k]. \end{aligned} \tag{3.122}$$

Equation (3.122) indicates that the cross-correlation function $R_{xy}[n]$ is the convolution of the impulse response $h[n]$ with the autocorrelation function $R_x[n]$ of the input. Note that if the input is white noise such that $R_x[l] = \sigma_x^2\delta[l]$, then (3.122) becomes

$$R_{xy}[l] = \sigma_x^2 h[l]. \tag{3.123}$$

This means that the cross-correlation function is proportional to the impulse response of the linear time-invariant system when the input random process is zero-mean white noise.

The power spectrum of the cross-correlation function $R_{xy}[l]$ is obtained by taking the discrete-time Fourier transform on both sides of (3.122),

$$P_{xy}(e^{j\omega}) = H(e^{j\omega})P_x(e^{j\omega}). \tag{3.124}$$

In a similar way, for a zero-mean white noise, (3.124) can be rewritten as follows:

$$P_{xy}(e^{j\omega}) = \sigma_x^2 H(e^{j\omega}). \tag{3.125}$$

As we can see, (3.125) is proportional to the frequency response of the linear time-invariant system. Therefore, (3.123) and (3.124) can be used for estimating the impulse response $h[n]$ and frequency response

$H(e^{j\omega})$ of a system when the input random process is a zero-mean white noise.

Example 3.4 The Output Power Spectrum for a Linear System Function

Consider a first-order difference equation

$$\mathbf{y}[n] = a\mathbf{x}[n] - b\mathbf{x}[n-1], \tag{3.126}$$

where $\mathbf{x}[n]$ is an input discrete-time random process and $\mathbf{y}[n]$ is the output process of the linear system. Assume that the input process $\mathbf{x}[n]$ is also white noise with $P_x(e^{j\omega}) = \sigma_x^2$.

The system function, or z-transform of the first-order difference equation, is

$$\begin{aligned} H(z) &= \frac{\mathbf{Y}(z)}{\mathbf{X}(z)} \\ &= a - bz^{-1}. \end{aligned} \tag{3.127}$$

Using (3.120), we obtain the output power spectrum in terms of the z-transform as follows:

$$\begin{aligned} P_y(z) &= (a - bz^{-1})(a - bz)P_x(z) \\ &= (a^2 + b^2 - abz - abz^{-1})\sigma_x^2. \end{aligned} \tag{3.128}$$

Thus, the output power spectrum can also be expressed in the frequency domain as follows:

$$\begin{aligned} P_y(e^{j\omega}) &= \sigma_x^2[a^2 + b^2 - ab(e^{j\omega} - e^{-j\omega})] \\ &= \sigma_x^2(a^2 + b^2 - 2abcos\omega). \end{aligned} \tag{3.129}$$

It follows that the output power spectrum in (3.129) is a real, even, and positive function, since the $cos\omega$ is real and even, as well as

$$\begin{aligned} P_y(e^{j\omega}) &= \sigma_x^2(a^2 + b^2 - 2abcos\omega) \\ &\geq \sigma_x^2(a - b)^2 \\ &\geq 0. \end{aligned} \tag{3.130}$$

3.4.6 Mean Square Estimation

Consider $\mathbf{x}_1[n], ..., \mathbf{x}_n[n]$ as discrete-time random variables. Let $T = t(\mathbf{x}_1[n], ..., \mathbf{x}_n[n])$ be an estimator of $\tau(\eta)$. Then,

$$E_\eta\{(T - \tau(\eta))^2\}, \tag{3.131}$$

is defined to be the *mean square error* (MSE) of the estimator $T = t(\mathbf{x}_1[n], ..., \mathbf{x}_n[n])$.

Note that the mean square error of an estimator in the above definition is a measure of goodness, with small values better than large values. We would like to think that $E_\eta\{(T-\tau(\eta))^2\}$ is a measure of the spread of function T value about $\tau(\eta)$ in such a way that the variance of a discrete-time random variable is a measure of its spread about its mean. Therefore, we would prefer the estimator with the smallest mean square error, or *minimum mean square error* (MMSE). However, in general, the mean square error of an estimator depends on η.

An estimator $T = t(\mathbf{x}_1[n], ..., \mathbf{x}_n[n])$ is called an *unbiased* estimator if and only if

$$\begin{aligned} E_\eta\{T\} &= E_\eta\{t(\mathbf{x}_1[n], ..., \mathbf{x}_n[n])\} \\ &= \tau(\eta). \end{aligned} \tag{3.132}$$

This means that an estimator is unbiased if its mean is equal to the parameter being estimated. Therefore, the mean square error in (3.131) can be rewritten as follows:

$$\begin{aligned} E_\eta\{(T - \tau(\eta))^2\} &= E_\eta\{[(T - E_\eta\{T\}) - (\tau(\eta) - E_\eta\{T\})]^2\} \\ &= E_\eta\{(T - E_\eta\{T\})^2\} \\ &\quad - 2(\tau(\eta) - E_\eta\{T\})E_\eta\{T - E_\eta\{T\}\} \\ &\quad + E_\eta\{(\tau(\eta) - E_\eta\{T\})^2\} \\ &= var\{T\} + [\tau(\eta) - E_\eta\{T\}]^2. \end{aligned} \tag{3.133}$$

If T is unbiased, then, (3.133) yields

$$E_\eta\{(T - \tau(\eta))^2\} = var\{T\}. \tag{3.134}$$

Note that the second-term $[\tau(\eta) - E_\eta\{T\}]$ in (3.133) is called the *bias* of the MSE estimator T and can be either positive, negative, or zero. Moreover, (3.133) indicates that the MSE is the sum of two positive quantities. It also provides the relationship among the MSE, variance, and bias of an estimator.

Example 3.5 Linear MSE Estimation

Consider the linear mean square estimator $\hat{\mathbf{y}}[n]$ of $\mathbf{y}[n]$ in terms of the discrete-time random variables, $\mathbf{x}_1[n], ..., \mathbf{x}_n[n]$. The estimator $\hat{\mathbf{y}}[n]$ is the sum as follows,

$$\hat{\mathbf{y}}[n] = \sum_{i=1}^{n} c_i \mathbf{x}_i[n], \tag{3.135}$$

where c_i is constant. Thus, the mean square error (MSE) is given by

$$\begin{aligned} MSE &= E\{(\hat{\mathbf{y}}[n] - \mathbf{y}[n])^2\} \\ &= E\left\{\left[\sum_{i=1}^{n} c_i \mathbf{x}_i[n] - \mathbf{y}[n]\right]^2\right\} \end{aligned} \tag{3.136}$$

Using (3.133), (3.136) can be written as

$$MSE = var\left\{\sum_{i=1}^{n} c_i \mathbf{x}_i[n]\right\} + \left[\mathbf{y}[n] - E_{\mathbf{y}[n]}\left\{\sum_{i=1}^{n} c_i \mathbf{x}_i[n]\right\}\right]^2. \tag{3.137}$$

Thus, the minimum mean square error can be achieved when the estimator $\hat{\mathbf{y}}[n]$ is an unbiased estimator of $\mathbf{y}[n]$. Then, (3.137) yields

$$MMSE = var\left\{\sum_{i=1}^{n} c_i \mathbf{x}_i[n]\right\}. \tag{3.138}$$

Based on (3.85), (3.138) yields

$$MMSE = \sum_{i=1}^{n} c_i^2 var\{\mathbf{x}_i[n]\} + \sum_{i} \sum_{j(j \neq i)} c_i c_j cov\{\mathbf{x}_i[n], \mathbf{x}_j[n]\}. \tag{3.139}$$

3.5 Summary

In this chapter, we presented an introduction to the fundamental theory of discrete-time random processes including probabilities and random variables, distribution and density functions, and stochastic processes. Beginning with a short review of probability consisting of intuitive, axiomatic, and conditional methods, we then introduced total probability and Bayes' theorems, the independence concept, and the definition of a random variable. These discussions of basic concepts and theories provide a necessary step for understanding random processes for discrete-time signal processing.

Another important topic introduced in this chapter is distribution and density functions and the role they play in the field of probability, random variables, and stochastic processes. First, we discussed the definitions of distribution and density functions, as well as their relationships. Second, we presented joint distributions and densities based on the two random variables. This led us to develop the generalization of joint distributions and densities for n random variables. Next, we defined the mean, variance, correlation, covariance, and the correlation coefficient of random processes. Gaussian and uniform random variables were also introduced, with emphasis on their density functions and properties. Additionally, we introduced chi-square and $\mathcal{F}$ distributions, which are useful in practical applications. Further, we developed results for a sum of n random variables that are useful in the analysis of random processes in terms of mean, variance, and covariance. In addition, the importance of the central limit theorem was discussed along with practical interpretations.

Next we looked at the stochastic processes of discrete-time random signals. We defined the concept of a discrete-time random process, and introduced autocorrelation and autocovariance as well as ensemble averages. We outlined properties of stationary random processes including strict-sense and wide-sense stationarity. Of particular importance in this chapter is the analysis of the mean, autocorrelation, and power spectrum in the context of stable linear time-invariant filters, along with the mathematic results in both the discrete-time and frequency domains. Finally, we concluded this chapter with the mean square error and estimation concepts as well as the relationships among the mean square error, variance, and bias of an estimator.

References

[1] Davenport, W. B., and W. L. Root, *Random Signals and Noise*, McGraw-Hill, New York, 1958.

[2] Davenport, W. B., *Probability and Random Processes*, McGraw-Hill, New York, 1970.

[3] Papoulis, A., *Probability, Random Variable, and Stochastic Processes*, McGraw-Hill, New York, 1991.

[4] Mood, A. M., F. A. Graybill, and D. C. Boes, *Introduction to the Theory of Statistics*, 3d ed., McGraw-Hill Book Company, New York, 1974.

[5] Bickel, P. J., and K. A. Doksum, *Mathematical Statistics: Basic Ideas and Selected Topics*, Holden-Day, Inc., Oakland, California, 1977.

4

Discrete-Time Transforms

4.1 Introduction

Discrete-time transforms are at the heart of discrete-time signal processing because they provide a frequency domain representation of the signal. This is useful as a design and analysis tool, and allows for a direct implementation of signal processing algorithms and systems. The fundamental advance in discrete-time transforms occurred when Cooley and Tukey [1], and Cooley, Lewis, and Welch [2] published their discovery of the fast Fourier transform (FFT) algorithm. Later, the FFT algorithms were introduced in the textbook *Digital Signal Processing*, by Oppenheim and Schafer [3]. The development of FFT algorithms launched modern digital signal processing because these algorithms made it possible to calculate discrete-time transforms efficiently.

In this chapter, we introduce the basic discrete-time transform pairs. We then discuss the various useful properties of these transforms. The transforms considered include the discrete Fourier transform (DFT) and its efficient algorithm, the FFT, the discrete cosine transform (DCT), the discrete sine transform (DST), the discrete Hartley transform (DHLT), the discrete Hilbert transform (DHT), the discrete fractional Hilbert transform (DFHT), the discrete-time wavelet transform (DTWT), the discrete Walsh transform (DWT), and the discrete Hadamard transform (DHAT). The relationships of the discrete-time transforms to the FFT are explored, emphasizing the computational savings.

This chapter is organized as follows. Section 4.1 has provided an introduction to discrete-time transformation. The DFT transform pairs, their properties, and frequency analysis using the DFT, as well as the relation of the FFT to the DFT are discussed in Sections 4.2 and 4.3, respectively. In Sections 4.4 and 4.5, we introduce the DCT and the DST, their properties and their relationships to the FFT. The DHLT, DHT, and DFHT are discussed in Sections 4.6, 4.7 and 4.8, respectively. Their relationships to the FFT are also explored. In Section 4.9, we describe the DTWT and its inverse DTWT. The DWT and DHAT are introduced in Sections 4.10 and 4.11, respectively. Finally, a brief summary is given in Section 4.12.

4.2 The Discrete Fourier Transform

The discrete Fourier transform (DFT) is defined as an operation for an N-point discrete-time signal $x[n]$ as follows:

$$X[k] = \sum_{n=0}^{N-1} x[n] W_N^{nk}, \qquad k = 0, 1, 2, ..., N-1, \tag{4.1}$$

where $W_N = e^{-j2\pi/N} = cos(2\pi/N) - jsin(2\pi/N)$. This operation in (4.1) is a transformation from an N-point of discrete-time signal $x[n]$ to an N-point of frequency-domain samples $X[k]$. From one point of view, (4.1) can be treated as two nested loops of a scalar operation containing a complex multiplication and a complex addition. It can also be viewed as a single loop that computes each point of frequency-domain $X[k]$ by an inner product of an N-point discrete-time signal $x[n]$ and an N-point basis factor W_N^{nk} for $n = 0, 1, ..., N-1$, and k fixed.

The definition of DFT in (4.1) can also be interpreted as a "frequency sampling" of the discrete-time Fourier transform (DTFT). It provides the frequency spectrum of a discrete-time signal $x[n]$,

$$X(f) = \sum_{n=-\infty}^{\infty} x[n] e^{\frac{-j2\pi f n}{F_s}}, \tag{4.2}$$

where F_s is called the sampling rate and f is a frequency value, both in hertz. Since the discrete-time signal $x[n]$ has N data points, the signal frequency range, from 0 to F_s, is divided into N discrete points with equally spaced frequencies, $f = \frac{kF_s}{N}$. Then, the DFT can be represented with a discrete variable k for frequency such that k is from 0 to $N-1$,

$$\begin{aligned} X[k] &= \sum_{n=0}^{N-1} x[n]e^{\frac{-j2\pi kn}{N}} \\ &= \sum_{n=0}^{N-1} x[n]W_N^{kn}, \end{aligned} \tag{4.3}$$

where $W_N = e^{-j(2\pi/N)}$. The region around each of the N points is known as a *frequency bin*, with a bandwidth of $\frac{1}{NT_s}$ or $\frac{F_s}{N}$ Hz.

The inverse discrete Fourier transform (IDFT) is given by

$$x[n] = \frac{1}{N}\sum_{k=0}^{N-1} X[k]W_N^{-kn}, \qquad n = 0, 1, ..., N-1. \tag{4.4}$$

In (4.1) and (4.4), both $x[n]$ and $X[k]$ may be complex valued.

4.2.1 Matrix Form of the DFT

Computation of the DFT can be viewed as a matrix operation. Consider the N-point DFT as a linear transformation of the discrete-time complex signal vector

$$\mathbf{x} = \{x[0] \;\; x[1] \;\; \cdots \;\; x[N-1]\}^T, \tag{4.5}$$

to another complex vector in the frequency domain

$$\mathbf{X} = \{X[0] \;\; X[1] \;\; \cdots \;\; X[N-1]\}^T. \tag{4.6}$$

The operation of the DFT in matrix notation is a multiplication as follows:

$$\mathbf{X} = \mathbf{W}\mathbf{x}, \tag{4.7}$$

where the matrix $\mathbf{W}$ contains the complex exponentials

$$\mathbf{W} = \begin{bmatrix} 1 & 1 & 1 & \cdots & 1 \\ 1 & W & W^2 & \cdots & W^{N-1} \\ 1 & W^2 & W^4 & \cdots & W^{2(N-1)} \\ \cdot & \cdot & \cdot & \cdots & \cdot \\ \cdot & \cdot & \cdot & \cdots & \cdot \\ \cdot & \cdot & \cdot & \cdots & \cdot \\ 1 & W^{N-1} & W^{2(N-1)} & \cdots & W^{(N-1)^2} \end{bmatrix} \tag{4.8}$$

where $W = e^{\frac{-j2\pi}{N}}$. The inverse DFT matrix can be obtained by using the inverse matrix $\mathbf{W}^{-1}$ as follows:

$$\mathbf{x} = \mathbf{W}^{-1}\mathbf{X}. \tag{4.9}$$

It should be pointed out that the column vectors $\mathbf{w}_i$ of matrix $\mathbf{W}$ are *orthonormal* because the inner product $\mathbf{w}_i^T\mathbf{w}_j = 0$ when $i \neq j$ and the inner product $\mathbf{w}_i^T\mathbf{w}_i = 1$ when $i = j$. This, in turn, implies that

$$\mathbf{W}\mathbf{W}^* = N\mathbf{I}, \tag{4.10}$$

where $\mathbf{W}^*$ denotes the complex conjugate of the matrix $\mathbf{W}$, and $\mathbf{I}$ is an $(N \times N)$ identity matrix. As a result, the inverse DFT matrix in (4.9) can be expressed as

$$\mathbf{x} = \frac{1}{N}\mathbf{W}^*\mathbf{X}. \tag{4.11}$$

This representation of the DFT in matrix notation will be useful when we introduce the uniform DFT filter bank in Section 7.7.

4.2.2 Computational Complexity

Note that N complex multiplications and $(N-1)$ complex additions are required to compute each value of the DFT in (4.3) since the discrete-time signal $x[n]$ may be complex. Therefore, in order to compute all N values directly, we require a total of N^2 complex multiplications and $N(N-1)$ complex additions. In other words, we need N^2 complex multiply accumulation and addition (MACs) operations. For a real discrete-time signal $x[n]$, the number of computations in (4.3) can be halved, because the DFT will be conjugate-symmetric with respect to the frequency value $\frac{N}{2}$.

The most useful approaches to reducing the computational complexity of the DFT exploit the symmetry and periodicity properties of W_N^{kn}. Consider:

1. Symmetry of the complex conjugate

$$W_n^{k(N-n)} = W_N^{-kn} = (W_N^{kn})^* \tag{4.12}$$

2. Periodicity in the index of n and k

$$W_N^{kn} = W_N^{k(n+N)} = W_N^{(k+N)n} \tag{4.13}$$

3. Compute factor $W_N^{q+\frac{N}{2}}$

$$W_N^{q+\frac{N}{2}} = W_N^{\frac{N}{2}} W_N^q = -W_N^q. \tag{4.14}$$

By exploiting the above observations, efficient computation of the DFT can be achieved. One such computation method is referred to as the fast Fourier transform (FFT) or the Cooley-Tukey algorithm. In Section 4.3, we present a detailed discussion of FFT algorithms in which the computation is proportional to $(Nlog_2N)$.

4.2.3 Properties of the DFT

The DFT exhibits many useful properties that are similar to the properties of the discrete-time Fourier transform. These DFT properties are briefly discussed below.

Linearity

The linearity property states that if

$$x_1[n] \stackrel{N}{\Longleftrightarrow} X_1[k], \tag{4.15}$$

and

$$x_2[n] \stackrel{N}{\Longleftrightarrow} X_2[k], \tag{4.16}$$

then

$$ax_1[n] + bx_2[n] \stackrel{N}{\Longleftrightarrow} aX_1[k] + bX_2[k], \tag{4.17}$$

for any complex number a and b. If $x_1[n]$ and $x_2[n]$ have the length N_1 and N_2, then their DFT must be computed at length of $max[N_1, N_2]$. This linearity property is also true in reverse. In other words, if two DFTs are added together, then the inverse DFT of the sum is the sum of the individual inverse DFTs.

Circular Shifting

The circular shifting property states that if

$$x[n] \overset{N}{\Longleftrightarrow} X[k], \tag{4.18}$$

then

$$x[((n-m))_N] \overset{N}{\Longleftrightarrow} e^{-j(2\pi k/N)m} X[k] \tag{4.19}$$

where $0 \leq n \leq N-1$ and is an integer, and $((i))_j$ means integer i modulo of integer j. If m is positive, the sequence $x[n]$ is shifted right, and if m is negative, then the discrete-time sequence $x[n]$ is shifted left.

Frequency Shifting or Modulation

The frequency shifting or modulation property states that if

$$x[n] \overset{N}{\Longleftrightarrow} X[k], \tag{4.20}$$

then

$$e^{j(2\pi l/N)n} x[n] \overset{N}{\Longleftrightarrow} X[((k-l))_N]. \tag{4.21}$$

In other words, if a discrete-time sequence $x[n]$ is multiplied by a complex exponential, its DFT is circularly shifted.

Duality

The duality property states that if

$$x[n] \overset{N}{\Longleftrightarrow} X[k], \tag{4.22}$$

then

$$X[n] \overset{N}{\Longleftrightarrow} Nx[((-k))_N], \quad 0 \leq k \leq N-1, \tag{4.23}$$

where $Nx[((-k))_N]$ is $Nx[k]$ index-reversed modulo of N. The duality property provides the solution to the DFT of $X[n]$ if $X[k]$ is the DFT of $x[n]$.

Reflection

The reflection property states that if

$$x[n] \overset{N}{\Longleftrightarrow} X[k], \tag{4.24}$$

then

$$x[((N-n))_N] \overset{N}{\Longleftrightarrow} X[((N-k))_N]. \tag{4.25}$$

Even Symmetry

If

$$x[n] \overset{N}{\Longleftrightarrow} X[k] = X_R[k] + jX_I[k], \tag{4.26}$$

where $X_R[k]$ and $X_I[k]$ are real valued, and if $x[n]$ is even, then

$$x[-n] = x[n] \overset{N}{\Longleftrightarrow} X[k] = X_R[k] = X_R[-k]. \tag{4.27}$$

Odd Symmetry

If

$$x[n] \overset{N}{\Longleftrightarrow} X[k] = X_R[k] + jX_I[k], \tag{4.28}$$

where $X_R[k]$ and $X_I[k]$ are real valued, and if $x[n]$ is odd, then

$$x[-n] = -x[n] \overset{N}{\Longleftrightarrow} X(k) = jX_I[k] = -jX_I[-k]. \tag{4.29}$$

Real and Imaginary Parts

If

$$x[n] \overset{N}{\Longleftrightarrow} X[k], \tag{4.30}$$

then

$$Re\{x[n]\} \overset{N}{\Longleftrightarrow} X_{ep}[k] = \frac{1}{2}\{X[((k))_N] + X^*[((-k))_N]\} \tag{4.31}$$

$$jIm\{x[n]\} \overset{N}{\Longleftrightarrow} X_{op}[k] = \frac{1}{2}\{X[((k))_N] - X^*[((-k))_N]\} \tag{4.32}$$

$$x_{ep}[n] = \frac{1}{2}\{x[((n))_N] + x^*[((-n))_N]\} \overset{N}{\Longleftrightarrow} Re\{X[k]\} \tag{4.33}$$

$$x_{op}[n] = \frac{1}{2}\{x[((n))_N] - x^*[((-n))_N]\} \stackrel{N}{\Longleftrightarrow} jIm\{X[k]\} \tag{4.34}$$

Time Reversal

The time reversal property states that if

$$x[n] \stackrel{N}{\Longleftrightarrow} X[k], \tag{4.35}$$

then

$$x[((-n))_N] \stackrel{N}{\Longleftrightarrow} X[((-k))_N]. \tag{4.36}$$

Parseval's Relation

Parseval's relation property states that if

$$x[n] \stackrel{N}{\Longleftrightarrow} X[k], \tag{4.37}$$

then

$$\sum_{n=0}^{N-1} |x[n]|^2 = \frac{1}{N} \sum_{k=0}^{N-1} |X(k)|^2. \tag{4.38}$$

Circular Convolution

The circular convolution property states that if

$$x[n] \stackrel{N}{\Longleftrightarrow} X[k], \tag{4.39}$$

and

$$y[n] \stackrel{N}{\Longleftrightarrow} Y[k], \tag{4.40}$$

then

$$x[n] \otimes y[n] = \sum_{m=0}^{N-1} x[m]y[((n-m))_N] \stackrel{N}{\Longleftrightarrow} X[k]Y[k]. \tag{4.41}$$

where the symbol $\otimes$ denotes the operation of circular convolution modulo N. Equation (4.41) is also referred to as an N-point circular convolution.

Modulation

The modulation property states that if

$$x[n] \stackrel{N}{\Longleftrightarrow} X[k], \tag{4.42}$$

and

$$y[n] \stackrel{N}{\Longleftrightarrow} Y[k], \tag{4.43}$$

then

$$x[n]y[n] \stackrel{N}{\Longleftrightarrow} X[k] \otimes Y[k] = \frac{1}{N}\sum_{m=0}^{N-1} X[m]Y[((k-m))_N]. \tag{4.44}$$

Equation (4.44) is a periodic convolution of two periodic functions over only one period of N-point.

4.2.4 Frequency Analysis of Discrete-Time Signal Using the DFT

In order to compute the magnitude spectrum $|X(e^{j\omega})|$ of a discrete-time signal $x[n]$, the value of the discrete-time signal $x[n]$ for all discrete-time is required. However, in practice, we only observe discrete-time signals for a finite duration. Therefore, in this section, we introduce the frequency analysis of a discrete-time signal with a finite data set by using the DFT.

Let $\{x[n]\}$ denote the sequence of the discrete-time signal to be analyzed using the DFT; $F_s/2$ is the highest frequency contained in the discrete-time signal, where F_s is the sampling frequency. Also assume that the finite duration of the discrete-time signal is L, where L is the number of samples in this duration. Then, this is equivalent to truncating $\{x[n]\}$ by using a rectangular window $w[n]$ of length L as follows:

$$\hat{x}[n] = x[n]w[n], \tag{4.45}$$

where the window function $w[n]$ is given by

$$w[n] = \begin{cases} 1 & 0 \le n \le L-1 \\ 0 & \text{otherwise} \end{cases} \tag{4.46}$$

Now suppose that the discrete-time signal $x[n]$ is a single sinusoid in the following form

$$x[n] = 2cos(n\omega_0), \tag{4.47}$$

where $\omega_0 = 0.45\pi$. Note that (4.47) can be expressed as

$$x[n] = e^{j\omega_0 n} + e^{-j\omega_0 n}. \tag{4.48}$$

Upon multiplying $w[n]$ by these two exponentials and using the frequency-shifting or modulation property described in Section 2.8.2, we obtain the DTFT of the finite-duration signal $x[n]$ as follows:

$$\hat{X}(e^{j\omega}) = W(e^{j(\omega-\omega_0)}) + W(e^{j(\omega+\omega_0)}), \tag{4.49}$$

where $W(e^{j\omega})$ is the DTFT of the window sequence $w[n]$ given by

$$W(e^{j\omega}) = e^{-j\omega(L-1)/2}\left[\frac{sin(\omega L/2)}{sin(\omega/2)}\right]. \tag{4.50}$$

To compute $\hat{X}(e^{j\omega})$, we use the N-point DFT for illustration, where $N > L$. Now, we can calculate the N-point DFT of the truncated discrete-time signal $\hat{x}[n]$ by padding the sequence $\hat{x}[n]$ with $N - L$ zeros. Figure 4.1 shows the result of the magnitude spectrum $|\hat{X}(k)| = |\hat{X}(e^{j\omega_k})|$ for $\omega_k = 2\pi k/N$, $k = 0, 1, 2, ..., N$, for $L = 100$ and $N = 512$. Note that the magnitude spectrum $|\hat{X}(e^{j\omega})|$ is symmetric at $\omega = \pi$. As can be seen, we can identify the peak frequency located at 0.45π. Also, we note that the magnitude spectrum $\hat{X}(e^{j\omega})$ is not completely localized to a signal frequency, but instead it is spread out over the entire frequency range, $0 \leq \omega \leq \pi$. This is because the power of the discrete-time signal $\{x[n]\}$, which was concentrated at a single frequency, has been spread by the rectangular window $w[n]$ into the whole frequency range. This phenomenon of windowing the signal is referred to as *leakage*.

Windowing not only distorts the spectral estimate because of the leakage effects but also reduces spectral resolution. So far, we have presented only the rectangular window with application to the discrete-time signals for frequency analysis. However, different window functions have different characteristics with respect to signal analysis. We will introduce other window functions in Section 5.4.1.

4.3 The Fast Fourier Transform

The DFT computational complexity can be reduced by exploiting both the symmetry and the periodicity of the complex sequence W_N^{kn}.

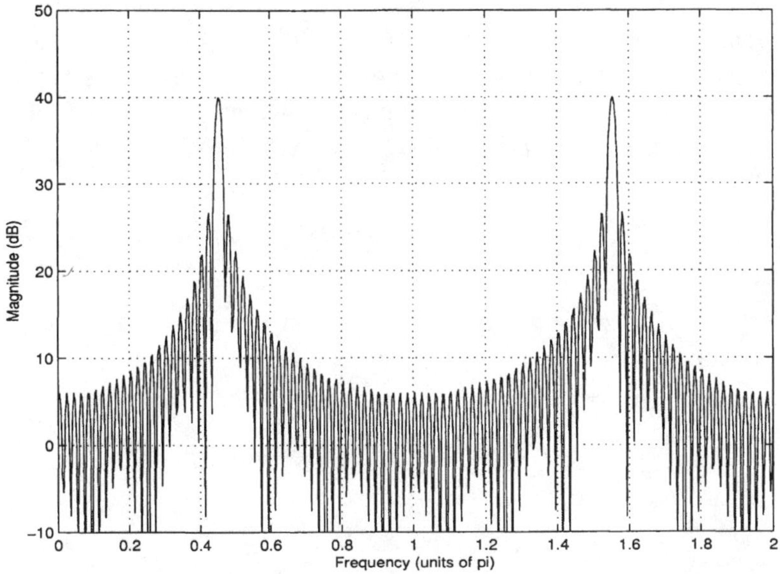

Figure 4.1 Magnitude spectrum for the truncated discrete-time signal $x[n]$ with $L = 100$ by using 512-point DFT.

The possibility of greatly reduced computation was popularized when Cooley and Tukey [1] published an algorithm for the computation of the DFT in 1965. This algorithm is often referred to as the fast Fourier transform (FFT).

FFT algorithms are based on the fundamental principle of decomposing a DFT into successively smaller DFTs. This method leads to a variety of different algorithms, all with comparable improvements in computational complexity. In this section, we introduce two basic classes of the FFT algorithms with *radix-2* that have been extensively used in practice. In the first class of FFT algorithms, the discrete-time sequence $x[n]$ is decomposed into successively smaller subsequences in which the computations are made on smaller transformations. This class is referred to as the *decimation-in-time FFT*. The second, called the *decimation-in-frequency FFT*, derives its name from the fact that the sequence $X[k]$ is decomposed into smaller subsequences.

4.3.1 Decimation-in-Time FFT

The principle of the decimation-in-time FFT algorithm is based on decomposing the discrete-time sequence $x[n]$ into successively smaller subsequences. It is most conveniently illustrated by considering a DFT of length N. The DFT in (4.3) is given by

$$X[k] = \sum_{n=0}^{N-1} x[n] W_N^{kn}. \tag{4.51}$$

We separate the sequence $x[n]$ into its even-numbered ($r = 2n$) and odd-numbered ($r = 2n + 1$) samples as follows:

$$\begin{aligned} X[k] &= \sum_{r=0}^{(N/2)-1} x[2r] W_N^{2rk} + \sum_{r=0}^{(N/2)-1} x[2r+1] W_N^{(2r+1)k} \\ &= \sum_{r=0}^{(N/2)-1} x[2r] (W_N^2)^{rk} \\ &\qquad + W_N^k \sum_{r=0}^{(N/2)-1} x[2r+1] (W_N^2)^{rk}. \end{aligned} \tag{4.52}$$

Note that $W_N^2 = W_{N/2}$ because

$$W_N^2 = e^{-2j(2\pi/N)} = e^{-j2\pi/(N/2)} = W_{N/2}. \tag{4.53}$$

Therefore, (4.52) can be expressed as

$$\begin{aligned} X[k] &= \sum_{r=0}^{(N/2)-1} x[2r] W_{N/2}^{rk} + W_N^k \sum_{r=0}^{(N/2)-1} x[2r+1] W_{N/2}^{rk} \\ &= E[k] + W_N^k O[k], \end{aligned} \tag{4.54}$$

where $E[k]$ and $O[k]$ are the $\left(\frac{N}{2}\right)$-point DFTs of $\frac{N}{2}$ length sequences $x[2r]$ and $x[2r+1]$, respectively. Even though a $\frac{N}{2}$-point DFT is only of length $\frac{N}{2}$, for values of k greater than $\frac{N}{2}$, we use the property that $E[k] = E[((k))_{\frac{N}{2}}]$ and $O[k] = O[((k))_{\frac{N}{2}}]$. If $\left(\frac{N}{2}\right)$ is still even, each of these can be broken into two $\left(\frac{N}{4}\right)$-point DFTs. Moreover, If N is equal to 2^v, then the decomposition process can be repeated until the

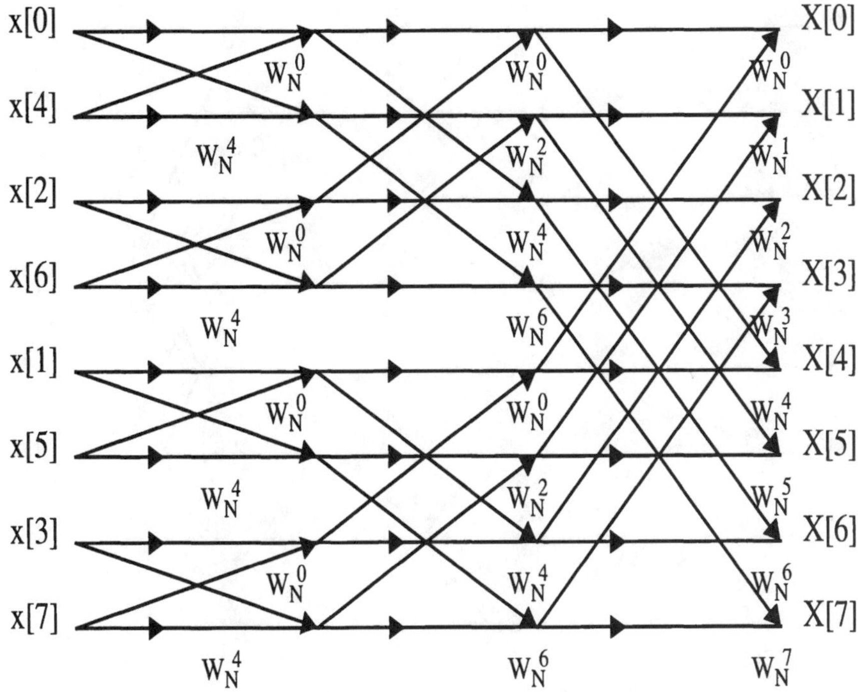

Figure 4.2 The flow graph of decimation-in-time FFT algorithm with radix-2 for an 8-point DFT calculation.

DFT can be calculated as a combination of $\frac{N}{2}$ 2-point DFTs. This algorithm is referred to as the *radix-2 decimation-in-time* FFT.

Figure 4.2 shows a flow graph for operations of the radix-2 decimation-in-time algorithm with $N = 8$. As we can see, each stage has 8 complex multiplications and 8 complex additions. The branches have transmittances of the form W_8^r. Since there are three stages, we have a total number of 24 complex multiplications and additions.

With observations $W_n^0 = 1$, $W_N^{-kN} = e^{j(\frac{2\pi}{N})Nk} = e^{j2\pi k} = 1$, and $W_N^{q+N/2} = -W_N^q$, Figure 4.2 can be simplified to the form shown in Figure 4.3, requiring one complex multiplication instead of two. In general, the number of complex multiplications and additions is equal to $\frac{N}{2}log_2N$ and $Nlog_2N$, respectively, because there are

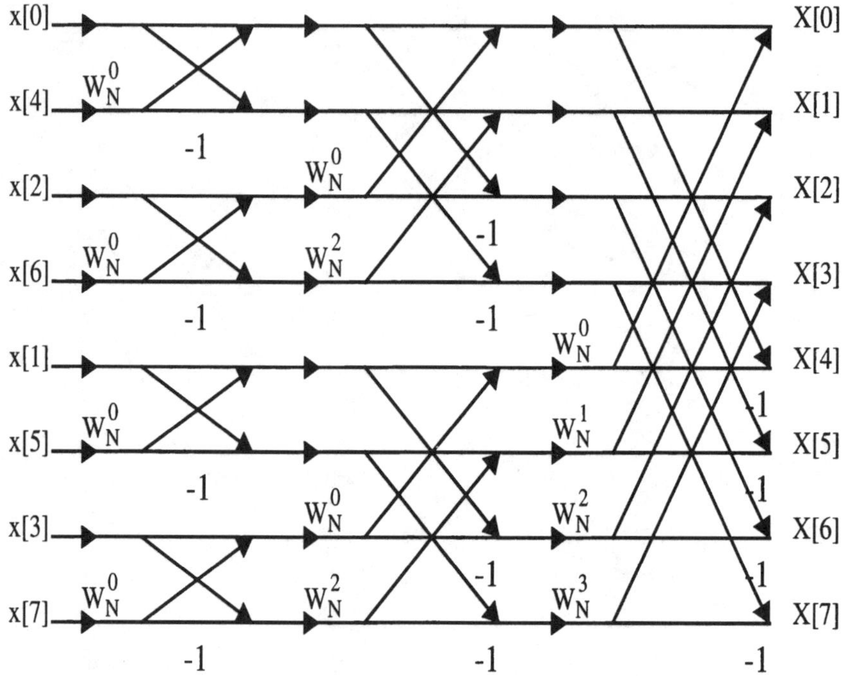

Figure 4.3 The flow graph of the decimation-in-time FFT algorithm with radix-2 for an 8-point DFT calculation using the simplification of W_N^r.

$v = log_2 N$ stages as well as $N/2$ complex multiplications and N complex additions for each stage.

4.3.2 Decimation-in-Frequency FFT

The decimation-in-frequency FFT algorithm is used to break up the output sequence $X[k]$ into smaller and smaller subsequences in the same manner that we did using the decimation-in-time FFT algorithm. Consider the DFT for an N-point sequence

$$X[k] = \sum_{n=0}^{N-1} x[n] W_N^{nk}, \qquad k = 0, 1, ..., N-1. \tag{4.55}$$

The output sequence $X[k]$ in (4.55) can be split into the even-numbered frequency samples $X[2q]$ and the odd-numbered frequency samples $X[2q+1]$. The even-numbered frequency samples $X[2q]$ can be calculated by

$$X[2q] = \sum_{n=0}^{N-1} x[n]W_N^{2nq}, \qquad q = 0, 1, ..., (N/2) - 1. \tag{4.56}$$

The summation in (4.56) can be expressed as

$$\begin{aligned} X[2q] &= \sum_{n=0}^{\frac{N}{2}-1} x[n]W_N^{2nq} + \sum_{n=N/2}^{N-1} x[n]W_N^{2nq} \\ &= \sum_{n=0}^{\frac{N}{2}-1} x[n]W_N^{2nq} \\ &\quad + \sum_{n=0}^{\frac{N}{2}-1} x\left[n+\frac{N}{2}\right] W_N^{2q[n+\frac{N}{2}]}. \end{aligned} \tag{4.57}$$

Note that the factor $W_N^{2q[n+(N/2)]} = W_N^{2qn}W_N^{qN} = W_N^{2qn}$ and $W_N^2 = W_{N/2}$. Therefore, (4.57) is equal to

$$\begin{aligned} X[2q] &= \sum_{n=0}^{\frac{N}{2}-1} x[n]W_N^{2nq} + \sum_{n=0}^{\frac{N}{2}-1} x\left[n+\frac{N}{2}\right] W_N^{2nq} \\ &= \sum_{n=0}^{\frac{N}{2}-1} \left(x[n] + x\left[n+\frac{N}{2}\right]\right) W_N^{2nq} \\ &= \sum_{n=0}^{\frac{N}{2}-1} \left(x[n] + x\left[n+\frac{N}{2}\right]\right) W_{N/2}^{nq} \end{aligned} \tag{4.58}$$

where $q = 0, 1, ..., (N/2) - 1$. As we can see, (4.58) is the $(N/2)$-point DFT of the $(N/2)$-point sequence $x[n]$ obtained by adding the first and the last half of the input sequence.

Now consider the specialization for odd-numbered frequency samples,

$$X[2q+1] = \sum_{n=0}^{N-1} x[n]W_N^{n(2q+1)}, \qquad q = 0, 1, ..., (N/2) - 1. \tag{4.59}$$

The summation in (4.59) can be broken into

$$\begin{aligned} X[2q+1] &= \sum_{n=0}^{\frac{N}{2}-1} x[n]W_N^{n(2q+1)} + \sum_{n=N/2}^{N-1} x[n]W_N^{n(2q+1)} \\ &= \sum_{n=0}^{\frac{N}{2}-1} x[n]W_N^{n(2q+1)} + \sum_{n=0}^{\frac{N}{2}-1} x\left[n+\frac{N}{2}\right] W_N^{(n+\frac{N}{2})(2q+1)} \\ &= \sum_{n=0}^{\frac{N}{2}-1} x[n]W_N^{n(2q+1)} \\ &\quad + W_N^{(N/2)(2q+1)} \sum_{n=0}^{\frac{N}{2}-1} x\left[n+\frac{N}{2}\right] W_N^{n(2q+1)}. \end{aligned} \quad (4.60)$$

Note that factors $W_N^{(N/2)2q} = 1$, $W_N^{(N/2)} = -1$, and $W_N^2 = W_{N/2}$. Therefore, (4.60) can be rewritten as

$$X[2q+1] = \sum_{n=0}^{(N/2)-1} \left(x[n] - x\left[n+\frac{N}{2}\right]\right) W_N^n W_{N/2}^{nq}, \quad (4.61)$$

where $q = 0, 1, ..., (N/2) - 1$. As we can see, (4.61) is the $\left(\frac{N}{2}\right)$-point DFT of the sequence $x[n]$ obtained by subtracting the second half of the input sequence from the first half and multiplying the resulting sequence by W_N^n.

Based on (4.58) and (4.61), we can define two new $\left(\frac{N}{2}\right)$-point discrete-time sequences $o[n]$ and $e[n]$ as follows

$$\begin{aligned} o[n] &= x[n] + x\left[n+\frac{N}{2}\right], \quad n = 0, 1, ..., (N/2) - 1 \\ e[n] &= x[n] - x\left[n+\frac{N}{2}\right], \quad n = 0, 1, ..., (N/2) - 1. \end{aligned}$$

Thus, we can compute the $\left(\frac{N}{2}\right)$-point DFT of the discrete-time sequence $o[n]$

$$X[2q] = \sum_{n=0}^{(N/2)-1} o[n]W_{N/2}^{nq}, \quad q = 0, 1, ..., (N/2) - 1, \quad (4.62)$$

and the $\left(\frac{N}{2}\right)$-point DFT of the discrete-time sequence $e[n]W_N^n$

$$X[2q+1] = \sum_{n=0}^{(N/2)-1} e[n]W_N^n W_{N/2}^{nq}, \quad q = 0, 1, ..., (N/2) - 1. \quad (4.63)$$

If the length $N = 2^v$, this decomposition process can be continued until the N-point DFT has been completely decomposed into $\frac{N}{2}$ 2-point DFTs. In this case, the algorithm is referred to as the radix-2 decimation-in-frequency FFT.

Figure 4.4 shows a flow graph for the radix-2 decimation-in-frequency FFT algorithm with $N = 8$. As can be seen, each stage has 4 complex multiplications and 8 complex additions. The branches have transmittances of the form W_8^r. Since there are three stages, we have a total number of 12 complex multiplications and 24 additions. In general, the number of complex multiplications and additions is equal to $\frac{N}{2}log_2N$ and $Nlog_2N$, respectively, because there are $v = log_2N$ stages as well as $N/2$ complex multiplications and N complex additions for each stage. Thus, the computational complexity is the same as that of decimation-in-time FFT.

4.3.3 Computational Complexity

In general, for the radix-2 FFT algorithm, a total of $\frac{N}{2}log_2N$ complex multiplications and $Nlog_2N$ additions are required. Some of the multiplications by the factors of W_N^0, $W_N^{\frac{N}{2}}$, $W_N^{kN/2}$, and $W_N^{kN/4}$ are 1, -1, $(-1)^k$, and $(-j)^k$, respectively. These operations are trivial and do not require actual multiplication. In addition, some multiplications by the factor of $W_N^{kN/8}$, which is $[(1-j)/\sqrt{2}]^k$, can be done by two real multiplications and two real additions. Moreover, general complex multiplications can be computed by three real multiplications and three real additions. Therefore, the total computational complexity of the radix-2 FFT algorithm is reduced to $\left[\frac{N}{2}(3log_2N - 10) + 8\right]$ multiplications and $\left[\frac{N}{2}(7log_2N - 10) + 8\right]$ additions.

Much research has been done over the past three decades on the DFT and FFT algorithms. For reference, we cite the books by Oppenheim and Schafer [3, 4], Dudgeon and Mersereau [5], Brigham [6], Chen [7], Lim [8], and Burrus et al. [9].

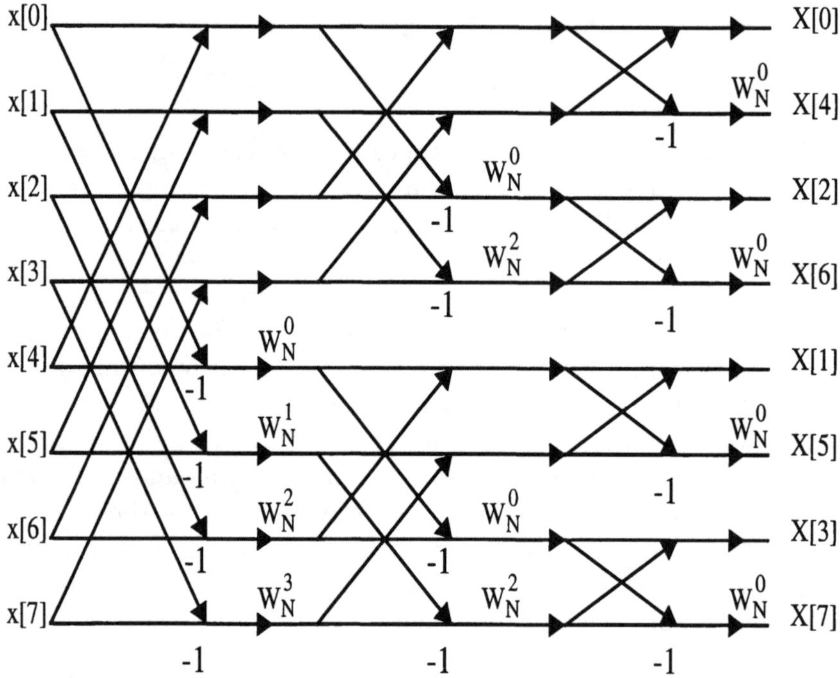

Figure 4.4 The flow graph of a radix-2 decimation-in-frequency 8-point FFT algorithm.

Example 4.1 Compute $2N$-Point DFT of Real Signal Using N-Point FFT

In this example, we develop a method to compute $2N$-point DFT of real signal using one N-point FFT. Consider a length of $2N$-point real signal $x[n]$ ($n = 0, 1, ..., 2N - 1$). We split the input sequence $x[n]$ into even-indexed points,

$$x_e[n] = x[2n], \tag{4.64}$$

and odd-indexed points,

$$x_o[n] = x[2n + 1], \tag{4.65}$$

for $n = 0, 1, 2, ..., N - 1$. Then we can write the $2N$-point FFT as follows:

$$\begin{aligned} X[k] &= \sum_{n=0}^{2N-1} x[n]W_{2N}^{nk} \\ &= \sum_{n=0}^{N-1} x[2n]W_{2N}^{2nk} + \sum_{n=0}^{N-1} x[2n+1]W_{2N}^{(2n+1)k} \\ &= \sum_{m=0}^{N-1} x_e[m]W_N^{mk} + W_{2N}^{k} \sum_{m=0}^{N-1} x_o[m]W_N^{mk} \\ &= DFT\{x_e[m]\} + W_{2N}^{k} DFT\{x_o[m]\}, \end{aligned} \tag{4.66}$$

for $k = 0, 1, ..., N-1$. Notice that the DFTs in (4.66) are N-point DFT. Therefore, we can calculate the $2N$-point FFTs by using two N-point DFTs of real signal.

Example 4.2 Compute $2N$-Point Inverse DFT of Conjugate-Symmetric Data Using N-Point Inverse FFT

Consider a length of $2N$-point conjugate-symmetric data $X[k]$, $k = 0, 1, ..., 2N-1$. We can develop a method to compute $2N$-point inverse DFT (IDFT) of $X[k]$ by using one N-point inverse FFT. We separate the discrete-time sequence $x[n]$ into the even-indexed sequence $x[2n]$,

$$\begin{aligned} x[2n] &= \frac{1}{2N} \sum_{k=0}^{2N-1} X[k]W_{2N}^{-2nk} \\ &= \frac{1}{2N} \sum_{k=0}^{N-1} \{X[k] + X[k+N]\}W_N^{-nk} \\ &= \frac{1}{2} IDFT\{X[k] + X[k+N]\}, \end{aligned} \tag{4.67}$$

and the odd-indexed sequence $x[2n+1]$,

$$\begin{aligned} x[2n+1] &= \frac{1}{2N} \sum_{k=0}^{2N-1} X[k]W_{2N}^{-(2n+1)k} \\ &= \frac{1}{2N} \sum_{k=0}^{N-1} W_{2N}^{-k}\{X[k] - X[k+N]\}W_N^{-nk} \\ &= \frac{1}{2} IDFT\{W_{2N}^{-k}(X[k] - X[k+N])\}, \end{aligned} \tag{4.68}$$

for $n = 0, 1, ..., N-1$. As we can see, the IDFTs are N-point IDFTs. Therefore, two N-point IFFTs can be computed simultaneously.

4.4 The Discrete Cosine Transform

The discrete cosine transform (DCT) is extensively used for speech and image transform coders, feature extraction, and in multirate systems known as filter banks. This is because the DCT can behave much like the theoretically optimal *Karhunen-Loève* transform [10-12]. The Karhunen-Loève transform is optimal in that it yields spectral values that are uncorrelated. The DFT and DCT are viable alternatives, although they are suboptimum. However, the DCT produces good performance compared with the Karhunen-Loève transform. To compute the DCT efficiently, fast algorithms have been intensively studied. In this section, we introduce the definition of the DCT as well as an efficient computation algorithm that is closely related to the FFT.

The N-point DCT of sequence $x[n]$ is defined by

$$C[k] = \sum_{n=0}^{N-1} 2x[n] cos\left[\frac{\pi}{2N} k(2n+1)\right], \quad k = 0, 1, ..., N-1. \qquad (4.69)$$

The inverse DCT is given by

$$x[n] = \frac{1}{N} \sum_{k=0}^{N-1} w[k] C[k] cos\left[\frac{\pi}{2N} k(2n+1)\right], \quad n = 0, 1, ..., N-1, \qquad (4.70)$$

where $w[k] = 0.5$ for $k = 0$ and $w[k] = 1$ for $k = 1, 2, ..., N-1$. The above DCT and inverse DCT are also referred to as the even symmetrical DCT.

4.4.1 Relationship of the DCT to the DFT

The DCT and inverse DCT can be computed by using the DFT. In order to derive the relationship between the DCT and the DFT, we first relate the N-point sequence $x[n]$ to a new $2N$-point sequence $y[n]$ that is then related to its $2N$-point DFT, $Y[k]$. Finally, we convert $Y[k]$ to $C[k]$, which is the N-point DCT of sequence $x[n]$.

The diagram below illustrates the relation:

$$\underset{x[n]}{N\text{-}points} \iff \underset{y[n]}{2N\text{-}points} \overset{DFT}{\iff} \underset{Y[k]}{2N\text{-}points} \iff \underset{C[k]}{N\text{-}points} \qquad (4.71)$$

Let $x[n]$ denote an N-point sequence that is zero outside $0 \leq n \leq N-1$. The sequence $y[n]$ is constructed by

$$\begin{aligned} y[n] &= x[n] + x[2N-1-n] \\ &= \begin{cases} x[n] & 0 \leq n \leq N-1 \\ x[2N-1-n], & N \leq n \leq 2N-1. \end{cases} \end{aligned} \qquad (4.72)$$

The sequence $y[n]$ in (4.72) is symmetric with respect to the value $N - \frac{1}{2}$. The $2N$-point DFT of $y[n]$ is obtained by

$$\begin{aligned} Y[k] &= \sum_{n=0}^{2N-1} y[n]W_{2N}^{kn} \\ &= \sum_{n=0}^{N-1} x[n]W_{2N}^{kn} + \sum_{n=N}^{2N-1} x[2N-1-n]W_{2N}^{kn} \\ &= W_{2N}^{-k/2} \sum_{n=0}^{N-1} 2x[n]cos\left[\frac{\pi}{2N}k(2n+1)\right], \end{aligned} \qquad (4.73)$$

where $0 \leq k \leq N-1$. The N-point DCT of sequence $x[n]$ is obtained from $Y[k]$ by multiplying by a factor of $W_{2N}^{k/2}$

$$\begin{aligned} C[k] &= W_{2N}^{k/2} Y[k] \\ &= \sum_{n=0}^{N-1} 2x[n]cos\left[\frac{\pi}{2N}k(2n+1)\right], \end{aligned} \qquad (4.74)$$

for $0 \leq k \leq N-1$. Note that the DCT $C[k]$ in (4.74) is N-point, so that the N values of sequence $x[n]$ are represented by the N values of the $C[k]$. If the sequence $x[n]$ is real, then $C[k]$ is real. If the sequence $x[n]$ is complex, then $C[k]$ is also complex.

4.4.2 Relationship Between the Inverse DCT and the Inverse DFT

To develop the relationship between the inverse DCT and the inverse DFT, we first relate the N-point DCT, $C[k]$, to the $2N$-point DFT,

$Y[k]$. Then we convert $Y[k]$ to $y[n]$, and transfer $y[n]$ to the N-point sequence $x[n]$ as shown in (4.71).

Note that the DFT $Y[k]$ is a $2N$-point sequence and the DCT $C[k]$ is an N-point sequence. However, the redundancy in $Y[k]$ can provide a way of determining $Y[k]$ from $C[k]$ because of its symmetry. Specifically, $Y[k]$ is obtained from $C[k]$ by

$$Y[k] = \begin{cases} W_{2N}^{-k/2}C[k] & 0 \leq k \leq N-1 \\ 0 & k = N \\ -W_{2N}^{-k/2}C[2N-k], & N+1 \leq k \leq 2N-1. \end{cases} \tag{4.75}$$

Then, the $2N$-point sequence $y[n]$ is the $2N$-point inverse DFT of $2N$-point $Y[k]$,

$$y[n] = \frac{1}{2N}\sum_{k=0}^{2N-1} Y[k]W_{2N}^{-kn}, \quad 0 \leq n \leq 2N-1. \tag{4.76}$$

Finally, we recover $x[n]$ from $y[n]$ by

$$x[n] = \begin{cases} y[n], & 0 \leq n \leq N-1 \\ 0, & \text{otherwise.} \end{cases} \tag{4.77}$$

4.4.3 Computational Complexity

Most of the computational complexity is due to the computation of the $2N$-point DFT and $2N$-point inverse DFT, which is required for the DCT and for inverse DCT computation. However, the DFT and inverse DFT can be calculated by using the FFT algorithms. Moreover, the $2N$-point DFT and the inverse $2N$-point DFT can be computed by using an N-point DFT and N-point inverse DFT, because $y[n]$ has symmetry. Therefore, the computational complexity is the same as that of the DFT. In general, the number of multiplications and additions for the DCT is $\left(\frac{N}{2}log_2N\right)$ and $\left(\frac{3N}{2}log_2N - N + 1\right)$, respectively.

In the context of the DCT, we suggest the books by Quackenbush, Barnwell, and Clements [13]; Rao and Yip [11]; Lim [8], and the reference paper by Miao [14].

4.5 The Discrete Sine Transform

The discrete sine transform (DST) was introduced into the field of digital signal processing by Jain [10] in 1976. Although the DST has not found as many applications as the DCT, various algorithms have been developed for the efficient computation of the DST [15]. In this section, we show that the N-point DST can be computed by using the N-point DCT algorithm. Therefore, the DST algorithm can be calculated by using the FFT algorithm.

The N-point DST of a sequence $x[n]$, $n = 0, 1, ..., N-1$, is defined by

$$S[k] = \frac{2}{N} W[k] \sum_{n=0}^{N-1} x[n] sin\left[(2n+1)\frac{k\pi}{2N}\right], \; k = 1, 2, ..., N, \quad (4.78)$$

where

$$W[k] = \begin{cases} \frac{1}{\sqrt{2}}, & \text{if } k = N \\ 1, & \text{otherwise.} \end{cases} \quad (4.79)$$

The inverse DST of $S[k]$ is given by

$$x[n] = \sum_{k=0}^{N-1} S[k] sin\left[(2n+1)\frac{k\pi}{2N}\right], \; n = 1, 2, ..., N. \quad (4.80)$$

Note that the definition of the inverse DST in (4.80) is the same as the forward DST in (4.78), except for a constant scaling, $\frac{2}{N}W[k]$. Therefore, only an algorithm for computing the forward DST must be derived.

4.5.1 Relation of DST and DCT

The N-point DST of $x[n]$ can be computed by using the N-point DCT algorithm. In order to prove the above statement, we first let $k = N - m$, then we obtain from (4.78) and (4.79)

$$S[N-m] = \frac{2}{N} W[N-m] \sum_{n=0}^{N-1} x[n] sin\left[(2n+1)\frac{(N-m)\pi}{2N}\right], \quad (4.81)$$

for $N - m = 1, 2, ..., N$, where

$$W[N-m] = \begin{cases} \frac{1}{\sqrt{2}}, & \text{if } N-m = N \\ 1, & \text{otherwise.} \end{cases} \quad (4.82)$$

Using the trigonometric identity,

$$sin(A + B) = sinAcosB - cosAsinB, \tag{4.83}$$

we let $A = [(2n+1)\frac{\pi}{2}]$, and $B = [-(2n+1)\frac{m\pi}{2N}]$. From (4.81), we have the equation

$$sin\left[(2n+1)\frac{(N-m)\pi}{2N}\right] = (-1)^n cos\left[(2n+1)\frac{m\pi}{2N}\right]. \tag{4.84}$$

If we also let $C[m] = S[N-m]$ and $D[m] = W[N-m]$, then we can rewrite (4.81) and (4.82) as follows:

$$\begin{aligned} C[m] &= S[N-m] \\ &= \frac{D[m]}{N}\sum_{n=0}^{N-1} 2y[n]cos\left[(2n+1)\frac{m\pi}{2N}\right], \end{aligned} \tag{4.85}$$

for $m = 0, 1, ..., N-1$, and where

$$D[m] = \begin{cases} \frac{1}{\sqrt{2}}, & \text{if } m = 0 \\ 1, & \text{otherwise,} \end{cases} \tag{4.86}$$

and

$$y[n] = (-1)^n x[n]. \tag{4.87}$$

As we can see, $C[m]$ in (4.85) is the DCT of the sequence $y[n]$, $n = 0, 1, ..., N-1$, except for a constant scaling $\frac{D[m]}{N}$. In other words, we can convert a DST computation to a DCT calculation where the FFT can be used for efficient computation.

4.5.2 Computational Complexity

The computational complexity of the DST is the same as that of the DCT. The number of multiplications and additions for the DST is $\frac{N}{2}log_2N$ and $\frac{3N}{2}log_2N - N + 1$, respectively. To compute the DST of a sequence, one can use any DCT algorithm or DCT-based architecture, ensuring that the computational complexity remains the same.

4.6 The Discrete Hartley Transform

The discrete Hartley transform (DHLT) was introduced by Bracewell [16, 17]. The DHLT is useful in many applications including filter bank design, signal representation, computing of convolutions, and fast computation for the DFT. The DHLT of a real-valued sequence $x[n]$ with length N is defined by

$$X[k] = \frac{1}{N}\sum_{n=0}^{N-1} x[n]cas\left(\frac{2\pi kn}{N}\right), \qquad n = 0, 1, ..., N-1, \tag{4.88}$$

where $cas\left(\frac{2\pi kn}{N}\right)$ is given by

$$cas\left(\frac{2\pi kn}{N}\right) = \left[cos\left(\frac{2\pi kn}{N}\right) + sin\left(\frac{2\pi kn}{N}\right)\right]. \tag{4.89}$$

The corresponding inverse DHLT is obtained by

$$\begin{aligned} x[n] &= \sum_{k=0}^{N-1} X[k]cas\left(\frac{2\pi kn}{N}\right) \\ &= \sum_{k=0}^{N-1} X[k]\left[cos\left(\frac{2\pi kn}{N}\right) + sin\left(\frac{2\pi kn}{N}\right)\right], \end{aligned} \tag{4.90}$$

for $n = 0, 1, ..., N-1$.

Note that the DHLT is real when the input $x[n]$ is real. The inverse DHLT defined in (4.90) is also identical to the DHLT given by (4.88), except for a scale factor of $\frac{1}{N}$. Thus, only one routine of computation is required for forward and inverse DHLTs. In addition, the real and imaginary parts of the DFT for a real-valued sequence are equal to the even and the negative odd parts of the corresponding DHLT components, respectively.

The fast DHLT algorithm for length $N = 2^m$ has a lower computational complexity than the corresponding DFT. However, for the length $N \neq 2^m$, we still need a fast DHLT algorithm in practice. Bi and Chen [18] and Vijayakumar and Prabhu [19], developed a fast DHLT algorithm for a general case with length $N = q2^m$ where q is an odd integer. They showed that the fast DHLT algorithm with length $N = 2^m$ is equivalent to the special case of $q = 1$.

If the length N of $x[n]$ is equal to $q2^m$, (4.88) can be decomposed into even-indexed transforms as

$$X[2k] = \frac{1}{N}\sum_{n=0}^{N/2-1} u[n]cas\left(\frac{2\pi kn}{N/2}\right), \tag{4.91}$$

for $0 \leq k \leq N/2-1$, where $u[n] = x[n] + x[n+N/2]$.

If $m = 1$ or $N = 2q$, the odd-indexed transform can be calculated by

$$\begin{aligned} X[((2k+q))_N] &= \frac{1}{N}\sum_{n=0}^{N-1} x[n]cas\left[\frac{2\pi n(2k+q)}{N}\right] \\ &= \frac{1}{N}\sum_{n=0}^{N-1} x[n](-1)^n cas\left(\frac{2\pi kn}{q}\right) \\ &= \frac{1}{N}\sum_{n=0}^{q-1}(-1)^n(x[n]-x[n+N/2])cas\left(\frac{2\pi kn}{q}\right) \\ &\quad \text{for } 0 \leq k \leq q-1. \end{aligned} \tag{4.92}$$

Both (4.91) and (4.92) are length q DHLTs.

If $m > 1$, or $N = q\cdot 2^m$, the odd-indexed transform with argument k by $4k \pm q$ is obtained by

$$\begin{aligned} X[4k \pm q] &= \frac{1}{N}\sum_{n=0}^{N/2-1} v[n]cas\left[\frac{2\pi n(4k\pm q)}{N}\right] \\ &= \frac{1}{N}\sum_{n=0}^{N/2-1} v[n]\left[cos\left(\frac{2\pi n}{N/q}\right)cas\left(\frac{2\pi kn}{N/4}\right)\right. \\ &\quad \left.\pm sin\left(\frac{2\pi n}{N/q}\right)cas\left(-\frac{2\pi kn}{N/4}\right)\right], \end{aligned} \tag{4.93}$$

for $0 \leq k \leq \frac{N}{4}-1$ and where $v[n] = x[n] - x[n+N/2]$. Now, we can create two new sequences $H[k]$ and $G[k]$ by using (4.93), as follows:

$$\begin{aligned} H[k] &= \frac{1}{2}\{X[4k+q] + X[4k-q]\} \\ &= \frac{1}{N}\sum_{n=0}^{N/2-1} v[n]cos\left(\frac{2\pi n}{N/q}\right)cas\left(\frac{2\pi nk}{N/4}\right) \end{aligned}$$

$$= \frac{1}{N}\sum_{n=0}^{N/4-1}\left[v[n]cos\left(\frac{2\pi n}{N/q}\right) + (-1)^{(q+1)/2}v[n+N/4]\right.$$
$$\left.\cdot\, sin\left(\frac{2\pi n}{N/q}\right)\right] cas\left(\frac{2\pi nk}{N/4}\right), \quad (4.94)$$

and

$$G[k] = \frac{1}{2}\{X[4k+q] - X[4k-q]\}$$
$$= \frac{1}{N}\sum_{n=0}^{N/2-1} v[n]sin\left(\frac{2\pi n}{N/q}\right) cas\left[\frac{2\pi nk}{N/4}\right]$$
$$= \frac{1}{N}\sum_{n=0}^{N/4-1}\left[v[n]sin\left(\frac{2\pi n}{N/q}\right) + (-1)^{(q-1)/2}v[n+N/4]\right.$$
$$\left.\cdot\, cos\left(\frac{2\pi n}{N/q}\right)\right] cas\left[\frac{2\pi nk}{N/4}\right]. \quad (4.95)$$

Note that the discrete-time input sequence $v[n]$ is rotated by using *twiddle factor* $cos\left(\frac{2\pi n}{N/q}\right)$ and $sin\left(\frac{2\pi n}{N/q}\right)$ in (4.94) and (4.95). Therefore, the odd-indexed transform outputs for $N/4$-point DHLTs are obtained by

$$X[((4k+q))_N] = H[k] + G[k] \quad (4.96)$$
$$X[((4k-q))_N] = H[k] - G[k], \quad (4.97)$$

for $0 \le k \le \frac{N}{4} - 1$. The decomposed procedure can be continued in the same way until the length-q and length-$2q$ DHLTs are achieved.

There are many technical sources with respect to fast algorithms for the DHLT. We recommend the reference papers by Chan and Siu [20], Hu, Chang, and Ersoy [21], Mether, Satapathy, and Panda [22], Bi and Lu [23], Bi, Chen, and Zeng [24], and Cheng [25].

4.7 The Discrete Hilbert Transform

The Hilbert transform is very important in signal processing and communication theory. It has been used in various signal processing systems, analytic signal evaluation, bandpass sampling, minimum-phase networks, spectral analysis, edge detection, digital filter design, and in many other applications (see Oppenheim and Schafer [4], and Venkata, Evans, and McCaslin [26]). The discrete Hilbert transform (DHT) has been developed so that the classical Hilbert transform can be used for digital signal processing. The DHT has become one of the most powerful tools in digital signal processing because the FFT algorithm can be used in computing the DHT.

The frequency response $H(e^{j\omega})$ of the DHT is defined by

$$H(e^{j\omega}) = \begin{cases} -j, & 0 \leq \omega < \pi \\ j, & -\pi \leq \omega < 0. \end{cases} \tag{4.98}$$

It is clear that (4.98) acts as a 90-degree phase shifter (converting sines into cosines and vice versa). The impulse response $h[n]$ corresponding to $H(e^{j\omega})$ is given by

$$h[n] = \frac{1}{2\pi}\int_{-\pi}^{0} je^{j\omega n}d\omega - \frac{1}{2\pi}\int_{0}^{\pi} je^{j\omega n}d\omega \tag{4.99}$$

$$= \begin{cases} \frac{2}{\pi}\left[\frac{sin^2(\pi n/2)}{n}\right], & n \neq 0 \\ 0, & n = 0. \end{cases} \tag{4.100}$$

The impulse response $h[n]$ of the DHT in (4.100) is plotted in Figure 4.5. The impulse response $h[n]$ of the discrete Hilbert transformer can be used to form a complex discrete-time signal as shown in Figure 4.6.

In order to compute the DHT by using the DFT algorithm, we first relate the N-point sequence $x[n]$ to the N-point DFT $X[k]$. We then multiply the weight sequence $W[k]$ by $X[k]$, and convert $W[k]X[K]$ to the sequence $\hat{x}[n]$. Specifically,

$$x[n] \overset{DFT}{\Longrightarrow} X[k] \overset{W[k]}{\Longrightarrow} W[k]X[k] \overset{IDFT}{\Longrightarrow} \hat{x}[n]. \tag{4.101}$$

Let $x[n]$ denote an N-point sequence that is zero outside $0 \leq n \leq N-1$. The N-point DFT $X[k]$ of sequence $x[n]$ is given by

$$X[k] = \sum_{n=0}^{N-1} x[n]W_N^{kn}, \ k = 0, 1, ..., N-1, \tag{4.102}$$

where $W_N = e^{-j(2\pi/N)}$. We now define the weight functions $W[k]$ as follows:

$$W[k] = [0, \overbrace{j, j, ..., j}^{(N/2)-1}, 0, \overbrace{-j, -j, ..., -j}^{(N/2)-1}], \quad \text{if } N \text{ is even,} \tag{4.103}$$

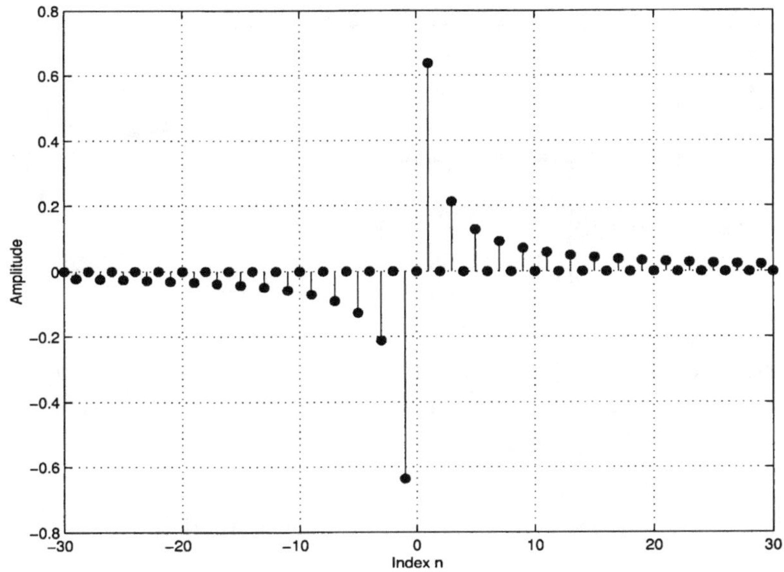

Figure 4.5 Impulse response $h[n]$ of the discrete Hilbert transform.

and

$$W[k] = [0, \overbrace{j, j, ..., j}^{(N-1)/2}, \overbrace{-j, -j, ..., -j}^{(N-1)/2}], \qquad \text{if } N \text{ is odd.} \qquad (4.104)$$

Finally, the output sequence $\hat{x}[n]$ is obtained by using the inverse DFT

$$\hat{x}[n] = \frac{1}{N} \sum_{k=0}^{N-1} \{W[k]X[k]\} W_N^{-kn}, \qquad n = 0, 1, ..., N-1. \qquad (4.105)$$

As a result, the sequence $\hat{x}[n]$ is the DHT of the sequence $x[n]$.

4.8 The Discrete Fractional Hilbert Transform

The discrete fractional Hilbert transform (DFHT) has only recently been developed (see Pei and Yeh [27], and Tseng and Pei [28]). So far, the DFHT has only been applied to digital communications, and edge and corner detections in images. The DFHT has much potential and needs further study.

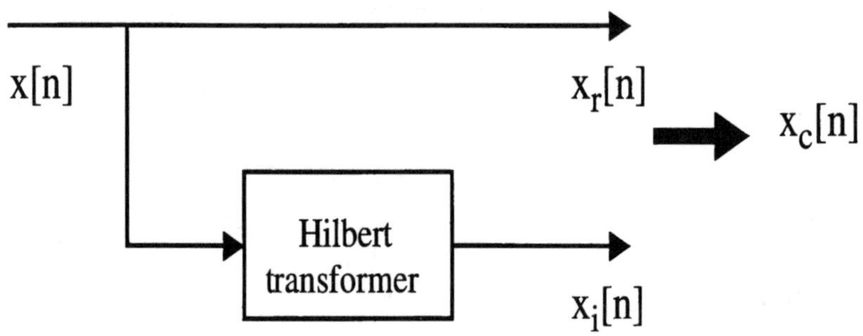

Figure 4.6 Block diagram representation of the creation of a complex sequence by using the DHT, where $x_c[n] = x_r[n] + jx_i[n]$.

The frequency response of the DFHT is defined by

$$H_v(e^{j\omega}) = \begin{cases} e^{-j\phi}, & 0 \leq \omega < \pi \\ e^{j\phi}, & -\pi \leq \omega < 0, \end{cases} \tag{4.106}$$

where v is its order and $\phi = \frac{v\pi}{2}$. Note that the DFHT in (4.106) is a $\frac{v\pi}{2}$-phase shifter rather than a 90-degree phase shifter. The impulse response of the DFHT is given by

$$h_v[n] = \begin{cases} cos(\phi), & n = 0 \\ sin(\phi)\left[\frac{2sin^2(\frac{n\pi}{2})}{n\pi}\right], & n \neq 0. \end{cases} \tag{4.107}$$

Note that the impulse response $h_0[n] = \delta[n]$ in (4.107) when $v = 0$, and $h_1[n] = \frac{2}{\pi}\left[\frac{sin^2(n\pi/2)}{n}\right]$ when $v = 1$. As can be seen, $h_1[n]$ is the DHT in (4.100). Therefore, using the functions of $h_0[n]$ and $h_1[n]$, (4.107) can be rewritten as

$$h_v[n] = cos(\phi)h_0[n] + sin(\phi)h_1[n]. \tag{4.108}$$

Taking the DTFT of (4.108), we obtain

$$H_v(e^{j\omega}) = cos(\phi)H_0(e^{j\omega}) + sin(\phi)H_1(e^{j\omega}). \tag{4.109}$$

Equation (4.109) indicates that the DFHT of $x[n]$ is a weighted sum of the original signal and its DHT.

4.8.1 Properties of the DFHT

Based on (4.106), (4.107), (4.108), and (4.109), the properties of the DFHT can be observed as follows.

Additivity

The DFHT satisfies the additive property of the order because $H_{v_1+v_2}(e^{j\omega}) = H_{v_1}(e^{j\omega})H_{v_2}(e^{j\omega})$. In other words, a signal passing through the cascaded systems $H_{v_1}(e^{j\omega})$ and $H_{v_2}(e^{j\omega})$ is equivalent to passing it through the system $H_{v_1+v_2}(e^{j\omega})$.

Unit Sample

When the order $v = 0$, the frequency response $H_0(e^{j\omega}) = 1$. In this case, a signal that passes through a DFHT is unchanged when the order $v = 0$; the impulse response $h_0[n]$ is equal to the unit sample $\delta[n]$.

Periodic Order

The frequency response $H_v(e^{j\omega})$ is periodic in order v. For example, the period of the DFHT is N, where N is an integer, because $H_{v+N}(e^{j\omega}) = H_v(e^{j\omega})$.

The DHT Property

The DFHT is equal to the DHT when the order $v = 1$. In this case, the impulse response $h_1[n] = \frac{2}{\pi}\left[\frac{sin^2(n\pi/2)}{n}\right]$. Therefore, this DFHT exhibits the DHT properties.

The Zero-Value

The impulse response $h_v[n]$ of the DFHT becomes zero-valued if and only if n is an even integer. This property of the DFHT enables us to design an FIR DFHT approximation by using the window design method.

4.8.2 The Sideband Signal Representation

Consider a real sequence $x[n]$. The corresponding complex analytic signal $x_v[n]$ is defined by

$$x_v[n] = x[n] - e^{-j\phi}\hat{x}_v[n], \tag{4.110}$$

where $\hat{x}_v[n]$ is the output of the DFHT with order v. Taking the DTFT for (4.110), we obtain

$$X_v(e^{j\omega}) = X(e^{j\omega}) - e^{-j\phi}\hat{X}_v(e^{j\omega}), \tag{4.111}$$

where $\hat{X}_v(e^{j\omega}) = H_v(e^{j\omega})X(e^{j\omega})$. For $0 \leq \omega < \pi$, substituting (4.106) into (4.111), (4.111) can be rewritten as

$$\begin{aligned} X_v(e^{j\omega}) &= X(e^{j\omega}) - e^{-j(2\phi)}X(e^{j\omega}) \\ &= \left(1 - e^{-j(2\phi)}\right)X(e^{j\omega}) \\ &= 2je^{-j(2\phi)}sin(\phi)X(e^{j\omega}). \end{aligned} \tag{4.112}$$

For $-\pi \leq \omega < 0$, substituting (4.106) into (4.111), we obtain the expression in the frequency domain as follows:

$$X_v(e^{j\omega}) = X(e^{j\omega}) - e^{j(\phi)}e^{-j(\phi)}X(e^{j\omega}) = 0. \tag{4.113}$$

As we can see, based on (4.113), the frequency content of $x_v[n]$ in the negative range from $-\pi$ to 0 is completely suppressed. However, the term $sin(\phi)$ in (4.112) is not zero if ϕ is not an integer multiple of π. In this case, the signal $x_v[n]$ will have the same information as $x[n]$. Therefore, this DFHT process can reduce the signal frequency bandwidth by half for the transmission of the signal in a communication channel. Often this is called single sideband.

4.9 The Discrete-Time Wavelet Transform

In Section 2.6.4, we introduced the z-domain discrete-time transfer function. A discrete-time LTI system can be expressed as

$$Y(z) = H(z)X(z), \tag{4.114}$$

where $H(z)$, $X(z)$, and $Y(z)$ are the system function, the input, and the output, respectively, all expressed in the z-domain. $H(z)$ is also called a *filter*.

For the development of the discrete-time wavelet transform (DTWT), we define filter $H(z) = \sum_{k=0}^{M-1} H_k(z)$. Then we define a discrete-time analysis filter as

$$H_k(z) = H(z^{2^k}), \tag{4.115}$$

where k is a positive integer. Therefore, the DTWT in the z-domain is obtained by

$$Y_k(z) = H_k(z)X(z). \tag{4.116}$$

The corresponding DTWT in the discrete-time domain is given by

$$y_k[n] = \sum_{m=-\infty}^{\infty} x[m]h_k[2^{k+1}n - m], \quad 0 \le k \le M-2, \tag{4.117}$$

and

$$y_{M-1}[n] = \sum_{m=-\infty}^{\infty} x[m]h_{M-1}[2^{M-1}n - m]. \tag{4.118}$$

Equations (4.117) and (4.118) are the definitions of the DTWT. The signal $y_k[n]$ is referred to as the set of *wavelet coefficients* of $x[n]$.

For the inverse DTWT, we first assume that a discrete-time synthesis filter $S_k(z)$ can be appropriately designed so that the reconstructed sequence $\hat{x}[n]$ is equal to the input sequence $x[n]$. Under this condition, we can express the inverse DTWT in the z-domain as

$$\begin{aligned}\hat{X}(z) &= S_0(z)Y_0(z^2) + S_1(z)Y_1(z^4) + \cdots + S_{M-2}(z)Y_{M-2}(z^{2^{M-2}}) \\ &\quad + S_{M-1}(z)Y_{M-1}(z^{2^{M-1}}).\end{aligned} \tag{4.119}$$

Therefore, the inverse DTWT in the discrete-time domain is obtained by

$$\begin{aligned}\hat{x}[n] &= \sum_{k=0}^{M-2} \sum_{m=-\infty}^{\infty} y_k[m]s_k[n - 2^{k+1}m] \\ &\quad + \sum_{m=-\infty}^{\infty} y_{M-1}[m]s_{M-1}[n - 2^{M}m],\end{aligned} \tag{4.120}$$

where the discrete-time synthesis filter $s_k[n - 2^{k+1}m]$ is referred to as the *shifted wavelet basis function*. The combination of the forward DTWT and inverse DTWT has many applications in the design of multirate filter banks [29, 30].

4.10 The Discrete Walsh Transform

The discrete Walsh transform (DWT) is useful in image processing, pattern recognition, digital filtering, and communication computation. The DWT can be calculated using algorithms similar to those used for the DFT, so that the arithmetic complexity is about $Nlog_2N$ addition operations.

For a sequence $x[n]$ with N samples, $N = 2^M$, the DWT of $x[n]$ is defined as

$$X[k] = \sum_{n=0}^{N-1} x[n] \left\{ \prod_{i=0}^{M-1} (-1)^{b_i[n] b_{M-1-i}[k]} \right\}, \quad k = 0, 1, ..., N-1, \tag{4.121}$$

where $b_i[n]$ is the ith bit in the binary representation of n. When $i = 0$, $b_0[n]$ is the least significant bit (LSB).

The corresponding inverse DWT is given by

$$x[n] = \frac{1}{N} \sum_{n=0}^{N-1} X[k] \left\{ \prod_{i=0}^{M-1} (-1)^{b_i[n] b_{M-1-i}[k]} \right\}, \quad n = 0, 1, ..., N-1. \tag{4.122}$$

Note that the definition of inverse DWT in (4.122) is the same as the forward DWT in (4.121) except for a constant scaling, $\frac{1}{N}$. Therefore, the algorithms for computing the forward DWT can be used for computing the inverse DWT as well.

A modified DWT, called the *plus-minus* (PM) DWT, was developed by Sundararajan and Ahmad [31, 32]. The PM algorithm is derived from the fact that the computation of the two-point transform of data values a and b is just a plus b or a minus b. The PM DWT algorithm provides a way to enhance performance in terms of the overhead operations of bit-reversal, data swapping, and array-index updating.

In order to develop the PM DWT algorithm, (4.121) is first rewritten in a form with $\frac{N}{2}$ operation as

$$\begin{aligned} X[k] &= \sum_{n=0}^{N/2-1} \left\{ \left(\sum_{d=0}^{1} x\left[n + d\frac{N}{2}\right] (-1)^{b_{M-1}(n+d\frac{N}{2}) b_0[k]} \right) \right. \\ &\quad \left. \times \prod_{i=0}^{M-2} (-1)^{b_i[n] b_{M-1-i}[k]} \right\}. \end{aligned} \tag{4.123}$$

As we can see, for each value of n, the inner summation produces a two-point DWT that will generate two distinct values for all of k. Then, we let a vector $\mathbf{v}[n]$ represent the nth input vector as given by

$$\begin{aligned}\mathbf{v}[n] &= \begin{bmatrix} v_0[n] \\ v_1[n] \end{bmatrix} \\ &= \begin{bmatrix} x[n] + x[n+N/2] \\ x[n] - x[n+N/2] \end{bmatrix},\end{aligned} \tag{4.124}$$

for $n = 0, 1, ..., \frac{N}{2} - 1$. Substituting (4.124) into (4.123), we obtain the PM DWT as

$$X[k] = \sum_{n=0}^{\frac{N}{2}-1} \left\{ \begin{bmatrix} x[n] + x[n+N/2] \\ x[n] - x[n+N/2] \end{bmatrix} \prod_{i=0}^{M-2} (-1)^{b_i[n]b_{M-1-i}[k]} \right\}. \tag{4.125}$$

The values of $\mathbf{v}[n]$ will repeat if $k \geq 2$ in (4.125), because the values of $\mathbf{v}[n]$ are a period of two. Therefore, replacing the argument k by using $k + p\frac{N}{2}$ (k mod 2) in (4.125) yields

$$\begin{aligned}\begin{bmatrix} X[k] \\ X[k+\frac{N}{2}] \end{bmatrix} &= \sum_{n=0}^{\frac{N}{2}-1} \left\{ \begin{bmatrix} x[n] + x[n+N/2] \\ x[n] - x[n+N/2] \end{bmatrix} \right. \\ &\quad \left. \cdot \prod_{i=0}^{M-2} (-1)^{b_i[n]b_{M-1-i}[k+p\frac{N}{2}]} \right\},\end{aligned} \tag{4.126}$$

for $k = 0, 1, ..., \frac{N}{2} - 1$. As a result, applying the vector form of input and output changes the basic definition of the DWT in (4.121). However, the new definition of the DWT requires the computation of 2-point transforms for each $\frac{N}{2}$ ordered input sequence. The operation of the new definition of the DWT is based on a two-dimensional (2-D) rather than a one-dimensional (1-D) interpretation.

4.11 The Discrete Hadamard Transform

Like the DWT, the discrete Hadamard transform (DHAT) has applications to image processing, pattern recognition, digital filtering, and wireless communication. The DHAT can be computed using the algorithms similar to those for the DFT. The arithmetic complexity of the DHAT is approximately $Nlog_2N$ addition operations.

For a sequence $x[n]$ with N samples, $N = 2^M$, the DHAT of sequence $x[n]$ is defined by

$$X[k] = \sum_{n=0}^{N-1} x[n]\left\{(-1)^{\left[\sum_{i=0}^{M-1} b_i[n]p_i[k]\right]}\right\}, \quad k = 0, 1, ..., N-1, \tag{4.127}$$

where $b_i[n]$ is the ith bit in the binary representation of n. When $i = 0$, $b_0[n]$ is the least significant bit (LSB). The relationship between $p_i(k)$ and $b_i(k)$ is as follows:

$$\begin{aligned} p_0[k] &= b_{M-1}[k] \\ p_1[k] &= b_{M-1}[k] + b_{M-2}[k] \\ p_2[k] &= b_{M-2}[k] + b_{M-3}[k] \\ &\vdots \\ p_{M-1}[k] &= b_1[k] + b_0[k]. \end{aligned}$$

The corresponding inverse DHAT is given by

$$x[n] = \frac{1}{N}\sum_{n=0}^{N-1} X[k]\left\{(-1)^{\left[\sum_{i=0}^{M-1} b_i[n]p_i[k]\right]}\right\}, \quad n = 0, 1, ..., N-1. \tag{4.128}$$

As we can see, the definition of the inverse DHAT in (4.128) is the same as the forward DHAT in (4.127) except for a constant scaling, $\frac{1}{N}$. Therefore, the algorithms for computing the forward DHAT can be used for computing the inverse DHAT as well.

In order to develop the PM DHAT algorithm, (4.127) can be rewritten in a form with $\left(\frac{N}{2}\right)$-point operations as

$$\begin{aligned} X[k] &= \sum_{n=0}^{\frac{N}{2}-1}\left\{(-1)^{\left[\sum_{i=0}^{M-2} b_{i+1}[2n]p_{i+1}[k]\right]}\right. \\ &\quad \left.\times \sum_{d=0}^{1} x[2n+d](-1)^{[b_0[2n+d]b_{M-1}[k]]}\right\}. \end{aligned} \tag{4.129}$$

As we can see, for each value of n, the inner summation produces a two-point DHAT that will generate two distinct values for all k.

Then, we can let a vector $\mathbf{u}[n]$ represent the nth input vector as given by

$$\mathbf{u}[n] = \begin{bmatrix} u_0[n] \\ u_1[n] \end{bmatrix} = \begin{bmatrix} x[2n] + x[2n+1] \\ x[2n] - x[2n+1] \end{bmatrix}, \tag{4.130}$$

for $n = 0, 1, ..., \frac{N}{2} - 1$. Substituting (4.130) into (4.129), we than obtain the PM DHAT as

$$X[k] = \sum_{n=0}^{\frac{N}{2}-1} \left\{ \begin{bmatrix} x[2n] + x[2n+1] \\ x[2n] - x[2n+1] \end{bmatrix} (-1)^{\left[\sum_{i=0}^{M-2} b_{i+1}[2n] p_{i+1}[k]\right]} \right\}. \tag{4.131}$$

The values of $\mathbf{u}[n]$ will repeat, because $b_{M-1}[k]$ is zero if $k < \frac{N}{2}$ and $b_{M-1}[k]$ is one if $k \leq \frac{N}{2}$. Therefore, replacing the argument k by using $2k + p$ and replacing the argument $b_{i+1}[2n]$ by using $br_i[2n]$ in (4.131) yields

$$\begin{bmatrix} X[2k] \\ X[2k+1] \end{bmatrix} = \sum_{n=0}^{\frac{N}{2}-1} \left\{ \begin{bmatrix} x[2n] + x[2n+1] \\ x[2n] - x[2n+1] \end{bmatrix} \cdot (-1)^{\left[\sum_{i=0}^{M-2} br_i[2n] p_{i+1}[2k+p]\right]} \right\} \tag{4.132}$$

for $k = 0, 1, ..., \frac{N}{2} - 1$, where $br_i[2n]$ denotes the bit-reversal operation on the index of n. As a result, applying the vector form of input and output changes the basic definition of the DHAT in (4.127). However, the new definition of the DHAT requires the computation of a 2-point transform for each $\frac{N}{2}$ ordered input sequence. As can be seen, the operation of the new definition of the DHAT is based on a two-dimensional (2-D) rather than a one-dimensional (1-D) representation.

4.12 Summary

In this chapter, we discussed many discrete-time transforms that are at the heart of discrete-time signal processing. We described their properties as well as their relationships among one another. The DFT is the most widely used transform, as the Fourier representation of finite-length sequences is equal to samples of the z-transform at equally spaced points on the unit circle. The DFT and its efficient implementation by FFT algorithms is among the most important transforms applied to digital signal processing.

The DCT and its relationship with the DFT were introduced. We showed that the DCT and its inverse can be computed by using the DFT and inverse DFT algorithms. We also described the DST. The relationships between the DCT and DST were explored, and it was shown that the DST can be calculated by using the DCT. Conversely, the DCT can be computed by using the DST.

The output sequences of the DHLT and its inverse are real when the input sequence is real. The inverse DHLT is identical to the DHLT except for a scaling factor. Thus, only one routine for computation is required for the DHLT and its inverse. Also, we introduced fast computation for the DHLT and its inverse for length $q \cdot 2^m$.

The DHT is a 90-degree phase shifter, while the DFHT is a $\frac{v\pi}{2}$ phase shifter. The DFHT is relatively new and needs to be explored further. An important application of the DFHT involves reducing the signal frequency bandwidth in a communication channel.

We presented the DTWT and its inverse based on the concept of a multirate filter bank, which will be introduced in Chapter 7. The DTWT was treated as the analysis filter while the inverse DTWT was treated as the synthesis filter.

Finally, the definitions of DWT and the DHAT were introduced. In addition, the alternative form of PM DWT and PM DHAT was shown in terms of 2-D representations of input and output rather than a 1-D representation, enabling reduced computational complexity.

References

[1] Cooley, J. W., and J. W. Tukey, "An Algorithm for the Machine Computation of Complex Fourier Series," *Mathematics of Computation*, Vol. 19, pp. 297-301, April 1965.

[2] Cooley, J. W., P. A. W. Lewis, and P. D. Welch, "Historical Notes on the Fast Fourier Transform," *IEEE Trans. on Audio Electroacoustics*, Vol. AU-15, pp. 76-79, June 1967.

[3] Oppenheim, A. V., and R. W. Schafer, *Digital Signal Processing*, Prentice Hall, Englewood Cliffs, New Jersey, 1975.

[4] Oppenheim, A. V., and R. W. Schafer, *Discrete-Time Signal Processing*, Prentice Hall, New Jersey, 1989.

[5] Dudgeon, D. E., and R. M. Mersereau, *Multidimensional Digital Signal Processing*, Prentice Hall, Englewood Cliffs, New Jersey, 1984.

[6] Brigham, E. O., *The Fast Fourier Transform and Its Applications*, Prentice Hall, Englewood Cliffs, New Jersey, 1988.

[7] Chen, C. H. (eds.), *Signal Processing Handbook*, Marcel Dekker, Inc., New York, 1988.

[8] Lim, J. E., *Two-Dimensional Signal and Image Processing*, Prentice Hall, Englewood Cliffs, New Jersey, 1990.

[9] Burrus, C. S., J. H. McClellan, A. V. Oppenheim, T. W. Parks, R. W. Schafer, and H. W. Schuessler, *Computer-Based Exercises for Signal Processing*, Prentice Hall, Englewood Cliffs, New Jersey, 1994.

[10] Jain, A. K., "A Fast Karhunen-Loève Transform for a Class of Stochastic Processes," *IEEE Trans. on Communications*, Vol. COM-24, pp. 1023-1029, 1976.

[11] Rao, K. R., and P. Yip, *Discrete Cosine Transform*, Academic Press, Inc., New Jersey, 1990.

[12] Miao, J., *Component Feature-Based Digital Waveform Analysis and Classification*, UMI, Ann Arbor, Michigan, 1996.

[13] Quackenbush, S. R., T. B. Barnwell, and M. A. Clements, *Objective Measures of Speech Quality*, Englewood Cliffs, Prentice Hall, New Jersey, 1988.

[14] Miao, J., "Vector Quantization of DCT Components for Speech Coding," *Proc. of Australian Inte. Conf. of Speech Science and Technology*, Roberto Togneri (eds.), Vol. 2, pp. 551-556, Perth, Australia, December 1994.

[15] Kar, D. C., and V. V. B. Rao, "On the Prime Factor Decomposition Algorithm for the Discrete Sine Transform," *IEEE Trans. on Signal Processing*, Vol. 42, No. 11, pp. 3258-3260, November 1994.

[16] Bracewell, R. N., "Discrete Hartley Transform," *Journal of Optical Society*, Vol. 73, No. 12, pp. 1832-1835, 1983.

[17] Bracewell, R. N., "The Fast Hartley Transform," *Proc. of IEEE*, Vol. 72, No. 8, pp. 1010-1018, 1984.

[18] Bi, G., and Y. Q. Chen, "Fast DHT Algorithms for Length $N = q * 2^m$," *IEEE Trans. on Signal Processing*, Vol. 47, No. 3, pp. 900-903, March 1999.

[19] Vijayakumar, N., and K. M. M. Prabhu, "FHT Algorithm for Length $N = q \cdot 2^m$," *Electric Letters*, Vol. 35, No. 12, pp. 966-968, June 1999.

[20] Chan, Y. H., and W. C. Siu, "New Fast Discrete Hartley Transform Algorithm," *Electronic Letters*, Vol. 27, No. 4, pp. 347-349, February 1991.

[21] Hu, N. C., H. I. Chang, and O. K. Ersoy, "Generalized Discrete Hartley Transforms," *IEEE Trans. on Signal Processing*, Vol. 40, No. 12, pp. 2931-2940, December 1992.

[22] Mether, P. K., J. K. Satapathy, and G. Panda, "New High-Speed Prime-Factor Algorithm for Discrete Hartley Transform," *IEE Proc.-F*, Vol. 140, No. 1, pp. 63-70, February 1993.

[23] Bi, G., and C. Lu, "Prime-Factor Algorithms for Generalised Discrete Hartley Transform," *Electronic Letters*, Vol. 35, No. 20, pp. 1708-1710, September 1999.

[24] Bi, G., Y. Q. Chen, and Y. Zeng, "Fast Algorithms for Generalized Discrete Hartley Transform of Composite Sequence Lengths," *IEEE Trans. on Circuits and Systems II: Analog and Digital Signal Processing*, Vol. 47, No. 9, pp. 893-901, September 2000.

[25] Cheng, L. Z., "Fast Computation of Discrete W Transform Through Discrete Hartley Transform (DHT)," *Electronic Letters*, Vol. 36, No. 5, pp. 474-476, March 2000.

[26] Venkata, N. D., B. L. Evans, and S. R. McCaslin, "Design of Optimal Minimum-Phase Digital FIR Filters Using Discrete Hilbert Transforms," *IEEE Trans. on Signal Processing*, Vol. 48, No. 5, pp. 1491-1495, May 2000.

[27] Pei, S. C., and M. H. Yeh, "Discrete Fractional Hilbert Transform," *IEEE Trans. on Circuits and Systems II: Analog and Digital Signal Processing*, Vol. 47, No. 11, pp. 1307-1311, November 2000.

[28] Tseng, C. C., and S. C. Pei, "Design and Application of Discrete-Time Fractional Hilbert Transformer," *IEEE Trans. on Circuits and Systems II: Analog and Digital Signal Processing*, Vol. 47, No. 12, pp. 1529-1533, December 2000.

[29] Gilbert, S., and T. Nguyen, *Wavelets and Filter Banks*, Wellesley-Cambridge Press, Wellesley, Massachusetts, 1996.

[30] Hanna, M. T., and S. A. Mansoori, "The Discrete Time Wavelet Transform: Its Discrete Time Fourier Transform and Filter Bank Implementation," *IEEE Trans. on Circuits and Systems II: Analog and Digital Signal Processing*, Vol. 48, No. 2, pp. 180-183, February 2001.

[31] Sundararajan, D., and M. O. Ahmad, "Fast Computation of the Discrete Walsh Transform," *IEEE 39th Midwest Symp. on Circuits and Systems*, Vol. 2, pp. 785-788, 1996.

[32] Sundararajan, D., and M. O. Ahmad, "Fast Computation of the Discrete Walsh and Hadamard Transforms," *IEEE Trans. on Image Processing*, Vol. 7, No. 6, pp. 898-904, 1998.

5

Digital Filtering

5.1 Introduction

In a variety of digital signal processing applications, it is important to change the relative amplitudes of the frequency components in digital signals or perhaps eliminate some frequency components entirely. This process is usually referred to as *digital filtering*. In the linear time-invariant (LTI) discrete-time systems, the output spectrum is that of the input signals multiplied by the frequency response of the discrete-time system. Consequently, digital filtering can be conveniently accomplished through the use of those discrete-time systems with an appropriately chosen frequency response.

An LTI discrete-time system can be completely characterized in the discrete-time domain by its impulse response $h[n]$, with the output $y[n]$ due to a given input $x[n]$ specified through the convolution sum

$$\begin{aligned} y[n] &= x[n] * h[n] \\ &= \sum_{k=-\infty}^{\infty} x[k]h[n-k]. \end{aligned} \tag{5.1}$$

The corresponding frequency responses of the LTI discrete-time system input and output are related by

$$Y(e^{j\omega}) = H(e^{j\omega})X(e^{j\omega}), \tag{5.2}$$

where $X(e^{j\omega})$ and $Y(e^{j\omega})$ are the Fourier transforms of the LTI discrete-time system input and output, respectively. Usually, the impulse response $h[n]$ or the frequency response $H(e^{j\omega})$ are referred to as a *digital filter*. As can be seen, digital filtering plays a very important role in the application of LTI discrete-time systems. Therefore, this chapter uses the theory of discrete-time signal processing to develop the design and implementation methods for both *finite impulse response* (FIR) and *infinite impulse response* (IIR) digital filters.

The design of digital filters involves three stages as follows:

1. Define the specifications or desired characteristics of the digital filter. (Of course, the specifications of digital filters depend on the intended applications.)
2. Realize the specifications in designing the digital filter.
3. Develop the digital signal processing (DSP) hardware or software to be used for implementation of the digital filter.

As we can see, these three steps are certainly interrelated. The first step is based on the application, the second step depends on the design methods of the digital filter, and the third step relies on the technology to be used for the real-time or off-line DSP implementation. In this chapter, we focus primarily on the first and second steps, developing the approximate digital filter satisfying the required specifications. For the third step, we introduce the flow-graph methods for DSP implementations.

Digital filters can be classified into two groups based on the length of the impulse response $h[n]$. In the first group, the impulse response $h[n]$ is a finite-extent sequence, so the filters in this group are called FIR filters. In the second group, the impulse response $h[n]$ is of infinite extent, so the filters in this group are called IIR filters. Usually, FIR filters differ considerably in design and implementation from IIR filters. The major reason for this is that the FIR filter does not have stability issues in design and implementation.

For the purpose of practical applications, we will restrict our discussion to a certain class of digital filter. One assumption is that the impulse response $h[n]$ must be real to ensure that the processed digital signals are real. Another assumption is that the impulse response $h[n]$ is stable; that is, $\sum_{n=-\infty}^{\infty} | h[n] |< \infty$. This condition corresponds to the stability type of the bounded-input bounded-output, usually known as BIBO stability.[1] Therefore, we will only discuss the class of digital filters whose impulse response $h[n]$ is real and stable.

This chapter is organized as follows. Section 5.1 provided an introduction to digital filter design. Section 5.2 gives the high-level filter specifications in the frequency domain. FIR filters with linear phase are discussed in Section 5.3. Symmetric and antisymmetric FIR filters are addressed. The designs of FIR and IIR filters are introduced in Sections 5.4 and 5.5, respectively, with emphasis on the window, equiripple, and least squares methods for FIR filters, and the bilinear and frequency transformation methods for IIR filters. Section 5.6 provides the implementation structures based on the FIR and IIR filter flow-graphs. The chapter is summarized in Section 5.7.

5.2 Filter Specifications

The filter specifications of the log magnitude response of a digital filter are usually given in terms of certain tolerances as shown in Figure 5.1 for the lowpass, highpass, bandpass, and bandstop cases. They are plotted as a function of normalized frequency ($\omega = \Omega T_s$), where Ω is continuous-time frequency in radians per second and T_s is the sampling period. The reciprocal function $F_s = \frac{1}{T_s}$ is the corresponding sampling frequency, in samples per second. We assume the filter coefficients to be real, so only the region $0 \leq \omega \leq \pi$ has to be specified corresponding to the range $0 \leq f \leq 0.5F_s$, because the remainder can be inferred from symmetry properties and periodicity of filter function $H(e^{j\omega})$ in the frequency domain.

For a lowpass filter in Figure 5.1(a), the frequency region $0 \leq \omega \leq$

1. The BIBO denotes the bounded-input bounded-output. A discrete-time system is stable in the BIBO sense if and only if every bounded input produces a bounded output.

ω_p is the passband and $\omega_s \leq \omega \leq \pi$ is the frequency stopband. The log magnitude response in the passband and stopband is required to lie within the tolerance regions. Between the frequency passband and stopband, we have to specify a transition band region, since the log magnitude response cannot change abruptly from unity gain to zero. The transition band is sometimes also called the "don't care" frequency region.

The standard definition of digital filter parameters for the lowpass case as shown in Figure 5.1(a) uses the following terminology:

1. δ_1 is called the peak passband ripple.
2. δ_2 is called the peak stopband ripple.
3. $A_p = 20\log_{10}(1+\delta_1)$ is the peak passband ripple in decibels.
4. $A_s = -20\log_{10}(\delta_2)$ is the minimum stopband attenuation in decibels.
5. ω_p is the frequency passband edge.
6. ω_s is the frequency stopband edge.
7. $\Delta\omega = \omega_s - \omega_p$ is defined as the transition bandwidth.

In a similar way, we can define the parameters for the cases of highpass, bandpass, and bandstop filters as shown in Figure 5.1(b), (c), and (d), respectively. However, for simplicity, we omit these parameter definitions.

In frequency-selective filters, the magnitude of the frequency response in the passband must approximate unity with an error of $\pm\delta_1$

$$(1-\delta_1) \leq | H(e^{j\omega}) | \leq (1+\delta_1), \quad | \omega | \leq \omega_p. \tag{5.3}$$

The other approximation band is the stopband, in which the magnitude response must approximate zero with an error less than

$$| H(e^{j\omega}) | \leq \delta_2, \quad \omega_s \leq | \omega | \leq \pi. \tag{5.4}$$

To approximate the ideal lowpass filter in such a way with a realizable discrete-time system, we must provide a transition band of nonzero width $(\omega_s - \omega_p)$ in which the magnitude of the frequency response changes smoothly from passband to stopband. Based on the above definition, the magnitude of the frequency response of a designed digital filter should meet the prescribed specification.

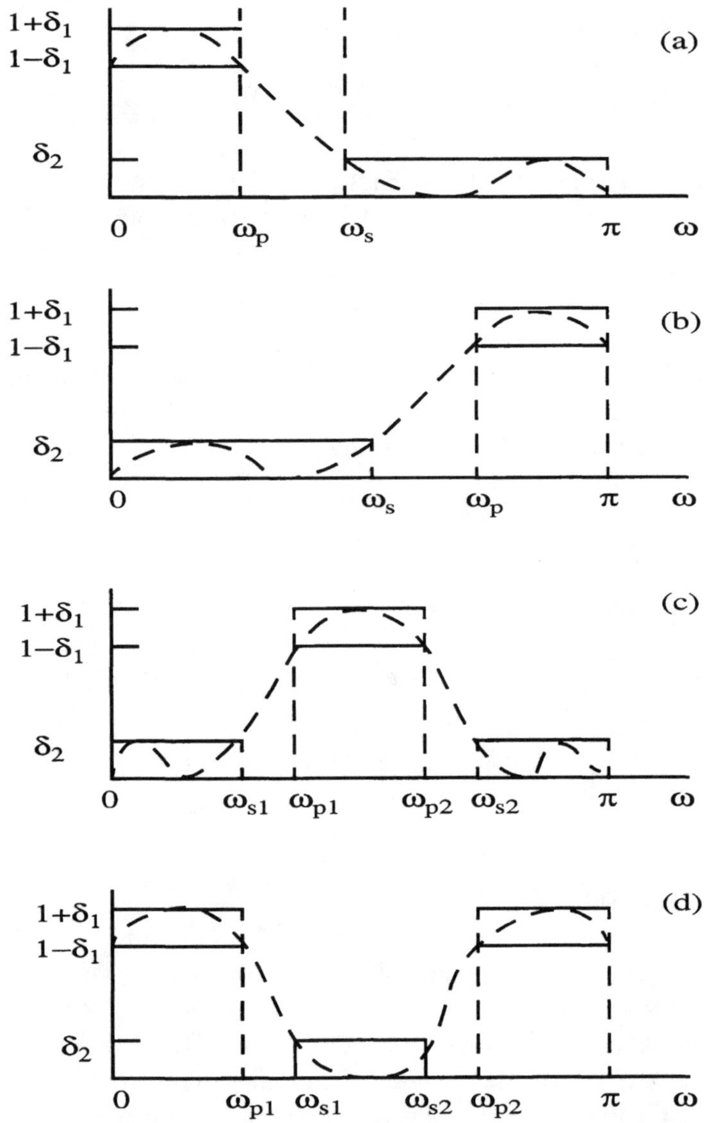

Figure 5.1 Specifications for digital filters: (a) Lowpass filter. (b) Highpass filter. (c) Bandpass filter. (d) Bandstop filter.

5.3 FIR Linear Phase

An FIR filter with input $x[n]$ and output $y[n]$ is given by the difference equation

$$y[n] = \sum_{k=0}^{M-1} b_k x[n-k], \tag{5.5}$$

where b_k is a set of filter coefficients, and M is the length of the filter. Let $h[k] = b_k$, $k = 0, 1, ..., M-1$. We can express (5.5) as follows:

$$y[n] = \sum_{k=0}^{M-1} h[k]x[n-k], \tag{5.6}$$

where the lower and upper limits in (5.6) are a reflection of causality and finite-extent sequence of the filter. The FIR filter can also be characterized by its system function

$$H(z) = \sum_{k=0}^{M-1} h[k]z^{-k}. \tag{5.7}$$

The FIR filter can be described by the zeros (roots) of the polynomial.

An FIR filter has *linear phase* if its impulse response $h[n]$ satisfies the following conditions:

$$h[n] = \begin{cases} h[M-1-n], & 0 \leq n \leq M-1 \\ 0, & \text{otherwise}, \end{cases} \tag{5.8}$$

or

$$h[n] = \begin{cases} -h[M-1-n], & 0 \leq n \leq M-1 \\ 0, & \text{otherwise}. \end{cases} \tag{5.9}$$

The conditions in (5.8) and (5.9) are sufficient to guarantee a causal and finite-duration system with linear phase. However, these requirements are not necessary conditions. For example, causal infinite-duration impulse responses can also have linear phase, as shown by Clements and Pease [1]. The linear-phase IIR filter can also be designed from the FIR specification by Fahmy et al. [2].

In general, FIR filters with linear phase can be classified as symmetric or antisymmetric filters. The significance of this property will be discussed in the following sections.

5.3.1 Symmetric FIR Filters

Symmetric FIR filters have a symmetric impulse response as follows:

$$h[n] = h[M-1-n], \quad 0 \le n \le M-1. \tag{5.10}$$

The frequency response, $H(e^{j\omega})$, can be expressed as

$$H(e^{j\omega}) = H_{sym}(e^{j\omega})e^{-j\omega(M-1)/2}, \tag{5.11}$$

where $H_{sym}(e^{j\omega})$ is a real function of ω, and the delay $(M-1)/2$ is an integer.

When M is an odd integer, by applying the symmetry condition in (5.10), $H_{sym}(e^{j\omega})$ can be expressed as

$$H_{sym}(e^{j\omega}) = h[(M-1)/2]+2\sum_{k=0}^{(M-3)/2} h[k]cos\omega[(M-1)/2-k], \tag{5.12}$$

where $H_{sym}(e^{j\omega})$ is a real, even, and periodic function of ω. In this case, $H(e^{j\omega})$ in (5.11) is referred to as a type I FIR filter with linear phase.

When M is an even integer, the symmetry condition in (5.10), $H_{sym}(e^{j\omega})$ can be expressed as

$$H_{sym}(e^{j\omega}) = 2\sum_{k=0}^{M/2-1} h[k]cos\omega[(M-1)/2-k], \tag{5.13}$$

where $H_{sym}(e^{j\omega})$ is also a real, even, and periodic function of ω. In this case, $H(e^{j\omega})$ in (5.11) is referred to as a type II FIR filter with linear phase.

5.3.2 Antisymmetric FIR Filters

If the FIR filter has an antisymmetric impulse response, then $h[n]$ is

$$h[n] = -h[M-1-n], \quad 0 \le n \le M-1. \tag{5.14}$$

It can be shown that the frequency response, $H(e^{j\omega})$, can be expressed as

$$H(e^{j\omega}) = H_{anti}(e^{j\omega})e^{-j\omega(M-1)/2+j\pi/2}, \tag{5.15}$$

where $H_{anti}(e^{j\omega})$ is a real, odd, and periodic function of ω, and the delay $(M-1)/2$ is an integer.

When M is equal to an odd integer, the center point of the antisymmetric impulse response $h[n]$ is $n = (M-1)/2$ so that $h[(M-1)/2] = 0$. By applying the antisymmetry condition in (5.14), $H_{anti}(e^{j\omega})$ can be expressed as

$$H_{anti}(e^{j\omega}) = 2 \sum_{k=0}^{(M-3)/2} h[k] sin\omega[(M-1)/2 - k]. \tag{5.16}$$

As can be seen, $H_{anti}(e^{j\omega})$ in (5.16) is a real, odd, and periodic function of ω. In this case, $H(e^{j\omega})$ in (5.15) is referred to as a type III FIR filter with linear phase.

When M is an even integer, each term in the impulse response $h[n]$ has a matching term with an opposite sign. Thus, by applying the antisymmetry condition in (5.14), $H_{anti}(e^{j\omega})$ can be expressed as

$$H_{anti}(e^{j\omega}) = 2 \sum_{k=0}^{M/2-1} h[k] sin\omega[(M-1)/2 - k], \tag{5.17}$$

where $H_{anti}(e^{j\omega})$ is also a real, odd, and periodic function of ω. In this case, $H(e^{j\omega})$ in (5.15) is known as a type IV FIR filter with linear phase.

So far, we have introduced symmetric and antisymmetric FIR filters and classified them into four types of FIR filters with linear phase. Note that the different types of FIR filters with linear phase have different applications in filter design. A symmetric FIR filter is suitable for some applications, while an antisymmetric FIR filter is suitable for other applications. Therefore, the choice of a symmetric or antisymmetric FIR filter depends on what problem is being solved. For example, in the antisymmetric FIR filter, (5.16) indicates that $H_{anti}(e^{j0}) = 0$ and $H_{anti}(e^{j\pi}) = 0$ when M is an odd integer. Consequently, in this case, the antisymmetric FIR filter in (5.15) is not suitable as either a lowpass filter or a highpass filter. In other words, we would not use the type III FIR filters with linear phase to design either a lowpass filter or a highpass filter. In a similar way, when M is equal to an even integer, (5.17) indicates that $H_{anti}(e^{j0}) = 0$. Consequently, in this case, the antisymmetric FIR filter in (5.15) is not suitable as a lowpass filter. As a result, we would not use a type IV FIR filter with linear phase to design a lowpass filter. On the other hand, when M is equal to an odd or even integer, the symmetric FIR filter in (5.11) yields $H(e^{j0}) \neq 0$, since $H_{sym}(e^{j0})$ in (5.12) and (5.13) is not equal to zero. In other words, we design a lowpass filter by using type I or II FIR filters.

In the following section, we present design methods for the linear phase FIR filters based on the filter specifications.

5.4 FIR Filter Design

An ideal lowpass digital filter with frequency response $H_i(e^{j\omega})$ is given by

$$H_i(e^{j\omega}) = \begin{cases} 1 & |\omega| < \omega_c \\ 0 & \text{otherwise,} \end{cases} \tag{5.18}$$

where ω_c is the cutoff frequency. The corresponding impulse response $h_i[n]$ of the lowpass digital filter is

$$\begin{aligned} h_i[n] &= \frac{1}{2\pi}\int_{-\pi}^{\pi} H_i(e^{j\omega})e^{j\omega n}d\omega \\ &= \frac{1}{2\pi}\int_{-\omega_c}^{\omega_c} e^{j\omega n}d\omega \\ &= \frac{\omega_c}{\pi}\left[\frac{sin(\omega_c n)}{\omega_c n}\right], \end{aligned} \tag{5.19}$$

where $-\infty \leq n \leq \infty$.

Note that (5.18) is a zero-phase digital filter with unit magnitude in the passband and zero in the stopband, since $H_i(e^{j\omega}) = H_i^*(e^{j\omega})$ and $H_i(e^{j\omega})$ is always positive in value. However, the ideal digital filter as shown in (5.19) is unstable because the impulse response $h_i[n]$ does not satisfy the BIBO stability requirement. Furthermore, this digital filter cannot be implemented since its impulse response $h_i[n]$ is infinite, noncausal, and nonrational (i.e., it cannot be implemented using any sort of finite-order difference equation).

The problem of designing an FIR digital filter is basically one of determining approximate $h[n]$ or $H(e^{j\omega})$ that meets the design specification based on the approximation of an ideal digital filter

$H_i(e^{j\omega})$ such that

$$\int_0^{2\pi} |H(e^{j\omega}) - H_i(e^{j\omega})|^2 d\omega \Longrightarrow 0, \tag{5.20}$$

for a given length of the digital filter, N. The five standard approaches for designing digital FIR filters are the window, the frequency sampling, the frequency transformation, the optimal Chebyshev equiripple design, and the least squares design methods. Our discussion of these methods will only focus on window, optimal (equiripple), and least squares methods. These are often used by systems engineers in practical applications.

5.4.1 Linear-Phase FIR Design Using Window Functions

In the window method, the desired frequency response $H_i(e^{j\omega})$ is assumed to be known. By inverse Fourier transforming $H_i(e^{j\omega})$, we can determine the desired impulse response of the filter, $h_i[n]$. In general, $h_i[n]$ is an infinite-extent sequence. With a sequence $w[n]$, the impulse response $h_i[n]$ of FIR filter is given by

$$h[n] = h_i[n]w[n]. \tag{5.21}$$

The sequence $w[n]$ is called a *window function.* If $h_i[n]$ and $w[n]$ are both symmetric with respect to the origin, then (5.21) is also symmetric with zero phase.

In the frequency domain, the response $H(e^{j\omega})$ is the periodic convolution of the ideal response $H_i(e^{j\omega})$ with window function $W(e^{j\omega})$ as follows,

$$\begin{aligned} H(e^{j\omega}) &= H_i(e^{j\omega}) * W(e^{j\omega}) \\ &= \frac{1}{2\pi}\int_{-\pi}^{\pi} H_i(e^{j\omega})W(e^{j(\omega-\theta)})d\theta. \end{aligned} \tag{5.22}$$

Based on (5.22), the effect of the window function in the frequency domain is to smooth $H_i(e^{j\omega})$. We like to have the mainlobe width of $W(e^{j\omega})$ small so that the transition width of $H(e^{j\omega})$ is small. We also want to have small sidelobe amplitude to ensure that the ripples in the passband and stopband regions will have small amplitudes. By appropriate choice of the window function, it is possible to control both the attenuation and the transition bandwidth of the frequency response $H(e^{j\omega})$.

Most of the effort in this area has been directed toward finding optimal window functions. An optimal window shape will taper off at the ends of the input record but still have a reasonably compact and narrow spectrum. This is important, because multiplying the time sequence by the window will cause the corresponding spectrum to be convolved with the spectrum of the window. These window functions are summarized as follows:

- Rectangular window

$$w[n] = \begin{cases} 1, & 0 \leq n \leq M \\ 0, & \text{otherwise}, \end{cases} \tag{5.23}$$

- Triangular window

$$w[n] = \begin{cases} 1 - \frac{|n|}{M}, & 0 \leq n \leq M \\ 0 & \text{otherwise}, \end{cases} \tag{5.24}$$

- Bartlett window

$$w[n] = \begin{cases} \frac{2n}{M}, & 0 \leq n \leq M/2 \\ 2 - \frac{2n}{M}, & M/2 \leq n \leq M \\ 0, & \text{otherwise}, \end{cases} \tag{5.25}$$

- Hanning window

$$w[n] = \begin{cases} 0.5\left[1 - cos(\frac{2\pi n}{M})\right], & 0 \leq n \leq M \\ 0, & \text{otherwise}, \end{cases} \tag{5.26}$$

- Hamming window

$$w[n] = \begin{cases} 0.54 - 0.46cos(\frac{2\pi n}{M}), & 0 \leq n \leq M \\ 0, & \text{otherwise}, \end{cases} \tag{5.27}$$

- Blackman window

$$w[n] = \begin{cases} 0.42 - 0.5cos(\frac{2\pi n}{M}) + 0.08cos(\frac{4\pi n}{M}), & 0 \leq n \leq M \\ 0, & \text{otherwise}, \end{cases} \tag{5.28}$$

- Kaiser window

$$w[n] = \begin{cases} \frac{I_o\left[\beta\sqrt{1-(2n/M)^2}\right]}{I_o(\beta)}, & -M/2 \le n \le M/2 \\ 0, & \text{otherwise,} \end{cases} \tag{5.29}$$

where $I_o(x)$ is the modified zero*th*-order Bessel function and is given by

$$I_o(x) = 1 + \sum_{m=1}^{\infty} \left[\frac{(0.5x)^m}{m!}\right]^2. \tag{5.30}$$

Kaiser determined empirically that the value of β needed to achieve a specified value of A is given by

$$\beta = \begin{cases} 0.1102(A - 8.7), & A > 50 \\ 0.5842(A - 21)^{0.4} + 0.07886(A - 21), & 21 \le A \le 50 \\ 0, & A < 21. \end{cases} \tag{5.31}$$

In addition, to achieve prescribed values of A and $\Delta\omega$, M must satisfy

$$M = \frac{A - 7.95}{2.285\Delta\omega}, \tag{5.32}$$

where $\Delta\omega = \omega_s - \omega_p$ is the transition region for lowpass filter approximation and $A_s = -20\log_{10}(\delta_2)$ is the minimum stopband attenuation.

- Dolph-Chebyshev window

$$W[k] = \begin{cases} \frac{cosh(Mcosh^{-1}x)}{cosh(Ncosh^{-1}\alpha)}, & |x| > 1 \text{ for mainlobe} \\ \frac{cos(Mcos^{-1}x)}{cosh(Ncosh^{-1}\alpha)}, & |x| \le 1 \text{ for sidelobes} \end{cases} \tag{5.33}$$

where $x = \alpha cos(\pi k/M)$, and $-(M-1) \le k \le (M-1)$. The sidelobe attenuation is obtained for a value of α given by

$$\alpha = cosh\left[\frac{1}{M}cosh^{-1}(10^{\beta})\right], \tag{5.34}$$

where α is an integer. Note that (5.33) is the frequency response of the Dolph-Chebyshev window. The corresponding coefficients $w[n]$ are obtained by evaluating (5.33) over the indicated range of k and then performing an inverse DFT on this sampled frequency response.

- Generalized cosine window

$$w[n] = A - B\left[cos\left(\frac{2\pi n}{M-1}\right)\right] + C\left[cos\left(\frac{4\pi n}{M-1}\right)\right], \quad (5.35)$$

where A, B, and C are constants that you defined. The concept behind this window is that by summing the individual terms to form the window, the low-frequency peaks in the frequency domain combine in such a way to decrease sidelobe height. This has the side effect of increasing the mainlobe width.

Note that Blackman, Hamming, Hanning, and rectangular windows are all special cases of the generalized cosine window. Eventually, the Hamming and Hanning windows are two-term generalized cosine windows, given by $A = 0.54$, $B = 0.46$, and $C = 0$ for Hamming and $A = B = 0.5$, and $C = 0$ for Hanning, respectively. When $A = 0.42$, $B = 0.5$, and $C = 0.08$, the generalized cosine window becomes the Blackman window.

The Dolph-Chebyshev window is an optimal window and has the narrowest mainlobe for a given sidelobe level. Theoretically, it can achieve extremely large values of sidelobe attenuation. In addition, the maximum sidelobe level of $|W(e^{j\omega})|$ is minimized with equiripple behavior in the region $\omega_s \leq \pi$. Consequently, this is also called a *minimax* window.

The rectangular window also has the narrowest mainlobe for a given length, M, but the first sidelobe is only slightly reduced below the main peak. The sidelobes in the Hamming, Hanning, and Blackman windows are greatly reduced; however, the mainlobes are wider and thus have wider transitions at discontinuities of $H_i(e^{j\omega})$. Oppenheim and Schafer have extensively studied the detailed properties of those commonly used windows [3].

Example 5.1 Linear-Phase Lowpass FIR Filter Using a Kaiser Window

Suppose the design specifications of a lowpass FIR filter are $\omega_p = 0.25\pi$, $\omega_s = 0.35\pi$, $\delta_1 = 0.01$, and $\delta_2 = 0.001$, with a sampling rate F_s of 8 kHz. Then, the cutoff frequency of the FIR filter is given by

$$\omega_c = \frac{\omega_p + \omega_s}{2} = 0.3\pi. \tag{5.36}$$

To determine the parameters of the Kaiser window, we first compute

$$\begin{aligned} \Delta\omega &= \omega_s - \omega_p = 0.1\pi && (5.37)\\ A &= -20\log_{10}\delta = 60, \ \text{(dB)} && (5.38) \end{aligned}$$

where $\delta = 0.001$, since the filter designed by using the window method has $\delta_1 = \delta_2$. Substituting these two quantities into (5.31) and (5.32), the formulas calculate

$$\beta = 5.6532, \tag{5.39}$$

and

$$M = 73. \tag{5.40}$$

The order M can be rounded off to the next even integer, 74. This makes $M/2$ integer for (5.29).

The impulse response $h[n]$ of the lowpass FIR filter is computed using (5.21),

$$\begin{aligned} h[n] &= h_i[n]w[n] \\ &= \left\{ \frac{sin[\omega_c(n - M/2)]}{\pi(n - M/2)} \right\} \frac{I_o\left[\beta\sqrt{1 - (\frac{2n}{M} - 1)^2}\right]}{I_o(\beta)}, && (5.41) \end{aligned}$$

where $0 \leq n \leq M$. Otherwise, the impulse response $h[n]$ is equal to zero. Figure 5.2(a), which shows the impulse response $h[n]$, displays the characteristic symmetry of the FIR filter system with sample numbers in the x-axis and the amplitude in the y-axis. Figure 5.2(b) shows the log magnitude response (dB) in the y-axis and indicates that $H(e^{j\omega})$ is zero at the frequency of 4 kHz.

The use of the windowing technique is not restricted to lowpass filters. Windows can be used to truncate any ideal impulse response $h_i[n]$ to obtain a causal FIR approximation. However, the windowing technique is not suitable for design of digital filters with more sophisticated specifications such as unequal ripple sizes, nonconstant passband, and stopband frequency responses.

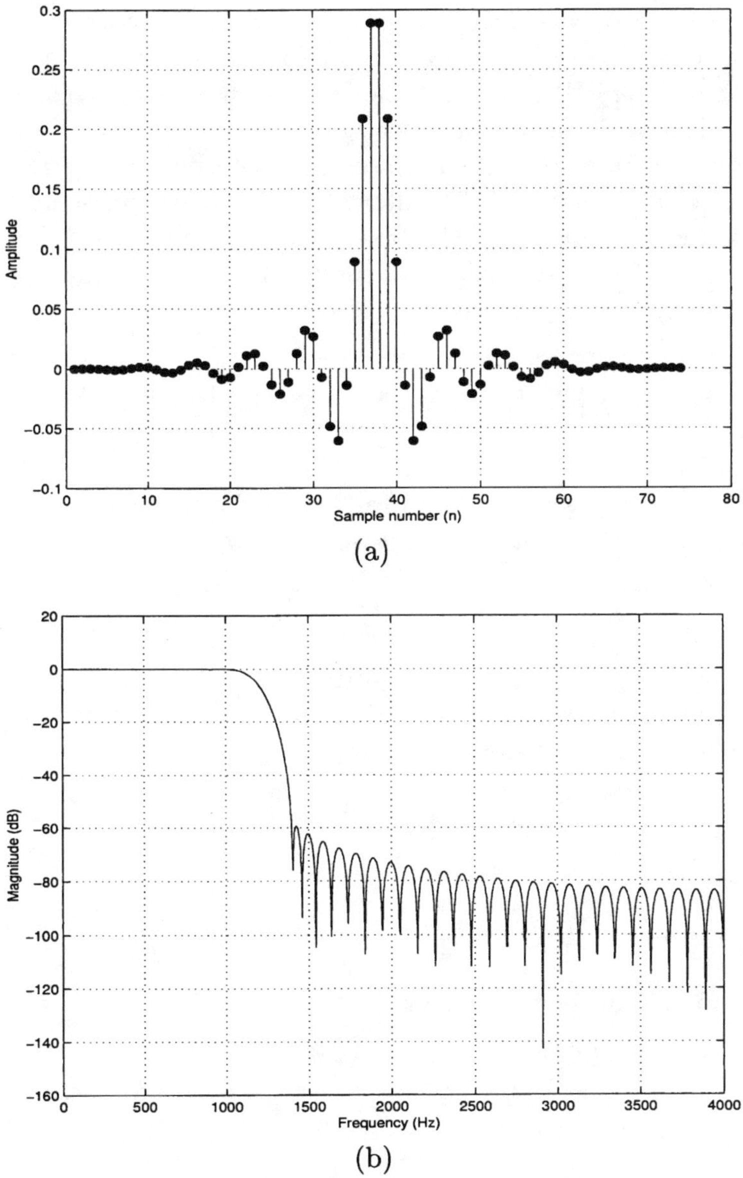

Figure 5.2 Design lowpass FIR filter using Kaiser window. (a) Impulse response when $M = 74$, and (b) log magnitude of FIR lowpass filter.

5.4.2 Linear-Phase FIR Using Equiripple Design

For a given set of specifications δ_1, δ_2, ω_1, and ω_2, it has been shown that an equiripple FIR filter has the smallest possible filter order, N. In addition, the FIR filter with the smallest weighted passband and stopband ripple is an equiripple design. The equiripple FIR design is often referred to as the *optimal filter* design technique.

Consider a symmetric FIR filter $h[n]$ with linear phase and filter impulse response $h[n] = h[M-1-n]$, for $0 \leq n \leq M-1$, with M odd integer. Recall that this is a type I FIR filter with linear phase. Thus, $H_{sym}(e^{j\omega})$ in (5.11) is expressed as

$$H_{sym}(e^{j\omega}) = h\left[\frac{M-1}{2}\right] + 2\sum_{k=0}^{(M-3)/2} h[k]cos\left[\omega\left(\frac{M-1}{2} - k\right)\right], \tag{5.42}$$

where $H_{sym}(e^{j\omega})$ is a real, even, and periodic function of ω. Then the frequency response, $H(e^{j\omega})$, is obtained by

$$H(e^{j\omega}) = H_{sym}(e^{j\omega})e^{-j\omega(M-1)/2}, \tag{5.43}$$

where the delay $(M-1)/2$ is an integer.

Suppose that we let $n = \frac{M-1}{2} - k$ and define a set of filter coefficients $\{b[n]\}$ as follows:

$$b[0] = h\left[\frac{M-1}{2}\right], \quad n = 0 \tag{5.44}$$

$$b[n] = 2h\left[\frac{M-1}{2} - n\right], \quad n = 1, 2, ..., (M-1)/2. \tag{5.45}$$

Then, (5.42) can be rewritten as

$$H_{sym}(e^{j\omega}) = \sum_{n=0}^{(M-1)/2} b[n]cos(\omega n). \tag{5.46}$$

Thus, (5.46) can be viewed as a linear combination of $[(M-1)/2+1]$ known basis functions, $cos(\omega n)$. The problem in designing an optimal FIR filter is to determine $b[n]$ such that δ_2 is minimized given the parameters ω_p, ω_s, $k = \frac{\delta_1}{\delta_2}$, and filter order $M-1$.

We must identify the coefficients $b[n]$ that minimize the maximum weighted error magnitude $|E(e^{j\omega})|$ over the range $0 \leq \omega \leq \pi$. The maximum error magnitude $|E(e^{j\omega})|$ is given by

$$\|E(e^{j\omega})\| = max_{0\leq\omega\leq\pi}\left[W(e^{j\omega})|H_i(e^{j\omega}) - H_{sym}(e^{j\omega})|\right], \qquad (5.47)$$

where $W(e^{j\omega})$ is a positive weighting function defined by

$$W(e^{j\omega}) = \begin{cases} 0, & \omega \text{ in the transition band} \\ \frac{\delta_2}{\delta_1}, & \omega \text{ in the passband} \\ 1, & \omega \text{ in the stopband,} \end{cases} \qquad (5.48)$$

and $H_i(e^{j\omega})$ is the ideal function that we wish to approximate with $H_{sym}(e^{j\omega})$. Equation (5.47) is called a *minimax* problem and is sometimes referred to as a *weighted Chebyshev approximation* problem. This equation is also called the *equiripple* FIR design criterion because the final error waveform contains many peaks or ripples, all of which have the same weighted magnitude.

A more precise statement of optimality is given by a powerful theorem known as the *alternation theorem* applied to the weighted Chebyshev approximation problem. The alternation theorem states that, for a lowpass or highpass filter, there is a unique solution to (5.47). The necessary and sufficient condition for this problem is that the maximum error magnitude $E(e^{j\omega})$ exhibits at least $[(M-1)/2+2]$ alternations in the range of $0 \leq \omega \leq \pi$. The frequencies at which alternation occurs are called *alternation frequencies*.

The Chebyshev approximation for designing FIR filters appears in the early literature. However, the Chebyshev approximation method did not enjoy widespread use until it was shown that the *Remez exchange* algorithm, which depends on the alternation theorem, could be used to design FIR filters with unequal ripple sizes in each of the frequency bands. This important algorithm for designing FIR filters is attributed to Parks and McClellan [4, 5], McClellan, Parks, and Rabiner [6], and McClellan and Parks [7]. The Parks-McClellan algorithm for designing FIR filters is the most popular and widely used technique in 1-D and 2-D FIR filter design. In 1995, Karam and McClellan published the complex Chebyshev approximation algorithm for FIR filter design. We skip its detailed discussion; interested readers are referred to [8].

In order to elaborate on the alternation theorem for equiripple FIR filter design, we consider designing a lowpass FIR filter with passband frequency range $0 \leq \omega \leq \omega_p$ and stopband frequency range $\omega_s \leq \omega \leq \pi$. Note that the ideal function $H_i(e^{j\omega})$ and the weighting function $W(e^{j\omega})$ are constant. Thus, we take the derivative of each side of (5.47),

$$\begin{aligned} \frac{dE(e^{j\omega})}{d\omega} &= \frac{d}{d\omega}\left\{W(e^{j\omega})[H_i(e^{j\omega}) - H_{sym}(e^{j\omega})]\right\} \\ &= -\frac{dH_{sym}(e^{j\omega})}{d\omega} \\ &= 0. \end{aligned} \tag{5.49}$$

Equation (5.49) means that a set of frequencies $\{\omega_i\}$ corresponding to the peaks of the function $E(e^{j\omega})$ also corresponds to the peaks in which the function $H_{sym}(e^{j\omega})$ meets the error tolerance. In addition, the terms $cos(\omega n)$ in (5.46) can be expressed as a sum of powers of $cos\omega$ as follows:

$$cos(\omega n) = T_n\{cos(\omega)\}, \tag{5.50}$$

where $T_n\{x\}$ is an nth-order polynomial. Therefore, (5.46) can be rewritten as

$$H_{sym}(e^{j\omega}) = \sum_{n=0}^{(M-1)/2} \beta[n]cos^n(\omega). \tag{5.51}$$

Note that the function $H_{sym}(e^{j\omega})$ can have $[(M-1)/2-1]$ local maximum and minimum in the frequency interval $0 < \omega < \pi$. In addition, there are alternations at $\omega = 0$ and $\omega = \pi$, as well as at $\omega = \omega_p$ and $\omega = \omega_s$, at all the frequency band edges of the functions $H_{sym}(e^{j\omega})$ and $E(e^{j\omega})$. Therefore, there are $[(M-1)/2+3]$ extremal frequencies in the function of $E(e^{j\omega})$, which is the best approximation of the ideal lowpass filter design. We refer to the filter design that contains more than $[(M-1)/2+2]$ alternations as an *extra ripple filter*. This condition does not exist for lowpass and highpass filters.

The alternation theorem provides necessary and sufficient conditions on the error for optimality in the Chebyshev sense. Thus, the optimum filter $H_{sym}(e^{j\omega})$ satisfies the following set of equations at the desired extremal frequencies $\{\omega_n\}$:

$$W(e^{j\omega_k})[H_i(e^{j\omega_k}) - H_{sym}(e^{j\omega_k})] = (-1)^k\delta, \tag{5.52}$$

for $k = 0, 1, ..., (M-1)/2 + 1$, where δ is the optimum error that is equal to δ_2. Note that (5.52) can be rearranged in an equivalent form,

$$\sum_{n=0}^{N} \beta[n] cos\omega_k n + \frac{(-1)^k \delta}{W(e^{j\omega_k})} = H_i(e^{j\omega_k}), \tag{5.53}$$

where $N = (M-1)/2$, for $k = 0, 1, ..., N$. Therefore, the parameters $\{\beta[n]\}$ and δ can be determined by expressing (5.53) in matrix form as

$$\mathbf{RP} = \mathbf{H}, \tag{5.54}$$

where the matrix $\mathbf{R}$ is

$$\mathbf{R} = \begin{bmatrix} 1 & cos\omega_0 & cos2\omega_0 & \cdots & cosN\omega_0 & \frac{1}{W(e^{j\omega_0})} \\ 1 & cos\omega_1 & cos2\omega_1 & \cdots & cosN\omega_1 & \frac{-1}{W(e^{j\omega_1})} \\ \cdot & \cdot & \cdot & \cdots & \cdot & \\ \cdot & \cdot & \cdot & \cdots & \cdot & \\ \cdot & \cdot & \cdot & \cdots & \cdot & \\ 1 & cos\omega_N & cos2\omega_N & \cdots & cosN\omega_N & \frac{(-1)^{N+1}}{W(e^{j\omega_{N+1}})} \end{bmatrix}, \tag{5.55}$$

the matrix $\mathbf{P}$ is

$$\mathbf{P} = [\beta[0], \beta[1], \cdots, \beta[N], \delta]^T, \tag{5.56}$$

and the matrix $\mathbf{H}$ is

$$\mathbf{H} = [H_{sym}(e^{j\omega_0}), H_{sym}(e^{j\omega_1}), \cdots, H_{sym}(e^{j\omega_{N-1}}), H_{sym}(e^{j\omega_N})]^T. \tag{5.57}$$

Equation (5.54) serves as the basis for an iterative algorithm for finding the optimum solution of $H_{sym}(e^{j\omega})$. The procedure starts by guessing a set of alternation frequencies ω_k, $k = 0, 1, ..., N$. Furthermore, the set of alternation frequencies should include ω_p and ω_s, because the values of ω_p and ω_s are fixed. Therefore, we can solve for the set of coefficients $\beta[n]$ and δ by using (5.54), which achieves the best approximation design for the ideal filter specification.

Example 5.2 Linear-Phase Lowpass FIR Filter Design Using the Equiripple Method

In this example, for the lowpass FIR filter case, we again use the specification parameters given in Example 5.1, so that we can compare all the design methods based on window and equiripple techniques. The design specifications of lowpass FIR filter are $\omega_p = 0.25\pi$, $\omega_s = 0.35\pi$, $\delta_1 = 0.01$, and $\delta_2 = 0.001$ with sampling rate F_s of 8 kHz.

For estimating the filter order of the equiripple lowpass FIR filter with specifications ω_p, ω_s, δ_1, and δ_2, the following formula, which was developed by Kaiser [9], can be used:

$$M = \frac{-20\log_{10}\sqrt{\delta_1\delta_2} - 13}{2.324\Delta\omega}, \tag{5.58}$$

where $\Delta\omega = \omega_s - \omega_p$. Substituting the specification parameters ω_p, ω_s, δ_1, and δ_2, into (5.58), the filter order, M, is equal to 53.

For the design results, Figure 5.3(a) shows the impulse response, and displays the characteristic symmetry of the FIR filter system with the sample number in the x-axis and the amplitude in the y-axis. Figure 5.3(b) shows the log magnitude response (dB) with equiripple in the passband and stopband.

Comparing the results of this example to the results of Example 5.1, we note that the lowpass FIR filter design requires filter order $M = 74$ using the Kaiser window method, while the Parks-McClellan design method only requires the filter order $M = 53$ to meet the same filter specifications. As we can see, the lowpass FIR filter order has up to 28.4% less computational complexity using the Parks-McClellan equiripple design method. This is because the window method can only produce approximately equal maximum errors in the passband and stopband. However, the Parks-McClellan design method can weight the error differently.

5.4.3 Linear-Phase FIR Using Least Squares Design

Consider a type I symmetric FIR filter $h[n]$ with linear phase and filter impulse response $h[n] = h[M-1-n]$, for $0 \le n \le M-1$, with M odd integer. Recall that $H_{sym}(e^{j\omega})$ in (5.11) is expressed as

$$H_{sym}(e^{j\omega}) = h\left[\frac{M-1}{2}\right] + 2\sum_{k=0}^{(M-3)/2} h[k]cos\left[\omega\left(\frac{M-1}{2} - k\right)\right], \tag{5.59}$$

where $H_{sym}(e^{j\omega})$ is a real, even, and periodic function of ω.

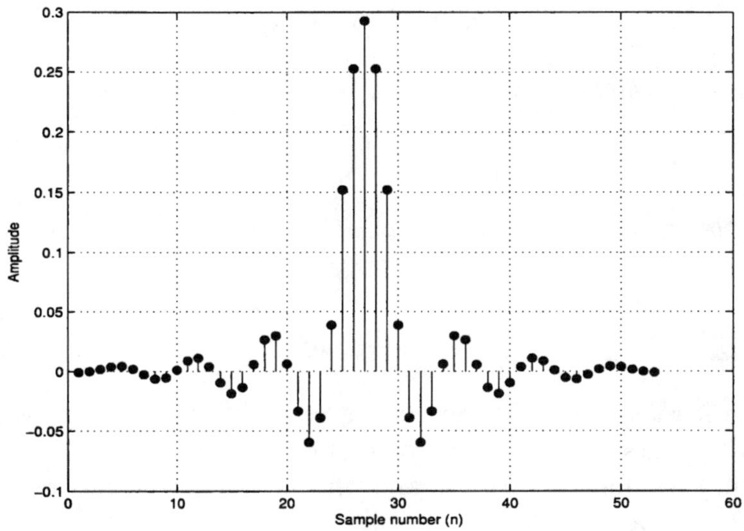

(a)

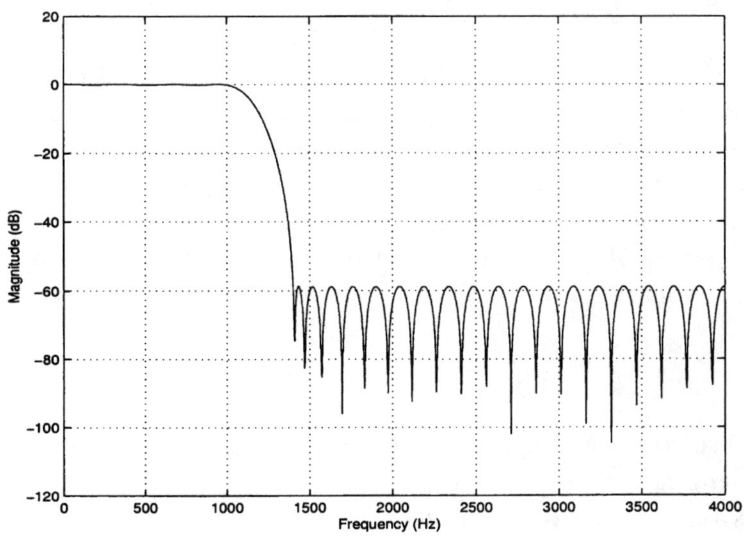

(b)

Figure 5.3 Design results of lowpass FIR digital filter using equiripple technique. (a) Impulse response when $M = 53$. (b) Log magnitude of FIR filter.

Define $n = [(M-1)/2-k]$ and filter coefficients $\{b[n]\}$ as follows:

$$b[0] = h[(M-1)/2], \quad n = 0, \tag{5.60}$$

$$b[n] = 2h[(M-1)/2-n], \quad n = 1, 2, ..., (M-1)/2, \tag{5.61}$$

and

$$b[n] = 0, \quad \text{otherwise}. \tag{5.62}$$

Then, (5.59) can be rewritten as

$$H_{sym}(e^{j\omega}) = \sum_{n=0}^{(M-1)/2} b[n]cos(\omega n). \tag{5.63}$$

The least squares FIR design approach minimizes the weighted integral of the square error approximation between an ideal piecewise frequency response, $H_i(e^{j\omega})$, and the actual frequency response, $H_{sym}(e^{j\omega})$. The scalar, $\mathcal{E}$, to be minimized is defined as

$$\mathcal{E} = \frac{1}{\pi}\int_0^{\pi} \left[W(e^{j\omega})|H_i(e^{j\omega}) - H_{sym}(e^{j\omega})|^2\right] d\omega, \tag{5.64}$$

where $W(e^{j\omega})$ is a weighting function. The ideal piecewise frequency response, $H_i(e^{j\omega})$, may be a linear function. For nonzero amplitude frequency bands, the weighted integrated square error can have a weight of $W(e^{j\omega})$ such that the error at low frequencies is much smaller than at high frequencies.

By setting $W(e^{j\omega}) = 1$, (5.64) becomes the full-band uniformly weighted square error,

$$\mathcal{E} = \frac{1}{\pi}\int_0^{\pi} |H_i(e^{j\omega}) - H_{sym}(e^{j\omega})|^2 d\omega. \tag{5.65}$$

Therefore, the least squares FIR filter design problem is to find a finite length $h[n]$ whose frequency response or amplitude response approximates a desired piecewise frequency response.

Using Parseval's theorem in (2.154), the error scalar defined in (5.65) can be expressed in the discrete-time domain as follows:

$$\mathcal{E} = 2\sum_{n=0}^{\infty} |h_i[n] - h_{sym}[n]|^2, \tag{5.66}$$

where $h_i[n]$ is symmetric at $n = 0$, is infinitely long, and is the inverse DTFT of $H_i(e^{j\omega})$ given by

$$h_i[n] = \frac{1}{2\pi}\int_{-\pi}^{\pi} H_i(e^{j\omega})cos(\omega n)d\omega. \tag{5.67}$$

Note that $h_{sym}[n]$ is the set of FIR filter coefficients being determined. Since $h_{sym}[n]$ is finite in duration, (5.66) can be separated into two terms,

$$\mathcal{E} = 2\sum_{n=0}^{(M-1)/2} |h_i[n] - h_{sym}[n]|^2 + 2\sum_{n=(M+1)/2}^{\infty} h_i^2[n]. \tag{5.68}$$

In order to minimize the scalar error $\mathcal{E}$, (5.68) clearly indicates that the $(M-1)/2$ values of impulse response $h_{sym}[n]$ should be selected equal to the corresponding $(M-1)/2$ values of $h_i[n]$.

Assume that a desired piecewise frequency response $H_i(e^{j\omega})$ is given by

$$H_i(e^{j\omega}) = \begin{cases} 1 & |\omega| < \omega_c \\ 0 & \text{otherwise,} \end{cases} \tag{5.69}$$

where ω_c is the cutoff frequency. The corresponding impulse response $h_i[n]$ of the frequency response $H_i(e^{j\omega})$ is

$$\begin{aligned} h_i[n] &= \frac{1}{2\pi}\int_{-\pi}^{\pi} H_i(e^{j\omega})e^{j\omega n}d\omega \\ &= \frac{1}{2\pi}\int_{-\omega_c}^{\omega_c} e^{j\omega n}d\omega \\ &= \frac{\omega_c}{\pi}\left[\frac{sin(\omega_c n)}{\omega_c n}\right], \end{aligned} \tag{5.70}$$

where $-\infty \leq n \leq \infty$. Now, (5.70) can be symmetrically truncated and shifted by $N = (M-1)/2$ to obtain the optimal and causal FIR filter coefficients as follows,

$$h_i[n] = \frac{\omega_c}{\pi}\left[\frac{sin\omega_c(n-N)}{\omega_c(n-N)}\right], \quad 0 \leq n \leq N-1, \tag{5.71}$$

and $h_i[n] = 0$ for otherwise.

Note that the minimization of the square error causes ripples in the desired piecewise frequency response $H_i(e^{j\omega})$ near the discontinuity of the actual frequency response $H_{sym}(e^{j\omega})$. This problem is referred to as the Gibb's phenomenon. The size of the ripple closest to the discontinuity is about 9% of the discontinuity (see [10]). Therefore, (5.71) does not provide a good solution to FIR filter design.

In order to eliminate the Gibb's phenomenon, a spline transition function that makes $H_i(e^{j\omega})$ continuous is used not only to remove the Gibb's phenomenon but also to asymptotically decrease as n increases. This method was reported in the least square FIR filter design by Burrus, Soewito, and Gopinath [10] and Burrus [11].

The first-order spline transition function for the desired frequency response $H_i(e^{j\omega})$ is defined by

$$H_i(e^{j\omega}) = \frac{\omega - \omega_s}{\omega_p - \omega_s}, \quad \omega_p < \omega < \omega_s \tag{5.72}$$

as well as $H_i(e^{j\omega}) = 1$ for $0 \leq \omega \leq \omega_p$, and $H_i(e^{j\omega}) = 0$ for $\omega_s \leq \omega \leq \pi$. Thus, the impulse response $h_i[n]$ can be calculated by using the inverse DTFT

$$\begin{aligned} h_i[n] &= \frac{1}{\pi}\left[\int_0^{\omega_p} cos(\omega n)d\omega + \int_{\omega_p}^{\omega_s}\left(\frac{\omega - \omega_s}{\omega_p - \omega_s}\right) cos(\omega n)d\omega\right] \\ &= \frac{\omega_c}{\pi}\left[\frac{sin(\omega_c n)}{\omega_c n}\right]\left[\frac{sin(\Delta n/2)}{\Delta n/2}\right], \end{aligned} \tag{5.73}$$

where $\omega_c = (\omega_p + \omega_s)/2$ is the average of the passband and stopband, and $\Delta = \omega_s - \omega_p$ is the transition bandwidth. As can be seen, (5.73) is equal to (5.70) multiplied by a sinc weighting function and asymptotically decreased by a factor of $\frac{1}{n^2}$.

In a similar way, if we use a qth-order spline as the transition function, the desired impulse response $h_i[n]$ is obtained by

$$h_i[n] = \frac{\omega_c}{\pi}\left[\frac{sin(\omega_c n)}{\omega_c n}\right]\left[\frac{sin(\Delta n/2)}{\Delta n/2}\right]^q, \tag{5.74}$$

where $\omega_c = (\omega_p + \omega_s)/2$ is the average of the passband and stopband, and $\Delta = \omega_s - \omega_p$ is the transition bandwidth. This impulse response $h_i[n]$ asymptotically drops off as $\frac{1}{n^{q+1}}$. The qth-order spline can produce a transition function that contains q segments of the qth-order polynomials connected together. Therefore, the $q-1$ derivatives are continuous at the junction.

Equation (5.74) is now truncated and shifted by $N = (M-1)/2$ to obtain the ideal and causal length-N FIR filter coefficients as follows:

$$h_i[n] = \frac{\omega_c}{\pi}\left[\frac{sin\omega_c(n-N)}{\omega_c(n-N)}\right]\left[\frac{sin(\Delta(n-N)/2)}{\Delta(n-N)/2}\right]^q, \quad 0 \le n \le N-1 \tag{5.75}$$

and $h_i[n] = 0$ otherwise.

Note that the exponent q in (5.74) is a parameter that can be used to control the least squares error approximation similar to the parameter β in the Kaiser window that we introduced in Section 5.4.1. Thus, an important question is how to choose this parameter so that we can minimize the least squares error approximation given $H_i(e^{j\omega})$ and filter length N. In practice, the optimal value of the power q is set to be

$$q = \frac{0.312}{\pi}(\omega_s - \omega_p)N, \tag{5.76}$$

where ω_p and ω_s are the frequency passband and stopband edges, and N is the FIR filter length. Hence, (5.76) provides the solution for minimizing the least squares approximation error. The mathematical formulation was contributed by Burrus, Soewito, and Gopinath [10].

The above least squares FIR filter method can also be applied to designing discrete-time Hilbert transforms as described in Section 4.7. From (4.100), the ideal impulse response $h_i[n]$ for the discrete-time Hilbert transform is given by

$$h_i[n] = \begin{cases} \frac{sin^2(\pi n/2)}{\pi n/2}, & n \neq 0 \\ 0, & n = 0. \end{cases} \tag{5.77}$$

If we use a qth-order spline as the transition function, then the desired impulse response $h_i[n]$ of the discrete-time Hilbert transform is obtained by

$$h_i[n] = \frac{sin^2(\pi n/2)}{\pi n/2}\left[\frac{sin(\Delta n/2)}{\Delta n/2}\right]^q, \quad n \neq 0 \tag{5.78}$$

and $h_i[n] = 0$ for $n = 0$. This can be truncated and shifted by N to obtain the causal length-N Hilbert transform coefficients as

$$h_i[n] = \frac{sin^2(\pi(n-N)/2)}{\pi(n-N)/2}\left[\frac{sin(\Delta(n-N)/2)}{\Delta(n-N)/2}\right]^q, \quad 0 < n \le N-1 \tag{5.79}$$

and $h_i[n] = 0$ for $n = 0$. Therefore, we can design the discrete-time Hilbert transform by using the least squares method.

One property of least-squares-designed FIR filters is that they have better frequency response over most of the passband and stopband. Also, the least squares FIR design can have a more uniform transition connecting the passband with the stopband. This property can help control frequency response magnitude in wide transition or "don't care" frequency regions.

The least squares FIR method can be applied to design lowpass, bandpass, bandstop, and highpass filters as well as Hilbert transforms. The theory of least squares FIR design was contributed by Parks and Burrus [12], Burrus, Soewito, and Gopinath [10], and Burrus [11]. The constrained least squares design of FIR filters without specified transition bands was reported by Selesnick, Lang, and Burrus [13]. For design software, see Burrus [11], Selesnick, Lang, and Burrus [13], and Matlab Signal Processing Toolbox [14].

Example 5.3 Linear-Phase Lowpass FIR Filter Design Using the Least Squares Method

In this example, we use the design specifications of lowpass FIR filter as $\omega_p = 0.25\pi$, $\omega_s = 0.35\pi$, the desired amplitudes with 1 and 0, and sampling rate F_s of 8 kHz. Figure 5.4 shows a designed FIR filter with frequency in the x-axis and magnitude response in the y-axis, where the dashed line is the least squares design and the solid line is the equiripple design. Now we can compare the design results with the two methods.

As we can see, the equiripple design shows the equiripple behavior at the passband and stopband. This indicates that the maximum error of the FIR filter over the passband and stopband is smaller. In other words, it is smallest for the band edge configuration and filter order. The filter designed using least squares shows a better response over most of the passband and stopband. Also note that the magnitude response in the stopband of the least squares filter tends to decrease as the frequency increases. This property is useful, for example in telecommunication applications, to reduce the adjacent channel power problem. However, for the band edge at $\omega_p = 0.25\pi$ and $\omega_s = 0.35\pi$, the frequency response is further away from

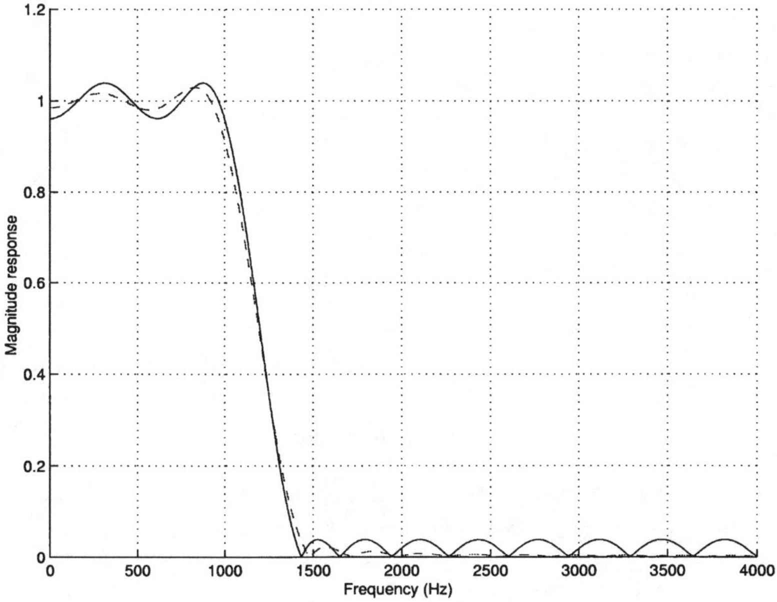

Figure 5.4 The plot of magnitude response using least squares and equiripple techniques, where the solid line denotes equiripple design method and the dashed line denotes the least squares method.

the desired frequency response than the designed filter using the equiripple technique.

In the next section, we will introduce the multiband least squares FIR design for linear-phase filter.

5.4.4 Multiband Linear-Phase FIR Using Least Squares Design

The general definition of the least squares error measure $\mathcal{E}$ for allowing multiple passband separated by transition bands is defined by

$$\mathcal{E} = \frac{1}{\pi} \sum_{k=0}^{N-1} \int_{\omega_k}^{\omega_{k+1}} \left[W(e^{j\omega}) |H_{ik}(e^{j\omega}) - H_{sym}(e^{j\omega})|^2 \right] d\omega, \qquad (5.80)$$

where the weights $W(e^{j\omega})$ are constant over each band defined by $\omega_k < \omega < \omega_{k+1}$. For notational simiplification, we let $W(e^{j\omega}) = W_k$

for the kth band. $H_{ik}(e^{j\omega})$ is the desired frequency response for the kth band.

In order to obtain the minimum in (5.80), we substitute (5.63) into (5.80):

$$\mathcal{E} = \frac{1}{\pi}\sum_{k=0}^{N-1}\int_{\omega_k}^{\omega_{k+1}}\left[W_k|H_{ik}(e^{j\omega}) - \sum_{n=0}^{(M-1)/2} b[n]cos(\omega n)|^2\right]d\omega. \tag{5.81}$$

We then compute the gradient of $\mathcal{E}$ and set it equal to zero,

$$\begin{aligned}\frac{\partial\mathcal{E}}{\partial b[m]} &= \frac{2}{\pi}\sum_{k=0}^{N-1}\int_{\omega_k}^{\omega_{k+1}}\left[W_k|H_{ik}(e^{j\omega}) - \sum_{n=0}^{(M-1)/2} b[n]cos(\omega n)|\right] \\ &\quad \cdot\ cos(\omega m)d\omega = 0.\end{aligned} \tag{5.82}$$

Note that (5.82) provides $(M+1)/2$ equations for $m = 0, 1, ..., (M-1)/2$. Thus, (5.82) can be rewritten in the following form:

$$\begin{aligned}&\sum_{k=0}^{N-1}\int_{\omega_k}^{\omega_{k+1}} W_k H_{ik}(e^{j\omega})cos(\omega m)d\omega \\ =&\sum_{n=0}^{(M-1)/2} b[n]\sum_{k=0}^{N-1}\int_{\omega_k}^{\omega_{k+1}} W_k cos(\omega n)cos(\omega m)d\omega.\end{aligned} \tag{5.83}$$

Also, we can express (5.83) in matrix form as follows:

$$\mathbf{h}_i = \mathbf{T}\mathbf{h}_{sym}, \tag{5.84}$$

where the ideal filter coefficients are

$$\mathbf{h}_i = \sum_{k=0}^{N-1}\left[W_k\int_{\omega_k}^{\omega_{k+1}} H_{ik}(e^{j\omega})cos(\omega m)d\omega\right], \tag{5.85}$$

and the weighted matrix is

$$\mathbf{T} = \sum_{k=0}^{N-1}\left[W_k\int_{\omega_k}^{\omega_{k+1}} cos(\omega n)cos(\omega m)d\omega\right]. \tag{5.86}$$

Therefore, the optimal filter coefficients $\mathbf{h}_{sym}$ that minimize the weighted least squared error measure in (5.80) are obtained by

$$\mathbf{h}_{sym} = \mathbf{T}^{-1}\mathbf{h}_i. \tag{5.87}$$

Note that the weighted matrix **T** in (5.86) has an analytical solution given by

$$T(m,n) = \frac{1}{2}\sum_{k=0}^{N-1} W_k \left[\frac{sin(n-m)\omega_{k+1} - sin(n-m)\omega_k}{n-m} + \frac{sin(n+m)\omega_{k+1} - sin(n+m)\omega_k}{n+m}\right]. \quad (5.88)$$

Also, the ideal filter coefficients $\mathbf{h}_i$ can be calculated numerically by using a truncated weighted sum of inverse DFTs from sampling $H_{ik}(e^{j\omega})$. As a result, this procedure provides a very powerful design method of allowing multiband weighted least squares design for linear-phase FIR filters. It is also possible that this method can be used to design arbitrarily shaped passbands by using an inverse DFT and introducing the spline function for transition bands to compute the ideal filter coefficients in (5.85).

The multiband least squares linear-phase FIR filter design method has contributions from Tufts, Rorabacher, and Mosier [15], Oethen, Parks, and Schüssler [16], Burrus, Soewito, and Gopinath [10], Burrus [11], and Selesnick, Lang, and Burrus [13].

Example 5.4 The Multiband Linear-Phase FIR Filter Design

In this example, we show the design results of the multiband linear-phase FIR filter by using the above described design approach. We define the multiband frequency response magnitude for the FIR filter as follows:

1. The magnitude of the frequency response in the first passband is $0.99 \leq |H(e^{j\omega})| \leq 1.01$, for $0 \leq \omega \leq 0.3\pi$.
2. The second passband response is $0.74 \leq |H(e^{j\omega})| \leq 0.76$, for $0.3\pi < \omega \leq 0.5\pi$.
3. The first stopband response is $-0.01 \leq |H(e^{j\omega})| \leq 0.01$, for $0.5\pi < \omega \leq 0.7\pi$.
4. The third passband response is $0.995 \leq |H(e^{j\omega})| \leq 1.005$, for $0.7\pi < \omega \leq 0.9\pi$.
5. The second stopband response is $-0.02 \leq |H(e^{j\omega})| \leq 0.02$, for $0.9\pi < \omega \leq \pi$.

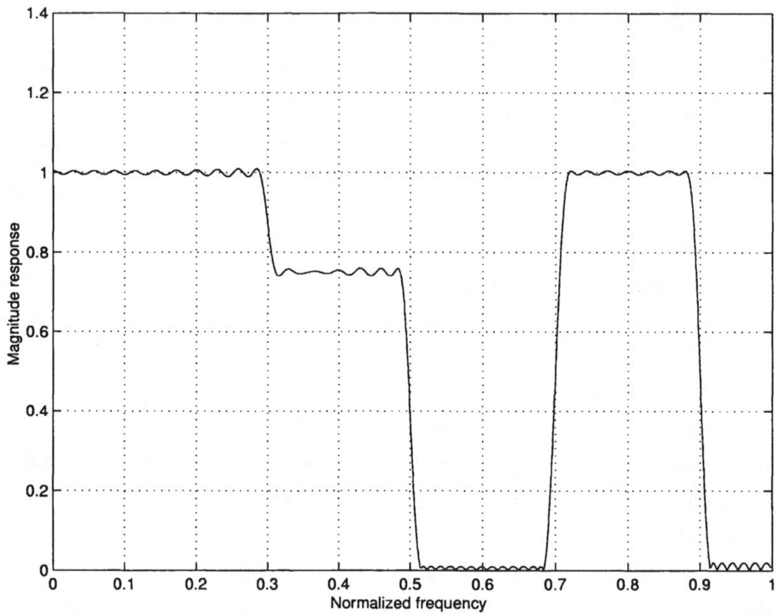

Figure 5.5 The multiband frequency response of the FIR filter with the filter length $M = 140$.

The designed multiband linear-phase FIR filter has filter length $M = 140$ with symmetric coefficients. The multiband magnitude of the frequency response plot is shown in Figure 5.5.

5.5 IIR Filter Design

The generalized higher-order difference equations of the infinite impulse response (IIR) filters are given by

$$y[n] = \sum_{k=0}^{N} b_k x[n-k] + \sum_{k=1}^{M} a_k y[n-k]. \tag{5.89}$$

The corresponding system transfer function of an IIR filter in the z-domain is

$$H(z) = \frac{Y(z)}{X(z)}$$

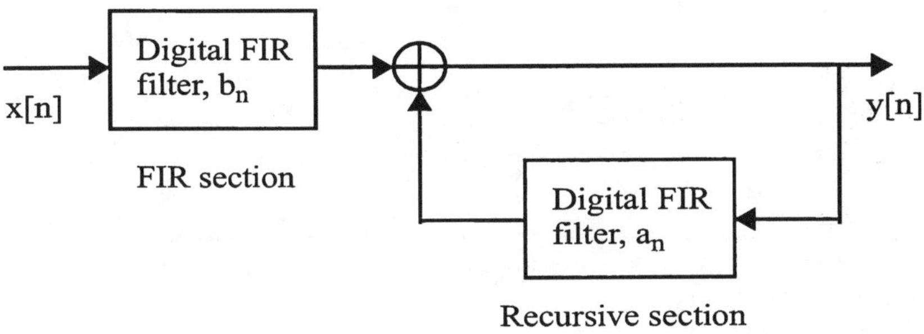

Figure 5.6 Digital IIR filter structure displaying two digital FIR filters.

$$= \frac{\sum_{k=0}^{N} b_k z^{-k}}{1 - \sum_{k=1}^{M} a_k z^{-k}}. \tag{5.90}$$

The IIR digital filters in (5.90) can be considered as having both a recursive and a nonrecursive section. Thus, an IIR digital filter can be viewed as two FIR digital filters, one of which is connected in a feedback loop as shown in Figure 5.6.

For a lowpass IIR filter with a transfer function of the form $H(z) = \frac{Y(z)}{X(z)}$, we may see that the magnitude of numerator $|Y(z)|$ has a lowpass FIR filter with a very poor passband response as shown in Figure 5.7(a). However, the magnitude of denominator $|X(z)|$ compensates for this problem as shown in Figure 5.7(b). Typically, the magnitude of the frequency response of $\frac{1}{X(z)}$ grows as ω increases in the passband frequency. Hence, the magnitude of the frequency response of $\frac{1}{X(z)}$ is large near the passband edge. As a result, the product of the magnitude $|Y(z)|$ and $\frac{1}{X(z)}$ tends to have the unit amplitude in the passband frequency, approximately.

A key design and implementation concern for an IIR digital filter is ensuring that the recursive section as shown in Figure 5.6 is stable. This is because a simple relationship does not exist between the IIR filter coefficients and the impulse response sequence $h[n]$. Another issue with regard to an IIR filter is that a causal and stable IIR filter may not have linear phase. However, in general, the IIR filter has an important advantage over the FIR filter; that is, the IIR filter requires fewer filter coefficients to approximate a given filter

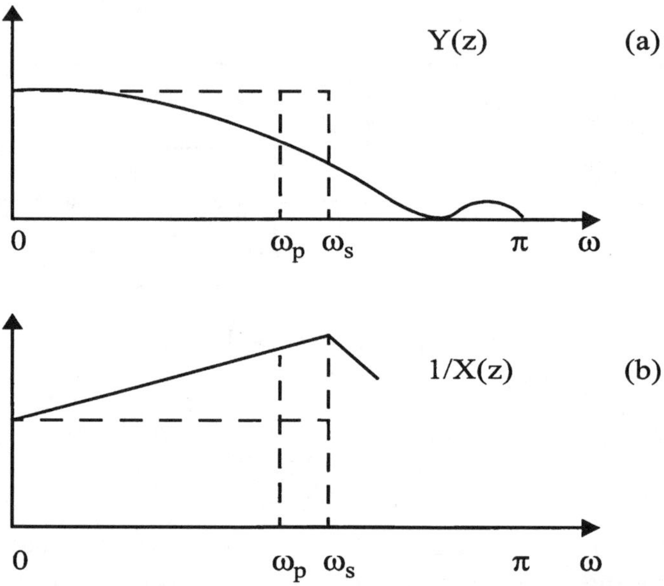

Figure 5.7 The separated plot for frequency response of IIR filter. (a) The frequency response plot by the numerator $Y(z)$. (b) The frequency response plot by the denominator $\frac{1}{X(z)}$.

frequency specification than the FIR filter does. This implies that the IIR filter is computationally efficient in terms of software and hardware implementation.

There are several methods that can be used to design an IIR filter. The techniques described in this section are all based on transferring an analog filter into a digital IIR filter, since the analog filter design is mature and well developed. Therefore, we begin designing a digital IIR filter in the analog domain and then convert the analog filter design into the digital IIR filter.

A system function of the analog filter is given by

$$H_a(s) = \frac{\sum_{i=0}^{M} a_i s^i}{s^N + \sum_{j=0}^{N-1} b_j s^j}, \tag{5.91}$$

where $\{a_i\}$ and $\{b_j\}$ are the analog filter coefficients, and N is the order of analog filter. Alternatively, analog impulse response $h(t)$ is related to $H_a(s)$ by the Laplace transform

$$H_a(s) = \int_{-\infty}^{\infty} h(t)e^{-st}dt. \tag{5.92}$$

Note that an analog filter with system function $H_a(s)$ in (5.91) is stable if all its poles lie in the left half of the s-plane. Consequently, if the conversion technique between an analog domain and a digital domain is to be effective, the conversion technique, which is sometimes referred to as a *frequency prewarping* transformation, should have the following properties:

1. The $j\Omega$ axis in the s-plane needs to map into the unit circle in the z-plane based on a relationship between the two frequency variables in the analog and digital domains.
2. The left-half plane in the s-plane needs to map into the inside of the unit circle in the z-plane. In such cases, if an analog filter is stable, a digital IIR filter is stable as well after mapping.

In the following section, we will introduce one of the simplest and most effective methods of designing digital IIR filters with prescribed magnitude response specifications based on the *bilinear transformation* method.

5.5.1 The Bilinear Transform

The IIR filter can be designed by using many methods, but by far the most common IIR filter design is the bilinear transformation. This method relies on the existence of a well-known *Laplace* transform of an analog filter and converts it into the z-domain transfer function of the digital filter. It can be shown that the z-domain transformed version $H(z)$ is stable if and only if the Laplace transform function $H_a(s)$ is stable.

With the continuous-time system function $H_a(s)$ and the discrete-time system function $H(z)$, the bilinear transformation is used to replace s defined by

$$s = \frac{2}{T_s}\left(\frac{1 - z^{-1}}{1 + z^{-1}}\right), \tag{5.93}$$

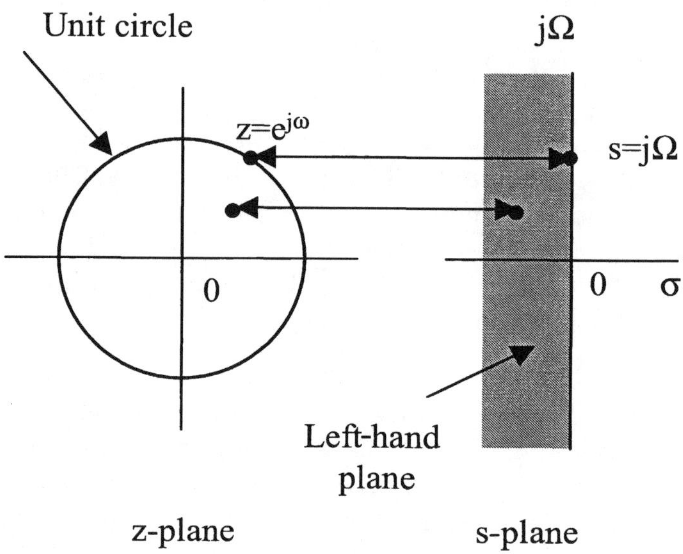

Figure 5.8 The mapping relationship between the unit circle in the z-plane and the left-hand plane in the s-plane.

and the result is a rational function $H(z)$ in the z-domain,

$$H(z) = H_a\left[\frac{2}{T_s}\left(\frac{1-z^{-1}}{1+z^{-1}}\right)\right], \tag{5.94}$$

where T_s is the sampling interval. The bilinear transformation in (5.93) maps the left-hand s-plane into the inside of the unit circle on the z-plane. Figure 5.8 illustrates this property.

Using (5.93), the relationship between the frequency variables ω and Ω can be developed after substituting $z = e^{j\omega}$,

$$\begin{aligned} s &= \sigma + j\Omega \\ &= \frac{2}{T_s}\left(\frac{1-e^{-j\omega}}{1+e^{-j\omega}}\right) \\ &= \frac{2}{T_s}\left[\frac{2e^{-j\omega/2}(j sin\omega/2)}{2e^{-j\omega/2}(cos\omega/2)}\right] \end{aligned}$$

$$= \frac{2j}{T_s} tan(\omega/2). \qquad (5.95)$$

Equating real and imaginary parts on both sides of (5.95) leads to the relations $\sigma = 0$ and

$$\Omega = \frac{2}{T_s} tan(\omega/2). \qquad (5.96)$$

This shows that the mapping transforms $\Omega = 0$ into $\omega = 0$, and $\Omega = \infty$ into $\omega = \pi$. As we can see, the entire range in the analog frequency Ω is only mapped into the digital frequency range $-\pi \leq \omega \leq \pi$. The mapping is highly nonlinear. So, (5.96) is sometimes referred to as a frequency prewarping transformation because the linear relationship between the digital frequency ω and the analog frequency Ω does not exist.

The passband edge ω_p and stopband edge ω_s of lowpass digital filter response $|H(e^{j\omega})|$ corresponding to the analog lowpass filter response are determined by (5.96) as follows,

$$\omega_p = 2\left[arctan\left(\frac{\Omega_p T_s}{2}\right)\right], \qquad (5.97)$$

and

$$\omega_s = 2\left[arctan\left(\frac{\Omega_s T_s}{2}\right)\right]. \qquad (5.98)$$

The size of the passband and stopband ripple δ_1 and δ_2 are unchanged. Therefore, given the digital filter frequency specifications, ω_p, ω_s, δ_1, and δ_2, the design procedure of digital IIR filter using the bilinear transformation is as follows:

1. Starting with the given frequency specification of the digital filter, the specifications in the digital domain are converted into the specifications of an equivalent analog lowpass filter by using the prewarping transformation in (5.96). Then, we can determine the analog passband edge Ω_p and stopband edge Ω_s of the analog lowpass filter.
2. We choose the type of classical analog filter and design the analog lowpass filter, which meets the specification Ω_p, Ω_s, δ_1, and δ_2.

3. Using the bilinear transformation in (5.93), we transfer the frequency response of analog filter $H_a(s)$ into the frequency response of the desired digital IIR filter $H(z)$.

Example 5.5 Digital IIR Filter Using Bilinear Transformation

In this example, we want to design a single-pole lowpass digital IIR filter with a 3-dB bandwidth at the cutoff frequency $\omega_c = 0.25\pi$ by using the bilinear transformation.

The frequency response of a single-pole analog lowpass filter in the s-domain is given by

$$H_a(s) = \frac{\Omega_c}{s + \Omega_c}, \tag{5.99}$$

where the cutoff frequency Ω_c is the 3-dB bandwidth of the analog lowpass filter.

Using (5.96) for the frequency prewarping transformation, the analog cutoff frequency Ω_c can be computed by

$$\begin{aligned} \Omega_c &= \frac{2}{T_s} tan(\omega_c/2) \\ &= \frac{2}{T_s} tan(0.125\pi) \\ &= \frac{0.8284}{T_s}. \end{aligned} \tag{5.100}$$

Thus, the analog lowpass filter has the system function

$$H_a(s) = \frac{0.8284/T_s}{s + 0.8284/T_s}. \tag{5.101}$$

Equation (5.101) represents the filter design in the analog domain and can be converted into the desired digital lowpass IIR filter by using the bilinear transformation in (5.93). As a result, we obtain the lowpass digital IIR filter in the z-domain,

$$H(z) = \frac{0.2929(1 + z^{-1})}{1 - 0.4142z^{-1}}, \tag{5.102}$$

where the sampling interval T_s has been divided out. Therefore, the frequency response of the lowpass digital IIR filter is

$$H(e^{j\omega}) = \frac{0.2929(1 + e^{-j\omega})}{1 - 0.4142e^{-j\omega}}. \tag{5.103}$$

As can be seen, at $\omega = 0$, $|H(e^0)| = 1$, and $\omega = 0.25\pi$, we obtain $|H(e^{j0.25\pi})| = 0.707$. This is equivalent to -3 dB, which is the desired frequency response.

5.5.2 Classical Analog Filters

We have shown that the design of digital IIR filters can easily be obtained by beginning with the analog filter in the s-domain, and then using the bilinear transformation to convert the s-plane into the z-plane. Thus, in designing an approximate analog filter, an important task is to preserve as much as possible the desired characteristics of the analog filter.

The analog filter is one of the very few areas of engineering for which a complete design theory exists, starting from specification requirements and ending with a circuit realization. In this section, we shall concentrate on a selection of topics that provides an introduction to the subject and the design method. Therefore, we introduce only three kinds of analog filters, including Butterworth, Chebyshev, and elliptic type filters.

The transfer function $H_a(s)$ of the analog filter can be written as the ratio of two polynomials as follows:

$$H_a(s) = \frac{\sum_{i=0}^{M} a_i s^i}{s^N + \sum_{j=0}^{N-1} b_j s^j}, \tag{5.104}$$

where N is the order of the analog filter. For this analog filter to be stable, the degree of the numerator must be less than or equal to that of the denominator.

In a special case, all the transmission zeros of this analog filter are at $s = \infty$. Under this situation, the transfer function $H_s(s)$ of the analog filter takes the form

$$H_a(s) = \frac{a_M}{s^N + \sum_{j=0}^{N-1} b_j s^j}. \tag{5.105}$$

The transfer function of the analog filter in (5.105) is known as an *all-pole* filter. Notice that Butterworth and Chebyshev type filters have this property, but not the elliptic type filter. We discuss the above three types of analog filters below.

The Butterworth Filter

The Butterworth filter shows a monotonically decreasing transmission with all the transmission zeros at $\Omega = 0$, making it an all-pole filter. The magnitude function for the Nth-order Butterworth filter in the frequency domain with passband Ω_p is given by

$$|H_a(j\Omega)| = \frac{1}{\sqrt{1+\varepsilon^2\left(\frac{\Omega}{\Omega_p}\right)}}. \tag{5.106}$$

As can be seen, when $\Omega = \Omega_p$, the passband magnitude function of the Butterworth filter is as follows:

$$|H_a(j\Omega)| = \frac{1}{\sqrt{1+\varepsilon^2}}. \tag{5.107}$$

Thus the parameter ε determines the maximum variation in the frequency of passband based on the following equation:

$$\delta_1 = 20log\sqrt{1+\varepsilon^2}. \tag{5.108}$$

At the edge of the stopband, when $\Omega = \Omega_s$, the attenuation of the Butterworth filter is given by

$$\delta_2 = -20log\left(\frac{1}{\sqrt{1+\varepsilon^2(\Omega_s/\Omega_p)^{2N}}}\right). \tag{5.109}$$

By setting $\delta_2 \geq \delta_{min}$, (5.109) can also be used to determine the filter order N.

One property of the Butterworth type filter is that it provides the best *Taylor* series approximation to the ideal lowpass filter response at $\Omega = 0$ and $\Omega = \infty$. For any order of filter N, the magnitude squared response has $2N-1$ zero derivatives at $\Omega = 0$ and $\Omega = \infty$. This property makes the Butterworth response very flat near $\Omega = 0$, yielding the famous description, *maximally flat* response. The frequency response is monotonic overall, decreasing smoothly from $\Omega = 0$ to $\Omega = \infty$.

The Chebyshev Filter

The Chebyshev type filter has an equiripple response in the passband and a monotonically decreasing transmission in the stopband. We refer to this property as a type I Chebyshev filter. While the odd-order Chebyshev filter has $|H(0)| = 1$, the even-order Chebyshev filter exhibits its maximum magnitude deviation at $\Omega = 0$. In both cases, one of the interesting things is that the total number of maxima and minima in the passband frequency equals the order of the Chebyshev filter. All the transmission zeros of the Chebyshev filter are at $\Omega = \infty$, making it an all-pole filter.

For $\Omega \leq \Omega_p$, the magnitude of the transfer function of an Nth-order Chebyshev filter with a passband edge (ripple bandwidth) Ω_p is given by

$$|H_a(j\Omega)| = \frac{1}{\sqrt{1 + \varepsilon^2 cos^2[N cos^{-1}(\Omega/\Omega_p)]}}. \tag{5.110}$$

For $\Omega \geq \Omega_s$, the magnitude of the transfer function of an Nth-order Chebyshev filter with a stopband edge (ripple bandwidth) Ω_s is given by

$$|H_a(j\Omega)| = \frac{1}{\sqrt{1 + \varepsilon^2 cosh^2[N cosh^{-1}(\Omega/\Omega_p)]}}. \tag{5.111}$$

Notice that at the passband edge, $\Omega = \Omega_p$, the magnitude function of a Chebyshev filter based on (5.111) is obtained by

$$|H_a(j\Omega)| = \frac{1}{\sqrt{1 + \varepsilon^2}}. \tag{5.112}$$

Thus, the parameter ε determines the passband ripple according to the following equation:

$$\delta_1 = 10log(1 + \varepsilon^2). \tag{5.113}$$

The attenuation achieved by Chebyshev filters at the stopband edge ($\Omega = \Omega_s$) is found using (5.113) as follows,

$$\delta_2 = 10log\left[1 + \varepsilon^2 cosh^2[N cosh^{-1}(\Omega_s/\Omega_p)]\right]. \tag{5.114}$$

Equation (5.114) can be used to determine the order of the Chebyshev filter by setting $\delta_2 \geq \delta_{min}$.

An important property of the type I Chebyshev filter is that it minimizes the absolute difference between the ideal and actual frequency response over the entire passband by incorporating an equal ripple in the passband. The stopband response of the Chebyshev filter is maximally flat. In addition, the transition from passband to stopband is more rapid than for the Butterworth filter.

In a type II Chebyshev filter, for $\Omega \leq \Omega_p$, the magnitude of the transfer function of an Nth-order Chebyshev filter with a passband edge Ω_p is given by

$$|H_a(j\Omega)| = \frac{1}{\sqrt{1 + \{\varepsilon^2 cos^2[N cos^{-1}(\Omega_p/\Omega)]\}^{-1}}}. \tag{5.115}$$

For $\Omega \geq \Omega_s$, the magnitude of the transfer function of an Nth-order Chebyshev filter with a stopband edge Ω_s is given by

$$|H_a(j\Omega)| = \frac{1}{\sqrt{1 + \{\varepsilon^2 cosh^2[N cosh^{-1}(\Omega_p/\Omega)]\}^{-1}}}. \tag{5.116}$$

In the type II Chebyshev filter, the magnitude of the frequency response is monotonic in the passband and equiripple in the stopband. One method of designing a type II Chebyshev filter is to first design a type I Chebyshev filter. We then convert it to the type II Chebyshev filter, since there are the reciprocal relations between (5.110) and (5.115), as well as (5.111) and (5.116).

The Elliptic Filter

The elliptic type filter distributes the error uniformly across the entire passband and stopband. To obtain equiripple error in both the passband and stopband, elliptic type filters must have both poles and zeros. The approximation of elliptic filters has the following form:

$$|H_a(j\Omega)| = \frac{1}{\sqrt{1 + \varepsilon^2 J_N^2(\Omega)}}, \tag{5.117}$$

where J_N^2 is a *Jacobian elliptic* function. A detailed description of Jacobian elliptic function is beyond the scope of this chapter. The interested reader is referred to the reference book by Sedra and Smith [17] for a more detailed discussion.

The important property of elliptic type filters is that they have equiripple behavior in both the passband and stopband. They generally meet filter requirements with the lowest order of any supported filter type. Given a filter order N with the passband ripple R_p in decibels, and stopband ripple R_s in decibels, elliptic type filters are able to minimize transition width.

5.5.3 Design Examples of a Digital IIR Filter

In this section, in order to illustrate the properties of the basic IIR approximation methods, we introduce some examples of realizations of a set of digital filter specifications by each of the design methods. We need the digital filter specification to determine the input parameters for the Butterworth, Chebyshev, and elliptic design formulas.

Suppose we want to design a lowpass IIR digital filter to meet the following specifications: $\omega_p = 0.30\pi$, $\omega_s = 0.70\pi$, passband ripple $R_p = 1$ dB, and stopband ripple $R_s = 80$ dB with frequency sampling rate F_s at 8 kHz. Note that the specifications of this digital filter are only given for the magnitudes of the frequency response.

Using the design procedure of bilinear transformation described in Section 5.5.1, the frequency specifications of the digital IIR filter need to be prewarped to the corresponding continuous-time frequencies. This is to ensure that the frequency distortion inherent in the bilinear transformation will map them back to the correct digital z-domain frequencies. We will introduce the design of digital IIR filters using Butterworth, Chebyshev, and elliptic as the examples. However, the details of the design computation are not presented herein since they are best carried out using computer programs.

Example 5.6 Digital IIR Filter Using Butterworth

The frequency response of the digital IIR filter is shown in Figure 5.9 after using the bilinear transformation of the appropriate prewarped Butterworth filter. The Butterworth approximation method requires an IIR filter with eighth order for the given specifications. Figure 5.9(a) shows the plot of log magnitude in decibels. The pole-zero plot for the eighth-order Butterworth filter is shown in Figure 5.9(b). As we can see from the plots, the Butterworth filter has flat response in the passband and decreases monotonically with frequency in the stopband.

Example 5.7 Digital IIR Filter Using Chebyshev

The frequency response of the digital IIR filter using Chebyshev types I and II approximation are shown in Figure 5.10 and Figure 5.11, respectively. Both methods require an IIR filter with sixth order rather than eighth-order for the Butterworth filter in Example 5.6, for the given frequency specification.

Note that the frequency response oscillates with equal error ripple on either side of the desired unity gain in the passband. Stopband response of the type I Chebyshev filter is maximally flat. As we can see from the plot in Figure 5.10(a), the transition from passband to stopband is more rapid than that of the Butterworth filter. However, the type I Chebyshev filter is similar to the Butterworth filter in that it has all zeros at $z = -1$.

The type II Chebyshev filter has the equiripple approximation behavior in the stopband as shown in Figure 5.11(a). The pole-zero plots for this case are shown in Figure 5.11(b). Note that the type II Chebyshev filter has its zeros arrayed on the unit circle to achieve the equiripple behavior in the stopband.

Example 5.8 Digital IIR Filter Using Elliptic Approximation

The frequency response of the digital IIR filter and pole-zero plot using elliptic approximation are shown in Figure 5.12(a) and (b), respectively. The design method only needs a fifth-order IIR filter for the given frequency specifications. The elliptic approximation has the error uniformly across the entire passband and stopband of frequencies. So, the elliptic filter must have both poles and zeros to obtain equiripple error in the frequency range of passband and stopband. Note that the elliptic filter, like the type II Chebyshev filter, has its zeros arrayed in the left range of the unit circle. Generally, the elliptic approximation can meet filter specifications with the lowest order.

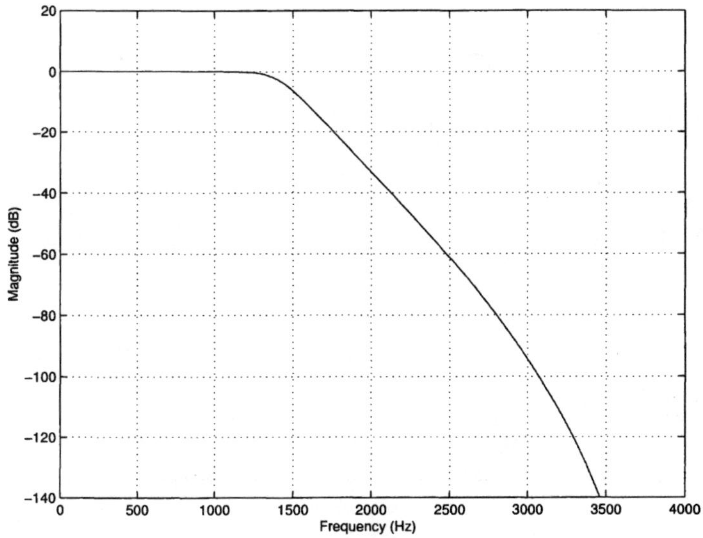

(a)

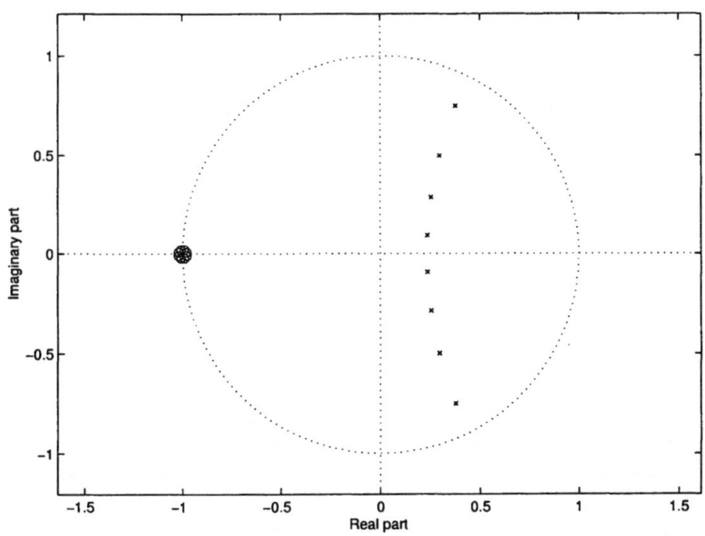

(b)

Figure 5.9 Frequency response of eighth-order Butterworth filter. (a) Plot log magnitude in dB. (b) Pole-zero plot.

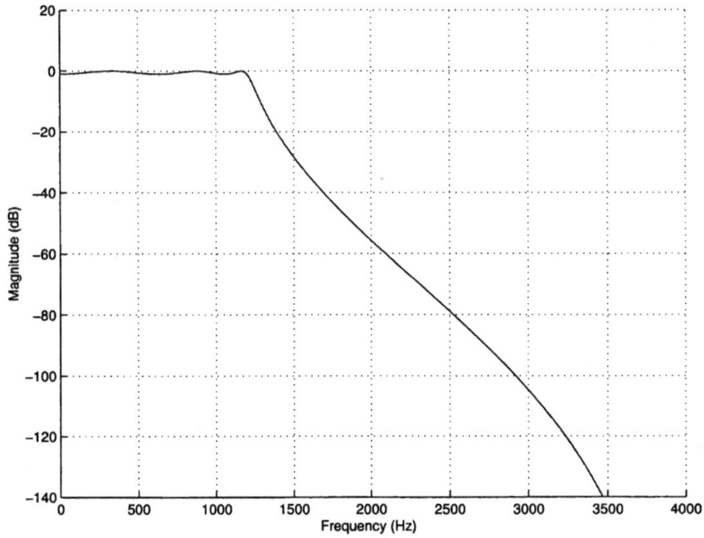

(a)

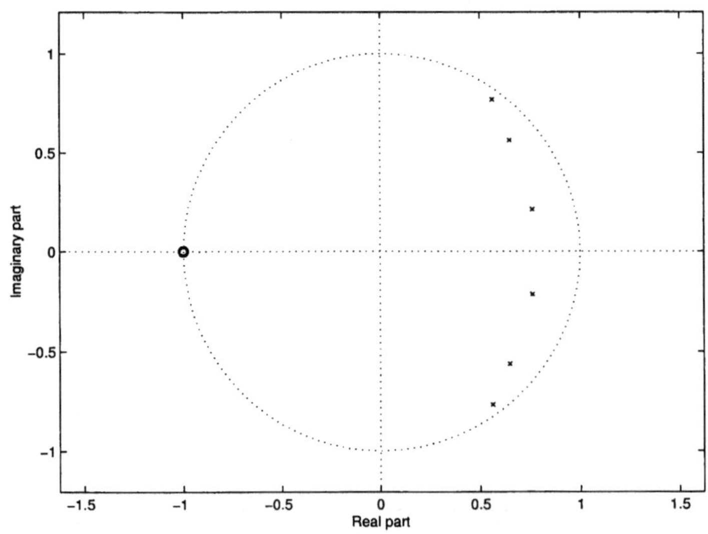

(b)

Figure 5.10 Frequency response of sixth-order type I Chebyshev filter. (a) Log magnitude plot in dB. (b) Pole-zero plot.

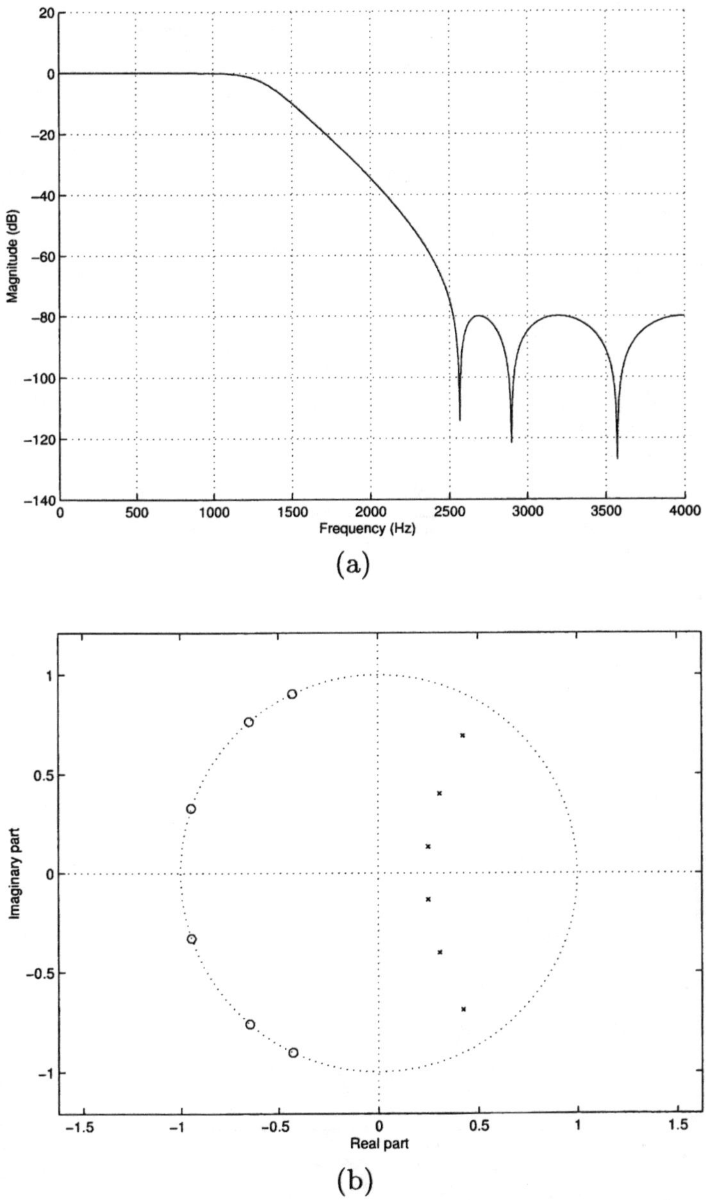

Figure 5.11 Frequency response of sixth-order type II Chebyshev filter. (a) Log magnitude plot in dB. (b) Pole-zero plot.

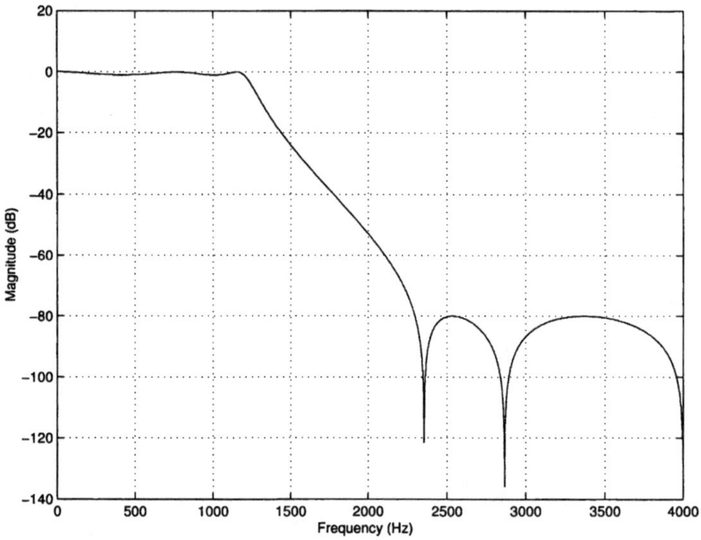

(a)

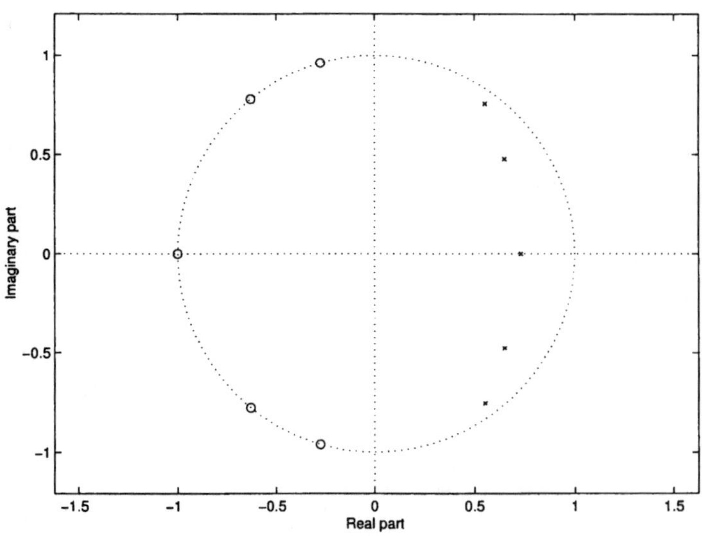

(b)

Figure 5.12 Frequency response of fifth-order elliptic filter. (a) Log magnitude plot in dB. (b) Pole-zero plot.

5.5.4 Frequency Transformations in Digital IIR Filters

So far, we have primarily discussed the design of lowpass digital IIR filters by using the bilinear transformation. In this section, we introduce the method of frequency transformation to design a highpass, bandpass, or bandstop digital IIR filter, based on a lowpass prototype digital IIR filter.

The approach of frequency transformation is first to convert the analog lowpass filter into a lowpass digital filter by using the bilinear transformation, and then to transform the lowpass digital filter into the desired filter, such as highpass, bandpass, or bandstop, by using a digital transformation. The digital transformation in the z-domain involves replacing the variable z^{-1} with a rational function $r(z^{-1})$. The $r(z^{-1})$ should satisfy the following properties:

1. The mapping z^{-1} replaced by a rational function $r(z^{-1})$ must convert points inside the unit circle into points also inside the unit circle in the z-plane.
2. A point on the unit circle in the z-plane must be mapped to another point also on the unit circle.

The above properties imply that a rational function $r(z^{-1})$ must be an all-pass function,

$$r(z^{-1}) = \pm \prod_{i=1}^{N} \frac{z^{-1} - \alpha_i}{1 - \alpha_i z^{-1}}, \tag{5.118}$$

where $|\alpha_i| < 1$. This condition ensures that a stable digital filter can be transformed into another stable digital IIR filter.

There are four types of rational functions $r(z^{-1})$ based on the requirements from the general form in (5.118), which can be used to transform a prototype digital lowpass filter into a highpass or a bandpass, bandstop, or other lowpass digital IIR filter. These rational functions $r(z^{-1})$ are discussed below.

Lowpass to Lowpass

If we wish to convert a lowpass digital filter into another lowpass digital IIR filter with passband edge frequency ω_P, the desired transformation is

$$z^{-1} \Longrightarrow \frac{z^{-1} - \alpha}{1 - \alpha z^{-1}}, \tag{5.119}$$

where the coefficient α in (5.119) is given by

$$\alpha = \frac{sin[(\omega_p - \omega_P)/2]}{sin[(\omega_p + \omega_P)/2]}, \tag{5.120}$$

where ω_p and ω_P are the passband edge frequency of old and new lowpass digital filters, respectively.

Lowpass to Highpass

If we wish to convert a lowpass digital filter into a highpass digital IIR filter with passband edge frequency ω_P, the desired transformation is obtained by

$$z^{-1} \Longrightarrow -\frac{z^{-1} + \alpha}{1 + \alpha z^{-1}}, \tag{5.121}$$

where the coefficient α in (5.121) is given by

$$\alpha = -\frac{cos[(\omega_p + \omega_P)/2]}{cos[(\omega_p - \omega_P)/2]}, \tag{5.122}$$

where ω_p and ω_P are the passband edge frequency of the old lowpass digital filter and the new highpass digital IIR filter, respectively.

Lowpass to Bandpass

If we wish to convert a lowpass digital filter into a bandpass digital IIR filter with passband edge frequencies ω_{p1} and ω_{p1}, the desired transformation is given by

$$z^{-1} \Longrightarrow -\frac{z^{-2} - \alpha_1 z^{-1} + \alpha_2}{\alpha_2 z^{-2} - \alpha_1 z^{-1} + 1}, \tag{5.123}$$

where the coefficient α_1 in (5.123) is obtained by

$$\alpha_1 = -\frac{2\beta K}{K + 1}, \tag{5.124}$$

and the coefficient α_2 in (5.123) is

$$\alpha_2 = \frac{K - 1}{K + 1}. \tag{5.125}$$

The parameter K is defined by

$$K = cot\left(\frac{\omega_{p2} - \omega_{p1}}{2}\right)\left[tan\left(\frac{\omega_p}{2}\right)\right], \tag{5.126}$$

and the parameter β is

$$\beta = \frac{cos[(\omega_{p2} + \omega_{p1})/2]}{cos[(\omega_{p2} - \omega_{p1})/2]}, \tag{5.127}$$

where ω_{p1} and ω_{p2} are the lower and upper passband edge frequencies of the bandpass digital IIR filter, respectively.

Lowpass to Bandstop

If we wish to convert a lowpass digital filter into a bandstop digital IIR filter with passband edge frequencies ω_{p1} and ω_{p2}, the desired transformation is given by

$$z^{-1} \Longrightarrow -\frac{z^{-2} - \alpha_1 z^{-1} + \alpha_2}{\alpha_2 z^{-2} - \alpha_1 z^{-1} + 1}, \tag{5.128}$$

where the coefficient α_1 in (5.128) is obtained by

$$\alpha_1 = -\frac{2\beta K}{K + 1}, \tag{5.129}$$

and the coefficient α_2 in (5.128) is

$$\alpha_2 = -\frac{K - 1}{K + 1}. \tag{5.130}$$

The parameter K is defined by

$$K = tan\left(\frac{\omega_{p2} - \omega_{p1}}{2}\right)\left[tan\left(\frac{\omega_p}{2}\right)\right], \tag{5.131}$$

and the parameter β is

$$\beta = \frac{cos[(\omega_{p2} + \omega_{p1})/2]}{cos[(\omega_{p2} - \omega_{p1})/2]}, \tag{5.132}$$

where ω_{p1} and ω_{p2} are the lower and upper passband edge frequencies of the bandpass digital IIR filter, respectively.

A frequency transformation can be implemented either in the analog domain or in the digital domain. We introduce only the frequency transformation in the digital domain. Using the bilinear transformation, or the frequency transformation, which is performed in the analog domain or in the digital domain, yields the same results in terms of identical digital IIR filters. In addition, in the case of the bilinear transformation, aliasing is not a problem for the frequency transformation either in the analog domain or in the digital domain.

The use of frequency transformations for the digital IIR filters was first proposed by Constantinides [18, 19]. Later, these frequency transformations were also described in the reference book by Proakis and Manolakis [20].

Example 5.9 Frequency Transformation for a Highpass Digital IIR Filter

Consider a lowpass digital IIR filter with system function

$$H(z) = \frac{0.2929(1 + z^{-1})}{1 - 0.4142z^{-1}}. \tag{5.133}$$

This lowpass digital IIR filter has 3-dB bandwidth at $\omega_p = 0.25\pi$. We want to convert this lowpass digital IIR filter into a highpass digital IIR filter by using the frequency transformation.

Using (5.121), the desired transformation is

$$z^{-1} \Longrightarrow -\frac{z^{-1} + \alpha}{1 + \alpha z^{-1}}, \tag{5.134}$$

where the coefficient α is given in (5.122). Substituting the above formula into the system function $H(z)$ yields

$$H(z) = \frac{0.2929\left(1 - \frac{z^{-1}+\alpha}{1+\alpha z^{-1}}\right)}{1 + 0.4142\left(\frac{z^{-1}+\alpha}{1+\alpha z^{-1}}\right)}. \tag{5.135}$$

Since $\omega_p = 0.25\pi$, the coefficient α in (5.122) is

$$\begin{aligned} \alpha &= -\frac{cos[(\omega_p + \omega_P)/2]}{cos[(\omega_p - \omega_P)/2]} \\ &= -\frac{cos[(0.25\pi + \omega_P)/2]}{cos[(0.25\pi - \omega_P)/2]}. \end{aligned} \tag{5.136}$$

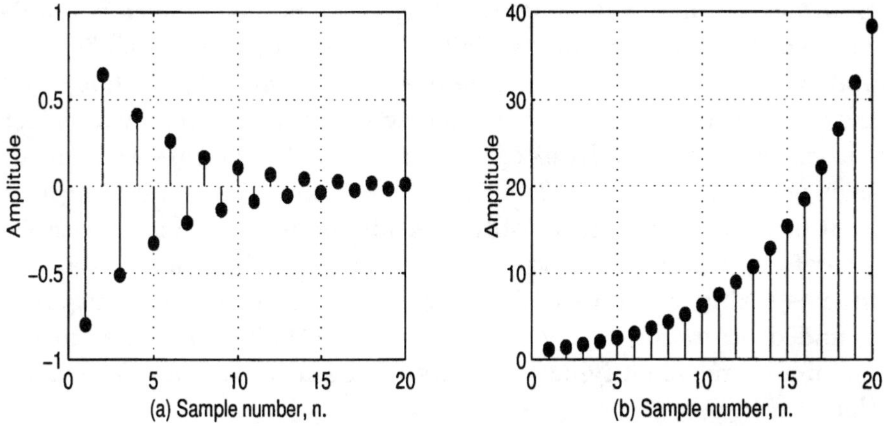

Figure 5.13 Impulse response of one-coefficient digital IIR filter. (a) The impulse response is very long in duration when $a_1 = -0.8$. (b) The impulse response diverges when $a_1 = 1.2$.

Suppose that $\omega_P = 0.75\pi$ for the highpass digital IIR filter. Then $\alpha = 0$. Therefore, we obtain the highpass digital IIR filter as follows,

$$H(z) = \frac{0.2929(1 - z^{-1})}{1 + 0.4142z^{-1}}. \tag{5.137}$$

This highpass digital IIR filter has a pole at $z = 0.4142$.

5.5.5 IIR Filter Stability

Consider a simple digital IIR filter having one feedback coefficient, $y[n] = x[n] + a_1 y[n-1]$. The impulse response $h[n]$ can be very long in duration for $a_1 < 1$, for example $a_1 = -0.8$, as shown in Figure 5.13(a). This is intuitively why the IIR digital filters can provide sharper cutoff with fewer filter coefficients. On the other hand, if $a_1 > 1$, the output of the IIR digital filter will diverge. This means that the digital IIR filter is unstable, as shown in Figure 5.13(b), if $a_1 = 1.2$ in this case. Therefore, a digital IIR filter with just one coefficient can have an impulse response $h[n]$ that is infinite in duration.

As can be seen, for a one-coefficient digital IIR filter, stability is guaranteed if that coefficient is less than one in magnitude. However, for more than one coefficient, it is no longer such a simple case. For example, consider a digital IIR filter with three coefficients, $y[n] = x[n] + 0.9y[n-1] + 0.9y[n-2] + 0.8y[n-3]$. Although all the coefficients are less than one in magnitude, their cumulative effect will cause the digital IIR filter to be unstable. In this case, the output sequence will be $\{1, 0.9, 3.149, ...\}$ for an input of a unit impulse $\{1, 0, 0, ...\}$. As a result, the output is unbounded and hence unstable. Therefore, extreme care is required to ensure that the digital IIR filter is stable. Some careful strategies to ensure that the output is bounded are required. By representing the digital IIR filter in the z-domain and identifying the poles of the filter we can do exactly this.

Identifying stability of a recursive filter requires that all poles of the digital IIR filter are less than one in magnitude so that none of the cascade of first-order sections will diverge. The poles may be complex numbers, but these complex numbers will always occur in conjugate pairs, given that the digital IIR filter coefficients are real. It is common practice to plot the pole-zero of a digital IIR filter on the z-plane. If all poles are within the unit magnitude circle, then the digital IIR filter is stable.

There are several computational procedures that help us in determining if any of the roots of the denominator polynomial $X(z)$ of $H(z) = \frac{Y(z)}{X(z)}$, namely,

$$X(z) = 1 + a_1 z^{-1} + a_2 z^{-1} + \cdots + a_N z^{-N}. \quad (5.138)$$

These procedures are called stability criteria. One of the test procedures for the stability is known as the Schur-Cohn test, which has been extensively used in practice. We cite the reference books by Kailath [21] and Brogan [22].

Example 5.10 Pole-Zero Plot of a Digital IIR Filter

Consider a simple fourth-order digital IIR filter with the sampling frequency 8,000 Hz in the following form,

$$H(z) = \frac{0.0102 + 0.0408z^{-1} + 0.0613z^{-2} + 0.0408z^{-3} + 0.0102z^{-4}}{1 - 1.9684z^{-1} + 1.7359z^{-2} - 0.7245z^{-3} + 0.1204z^{-4}}. \quad (5.139)$$

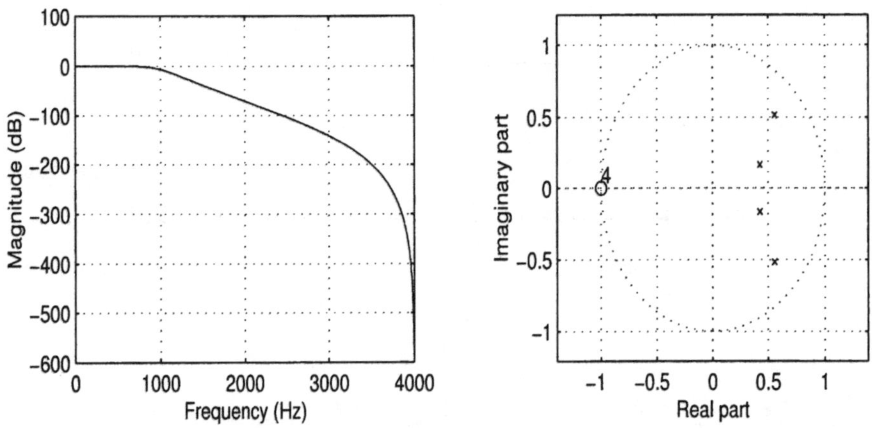

Figure 5.14 Frequency response for IIR digital filter. (a) Log magnitude. (b) Pole-zero plot, where "×" indicates the poles and "o" is the zeros of IIR digital filter.

The log magnitude and pole-zero plot are shown in Figure 5.14(a) and (b), respectively. This example is a Butterworth digital IIR filter with all the poles within the unit circle. This example also illustrates that the poles are inside the unit circle for stable digital IIR filters.

5.6 Implementation of Filter Structures

The signal flow-graph representation is useful in implementing LTI discrete-time systems, where specified either by difference equation or system function. The basic elements in the signal flow-graph representation required for implementation of the system transfer function, $H(z)$, are adders, multipliers, and memory for storing delayed sequence values. Figure 5.15 shows the basic elements used in the signal flow-graph representation of a discrete-time filter system.

In any digital implementation, the structures with the fewest constant multipliers and the fewest delay branches are often most desirable. This is because multiplication is generally a time-consuming and costly operation in the implementation of digital hardware or in DSPs and because each delay element corresponds to a memory register. Therefore, the reduction of the number of

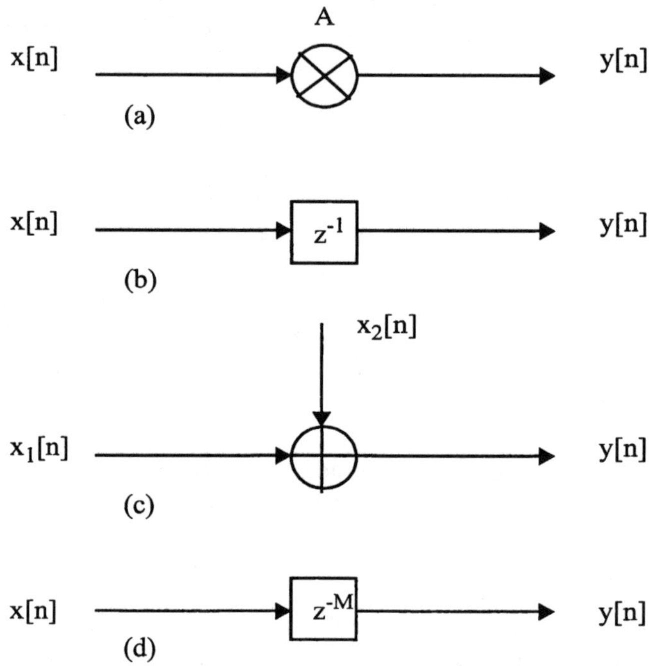

Figure 5.15 Basic elements in the signal flow-graph representation of a discrete-time filter system. (a) Multiplication of a discrete-time sequence with a constant, $y[n] = Ax[n]$. (b) Sample unit delay, $y[n] = x[n-1]$. (c) Two discrete-time sequence additions, $y[n] = x_1[n] + x_2[n]$. (d) Unit delay of M samples, $y[n] = x[n-M]$.

multipliers can increase the implementation speed, and a reduction in the number of delay elements can reduce the memory requirements.

In this section, we develop several of the most practically used forms along with their flow-graph representations for implementing LTI digital FIR and IIR filter systems.

5.6.1 FIR Filter Structures

For a discrete-time FIR system, the filter performs a weighted average on a window of M samples,

$$y[n] = \sum_{k=0}^{M-1} b[k]x[n-k], \tag{5.140}$$

with a corresponding z-transform system function,

$$H(z) = \sum_{k=0}^{M-1} b[k]z^{-k}. \tag{5.141}$$

The FIR filter transfer function $H(z)$ will always be a polynomial in the z-domain. The roots of this polynomial are the zeros of the FIR filter. Note that the zeros may be complex numbers, and if they are not real, they will occur in conjugate pairs.

Direct-Form Structure

The direct-form structure of FIR filter implementation is shown in Figure 5.16(a). This implementation structure is referred to as a *tapped delay line* filter or a *transversal* filter that consists of three basic elements: (a) unit-delay, (b) multiplier, and (c) adder. The number of delay elements that is commonly called the *FIR filter order* used in the filter determines the finite duration of its impulse response. As can be seen from Figure 5.16(a), at each tap, the signal along this chain is weighted by the appropriate filter coefficient and the resulting products are summed to form the discrete-time output $y[n]$. Note that the implementation structure of a transversal filter requires M multiplications and $M-1$ additions *per unit time*.

In addition to the transversal implementation form, the *transposed* direct form for the FIR filter is shown in Figure 5.16(b). Computationally, the transposed form is equivalent to the transversal and needs to have M multiplications and $M-1$ additions per unit time for implementation.

Cascade-Form Structure

The transfer system function $H(z)$ for an FIR filter can be

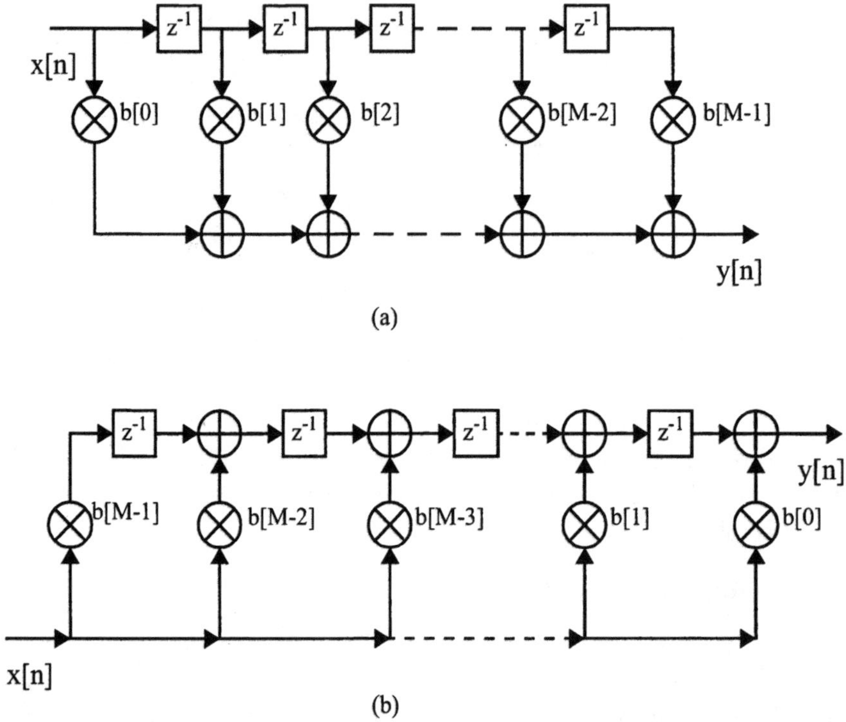

Figure 5.16 Direct-form structure of an FIR filter. (a) Transversal FIR filter form. (b) Transposition FIR filter form.

factored into a product of polynomials as follows:

$$\begin{aligned} H(z) &= \sum_{k=0}^{M-1} b[k]z^k \\ &= \sum_{k=0}^{M-1} b_k z^k \\ &= \prod_{k=1}^{M/2} (b_{0k} + b_{1k}z^{-1} + b_{2k}z^{-2}). \end{aligned} \quad (5.142)$$

If M is even, one of the coefficients b_{2k} will be zero because $H(z)$ would have an odd number of real zeros. Note that the pairs of

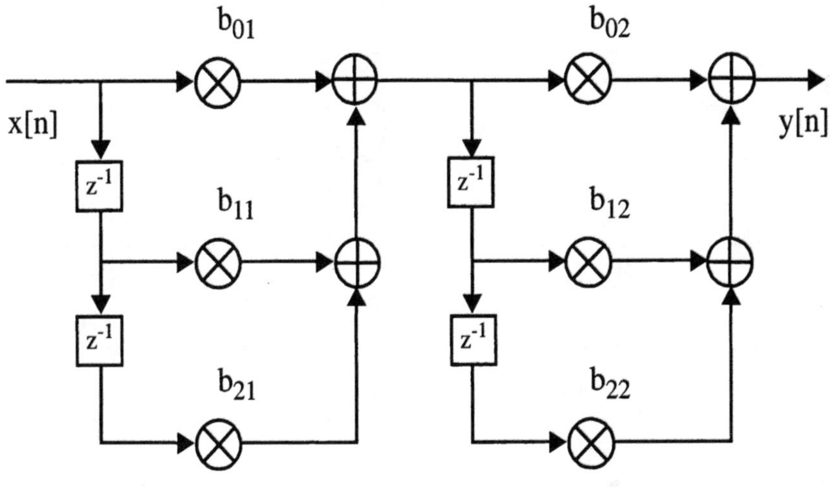

Figure 5.17 Cascade implementation form of an FIR filter.

complex-conjugate roots can always be formed so that the coefficients $\{b_{ik}\}$ in (5.142) are real valued. On the other hand, real-valued roots in (5.142) can be paired in any arbitrary way. Figure 5.17 shows the cascade implementation form of an FIR filter system when $M = 4$.

Symmetry-Form Structure

The causal FIR filter system has generalized linear phase if the impulse response $h[n]$ satisfies certain symmetry conditions as described in Section 5.3. With such symmetry conditions of FIR filter coefficients, the number of coefficient multipliers can be halved. The four different types of linear-phase FIR filter are shown in Figure 5.18. Figure 5.18(a) depicts the impulse response for a type I linear-phase FIR filter with even coefficient symmetry about $n = 5$. Figure 5.18(b) shows the impulse response for a type II linear-phase FIR filter with even coefficient symmetry about 4.5. Figure 5.18(c) illustrates the impulse response for a type III linear-phase FIR filter with odd coefficient symmetry at $n = 5$. Figure 5.18(d) presents the impulse response for a type IV linear-phase FIR filter with coefficient odd symmetry about 4.5.

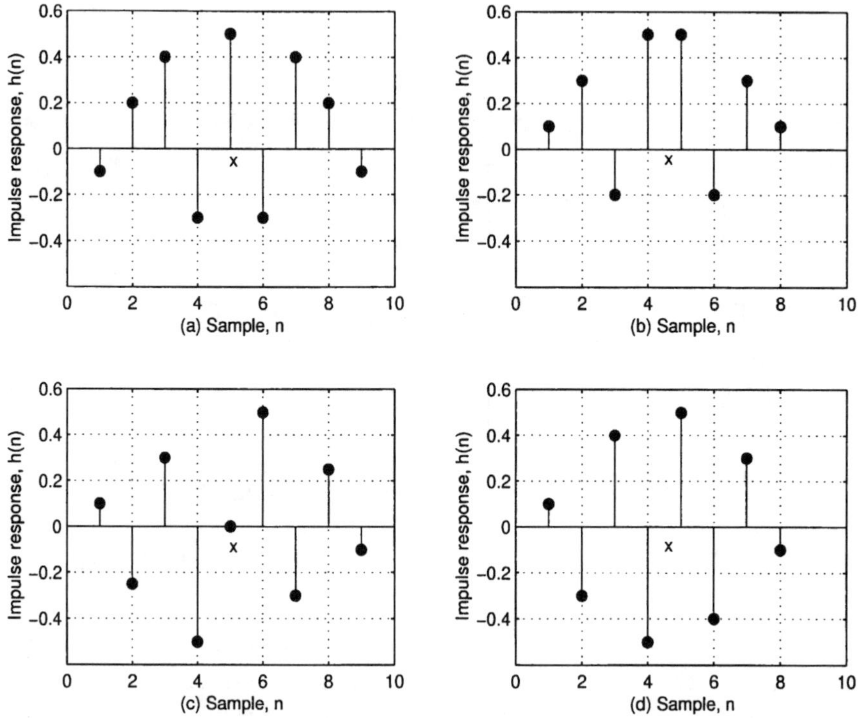

Figure 5.18 Impulse response for linear-phase FIR filter, where "×" in the figure indicates the symmetry point. (a) Type I symmetry coefficients. (b) Type II symmetry coefficients. (c) Type III symmetry coefficients. (d) Type IV symmetry coefficients.

Given the conditions of symmetry, $h[M-n] = \pm h[n]$, for $n = 0, 1, ..., M$, the difference equation for type 1, with M an even integer, can be written as

$$\begin{aligned} y[n] &= \sum_{k=0}^{M} h[k]x[n-k] \\ &= \sum_{k=0}^{M/2-1} h[k](x[n-k] + x[n-M-k]) \\ &\quad + h[M/2]x[n-M/2]. \end{aligned} \tag{5.143}$$

For type III with M an even integer, the difference equation relating output to input is

$$y[n] = \sum_{k=0}^{M/2-1} h[k](x[n-k] - x[n-M+k]). \tag{5.144}$$

For type II with M an odd integer, the corresponding equation is

$$y[n] = \sum_{k=0}^{(M-1)/2} h[k](x[n-k] + x[n-M+k]). \tag{5.145}$$

For the case of M, an odd integer, the type IV system equation is

$$y[n] = \sum_{k=0}^{(M-1)/2} h[k](x[n-k] - x[n-M+k]). \tag{5.146}$$

Note that equations (5.143), (5.144), (5.145), and (5.146) show the implementation structures with either $M/2+1$, $M/2$, or $(M+1)/2$ multipliers rather than the M multipliers of the general direct form structure in Figure 5.16. The correspondingly efficient implementations of a linear-phase FIR filter with type I and type II structures are shown in Figure 5.19(a) and (b), respectively. The efficient implementation forms of a linear-phase FIR filter with type III and type IV structures are shown in Figure 5.20(a) and (b), respectively.

5.6.2 Digital IIR Filter Structures

For the discrete-time IIR filter system, the input and output satisfy a difference equation of the form

$$y[n] = \sum_{k=0}^{M} b_k x[n-k] + \sum_{k=1}^{N} a_k y[n-k], \tag{5.147}$$

and corresponding rational system function,

$$H(z) = \frac{\sum_{k=0}^{M} b_k z^{-k}}{1 - \sum_{k=1}^{N} a_k z^{-k}}. \tag{5.148}$$

The implementation forms of the flow-graph are presented below.

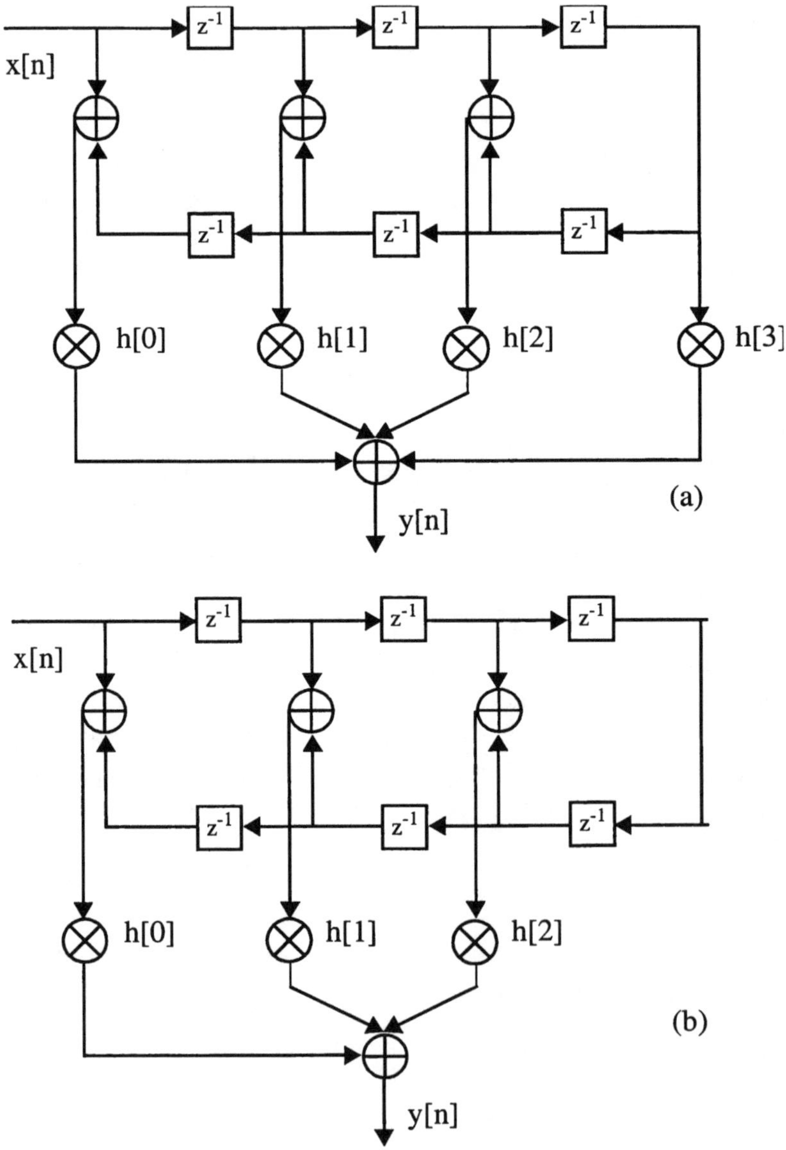

Figure 5.19 Symmetry-form structure of FIR filter. (a) Efficient implementation form of a linear-phase FIR type I filter. (b) Efficient implementation form of a linear-phase FIR type II filter.

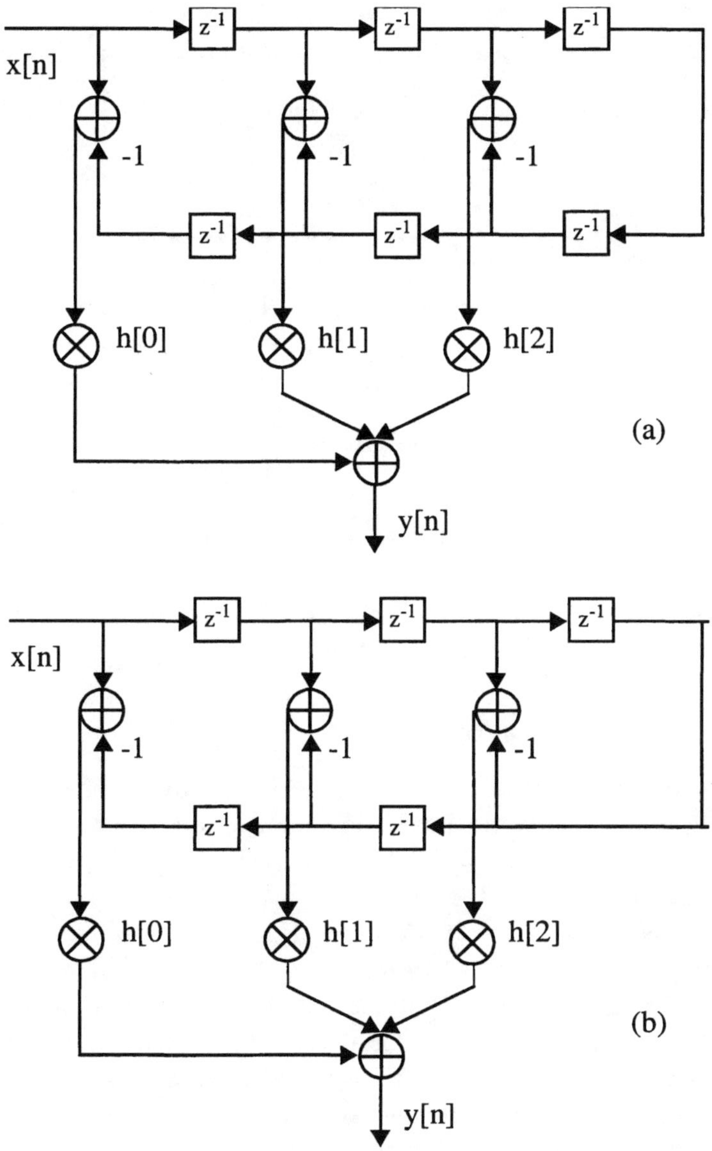

Figure 5.20 Symmetry-form structure of FIR filter. (a) Efficient implementation form of a linear-phase FIR type III filter. (b) Efficient implementation form of a linear-phase FIR type IV filter.

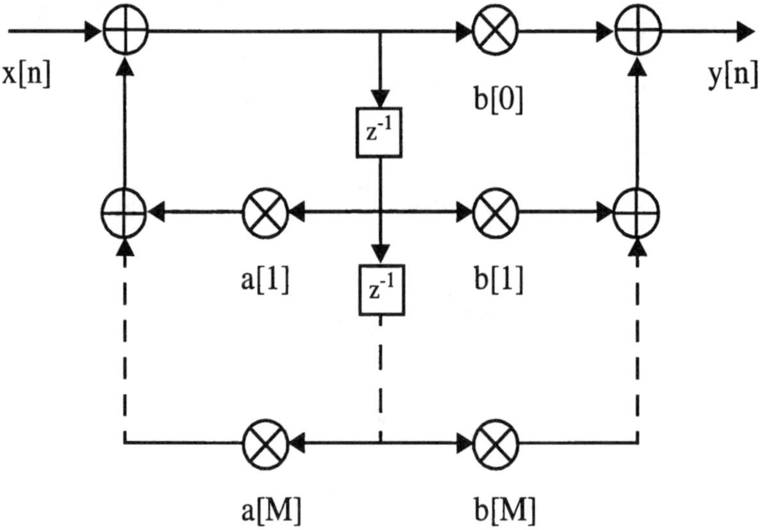

Figure 5.21 Direct form II or canonic direct form for digital IIR filter implementation.

Direct-Form Structure

In the implementation of IIR digital filters, unit delays in parallel, running down the center based on (5.148), can be paired such that within each pair the two delays take the same input signal, as shown in Figure 5.21. The structure in Figure 5.21 is known as the *direct form II IIR filter realization.* It uses the minimum number of multiplications and delays for implementing a IIR filter system. Specifically, the minimum number of delays required in general is $max\{N, M\}$. Such an implementation with the minimum number of delay elements is also commonly referred to as a *canonic direct form* implementation. This implementation structure requires $M + N + 1$ multiplications, $M + N$ additions, and the maximum of $\{N, M\}$ memory locations.

Cascade-Form Structure

The implementation structure of the cascade form can be

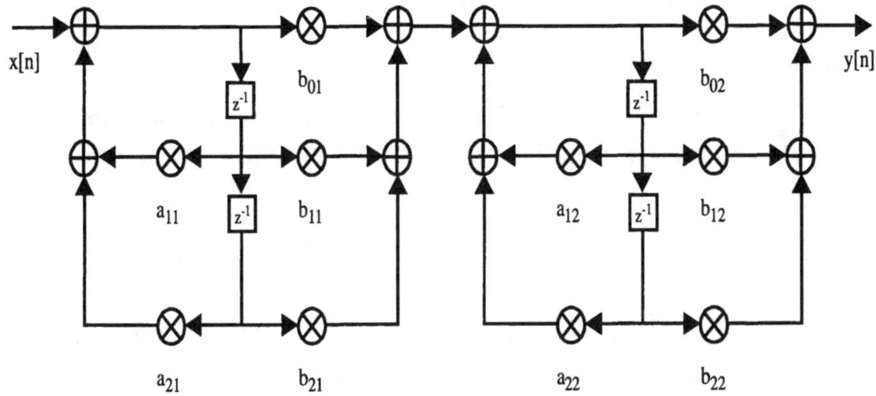

Figure 5.22 Cascade-form structure of a fourth-order system with a canonic implementation form for each second-order IIR subsystem.

obtained from the system function $H(z)$ with a ratio of polynomials as in (5.148). By combining pairs of real factors and complex conjugate pairs into second-order systems, (5.148) can be expressed as

$$H(z) = \prod_{k=1}^{N_s} \frac{b_{0k} + b_{1k}z^{-1} + b_{2k}z^{-2}}{1 - a_{1k}z^{-1} - a_{2k}z^{-2}}, \tag{5.149}$$

where N_s is the largest integer for $(N+1)/2$. The individual second-order sections in the cascade form can be implemented with a minimum number of multiplications and a minimum number of delay elements if we use the canonic direct form. A cascade form for a fourth-order system using two canonic direct forms is shown in Figure 5.22.

Note that the coefficients $\{a_{ik}\}$ and $\{b_{ik}\}$ in (5.149) are real valued. This indicates that we need to group together a pair of complex-conjugate zeros, and a pair of complex-conjugate poles to form the second-order subsystem. Moreover, any two real-valued zeros, and any two real-valued poles can be paired to form a second-order IIR subsystem. In addition, the second-order IIR subsystem in the numerator of (5.149) or the second-order IIR subsystem in the denominator of (5.149) may contain either a pair of real roots or a pair of complex-conjugate roots.

Parallel-Form Structure

The parallel form is realized by transforming the polynomials of the rational system function, $H(z)$, into a partial fraction expansion:

$$H(z) = \sum_{k=0}^{N_p} A_k z^{-k} + \sum_{k=1}^{N_s} \frac{e_{0k} + e_{1k}z^{-1}}{1 - a_{1k}z^{-1} - a_{2k}z^{-2}}, \tag{5.150}$$

where N_s is the largest integer for $(N+1)/2$ and the first sum is not included in (5.150) if $N_p = M - N$ is negative.

Note that the coefficients $\{e_{ik}\}$ and $\{a_{ik}\}$ in (5.150) are real valued. This means that we can combine pairs of complex-conjugate poles to form a two-pole subsystem. Furthermore, real-valued poles can form two-pole subsystems in an arbitrary manner. Figure 5.23 shows the parallel form for a fourth-order system with real and complex poles grouped in pairs.

So far, we have discussed the direct-form, cascade-form, and symmetry-form structures for the FIR filter. For the IIR filter, we have presented the direct-form, cascade-form, and parallel-form structures. However, there is another FIR and IIR filter structure, which is called the *lattice filter*. This type of filter structure is very attractive in some applications because the filter demonstrates stability and robustness to finite-word-length effects. For example, the latter structure is used extensively in digital speech processing, adaptive filtering, and geophysical signal processing. We cite the references on these structures and their efficient implementation methods by Rabiner and Schafer [23], Hayes and Clements [24], and Proakis and Manolakis [20].

5.7 Summary

In this chapter, we introduced digital filtering for linear time-invariant discrete-time systems. We discussed a wide range of methods for designing FIR and IIR filters as well as filter structures for implementation.

The selection of an FIR or an IIR filter in practical applications

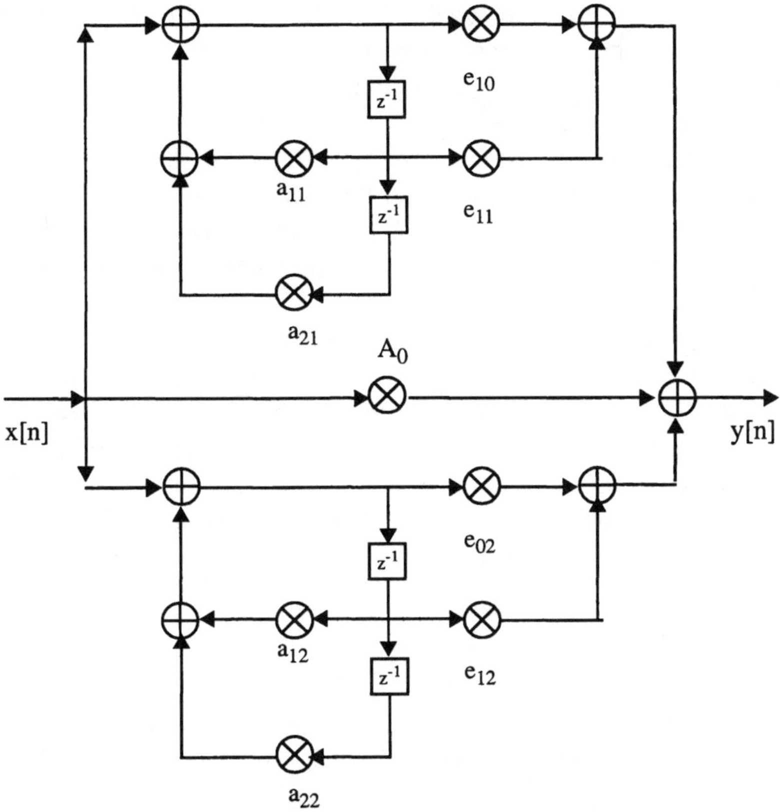

Figure 5.23 Implementation structure of parallel-form for fourth-order IIR digital filter system.

depends on the importance to the design problem of the advantages of each type. The FIR filter does not have closed-form equations for design. The window method can be straightforwardly applied with some iterations to meet the required filter specifications. The least squares method and the Parks-McClellan algorithm lead to FIR filters of lower order than those of the window method. These design methods can be easily implemented in special hardware or in DSPs. FIR filters can have precisely linear phase. Also, FIR filters do not have instability problems. Therefore, the quantization problem of filter coefficients will not lead to unstable filters, since there are no poles in FIR filters. In contrast, IIR filters have the advantage that a variety of frequency-selective filters can be designed using closed-form design formulas based on the Butterworth, Chebyshev, or elliptic descriptions. IIR filters can satisfy a magnitude response specification more efficiently, with lower filter order. However, IIR filters are not able to achieve precisely linear phase. IIR filters are not guaranteed to be stable. Quantization of IIR filter coefficients is a problem when the poles of the IIR filter are close to the unit circle, since small perturbations can cause them to go outside the unit circle.

The practical concerns are usually measured in terms of hardware complexity, chip area, processing comsumption, computation speed, and cost. These factors are directly related to the filter order and which implementation architecture is used. The computational efficiency of the implementation architecture is very important for digital filter operations. We will introduce efficient implementation architectures based on multirate signal processing systems and filter banks in Chapter 7.

References

[1] Clements, M. A., and J. Pease, "On Causal Linear Phase IIR Digital Filters," *IEEE Trans. on Acoustic, Speech, and Signal Processing*, Vol. ASSP-3, No. 4, pp. 479-484, March 1989.

[2] Fahmy, M. F., Y. M. Y. Hasan, G. Abdel-Raheem, and N. EI-Gayed, "Design of Linear-Phase IIR Filters from FIR Specifications," *IEEE Trans. on Signal Processing*, Vol. 42, No. 2, pp. 437-440, February 1994.

[3] Oppenheim, A. V., and R. W. Schafer, *Discrete-Time Signal Processing*, Prentice Hall, Englewood Cliffs, New Jersey, 1989.

[4] Parks, T. W., and J. H. McClellan, "Chebyshev Approximation for Nonrecursive Digital Filters with Linear Phase," *IEEE Trans. on Circuit Theory*, Vol. CT-19, pp. 189-194, March 1972.

[5] Parks, T. W., and J. H. McClellan, "A Program for the Design of Linear Phase Finite Impulse Response Filters," *IEEE Trans. on Audio Electroacoustics*, Vol. AU-20, No. 3, pp. 195-199, August 1972.

[6] McClellan, J. H., T. R. Parks, and L. R. Rabiner, "A Computer Program for Designing Optimum FIR Linear Phase Digital Filters," *IEEE Trans. on Audio Electroacoustics*, Vol. AU-21, pp. 506-526, December 1973.

[7] McClellan, J. H., and T. R. Parks, "A Unified Approach to the Design of Optimum FIR Linear Phase Digital Filters," *IEEE Trans. on Circuit Theory*, Vol. CT-20, pp. 697-701, November 1973.

[8] Karam, L. J., and J. H. McClellan, "Complex Chebyshev Approximation for FIR Filter Design," *IEEE Trans. on Circuits and Systems II*, March 1995.

[9] Kaiser, J. F., "Nonrecursive Digital Filter Design Using the I_0-Sinh Window Function," *Proc. of IEEE Int. Symp. Circuits and Systems*, pp. 20-23, San Francisco, California, April 1974.

[10] Burrus, C. S., A. W. Soewito and R. A. Gopinath, "Least Squared Error FIR Filter Design with Transition Bands," *IEEE Trans. on Signal Processing*, Vol. 40, No. 6, pp. 1327-1340, June 1992.

[11] Burrus, C. S., "Multiband Least Squares FIR Filter Design," *IEEE Trans. on Signal Processing*, Vol. 43, No. 2, pp. 412-421, February 1995.

[12] Parks, T. W., and C. S. Burrus, *Digital Filter Design*, John Wiley & Sons, New York, 1987.

[13] Selesnick, I. W., M. Lang, and C. S. Burrus, "Constrained Least Square Design of FIR Filter Without Specified Transition Bands," *IEEE Trans. on Signal Processing*, Vol. 44, No. 8, August 1996.

[14] The MathWorks, *Matlab Signal Processing Toolbox*, User's Guide, Version 4, 1996.

[15] Tufts, D. W., D. W. Rorabacher, and W. E. Mosier, "Designing Simple Effective Digital Filters," *IEEE Trans. on Audio and Electroacoustics*, Vol. 8, No. 2, pp. 142-158, June 1970.

[16] Oethen, G., T. W. Parks, and H. W. Schüssler, "New Results in the design of Digital Interpolators," *IEEE Trans. on Acoustic, Speech, and Signal Processing*, Vol. 23, pp. 301-309, June 1975.

[17] Sedra, A. S., and K. C. Smith, *Microelectronic Circuits*, 3d ed., Saunders College Publishing, Chicago, Illinois, 1991.

[18] Constantinides, A. G., "Frequency Transformations for Digital Filters," *Electronic Letter*, Vol. 3, pp. 487-489, November 1967.

[19] Constantinides, A. G., "Spectral Transformations for Digital Filters," *Proc. of IEEE*, Vol. 117, pp. 1585-1590, August, 1970.

[20] Proakis, J. G., and D. G. Manolakis, *Digital Signal Processing: Principles, Algorithms, and Applications*, 3d ed., Prentice-Hall, Upper Saddle River, New Jersey, 1996.

[21] Kailath, T., *Linear Systems*, Prentice Hall, Englewood Cliffs, New Jersey 1980.

[22] Brogan, W. L., *Modern Control Theory*, 3d ed., Prentice Hall, Englewood Cliffs, New Jersey 1991.

[23] Rabiner, L. R., and R. W. Schafer, *Digital Processing of Speech Signals*, Prentice Hall, Englewood Cliffs, New Jersey, 1978.

[24] Hayes, M. H., and M. A. Clements, "An Efficient Algorithm for Computing Pisarenko's Harmonic Decomposition using Levinson's Recursion," *IEEE Trans. on Acoustic, Speech, and Signal Processing*, Vol. ASSP-34, No. 3, pp. 485-491, June 1986.

6

Adaptive Filters

6.1 Introduction

The subject of *adaptive digital filters* is an important part of statistical digital signal processing. The use of an adaptive digital filter provides an attractive option when signals result from operations in an environment of unknown statistics. Usually, the adaptive digital filter offers new capabilities for digital signal processing that would not be possible with conventional methods.

The adaptive digital filter is fundamentally based on the minimization theory of least squares that was initially contributed by the German mathematician Gauss in the nineteenth century. The theory of least squares is widely used in statistical analysis and in many fields of science and engineering. In 1960, Widrow and Hoff published a paper to address adaptive switching circuits [1] and introduced the algorithm of the least mean squares (LMS). Subsequently, Widrow et al. contributed two classic papers entitled "Adaptive Noise Cancelling: Principles and Applications" and "Stationary and Non-Stationary Learning Characteristics of the LMS Adaptive Filter" in 1975 and 1976, respectively [2, 3]. These were germinal contributions to adaptive DSP development.

In the area of digital signal processing, the problem of least squares minimization is often applied to sampled data in real-

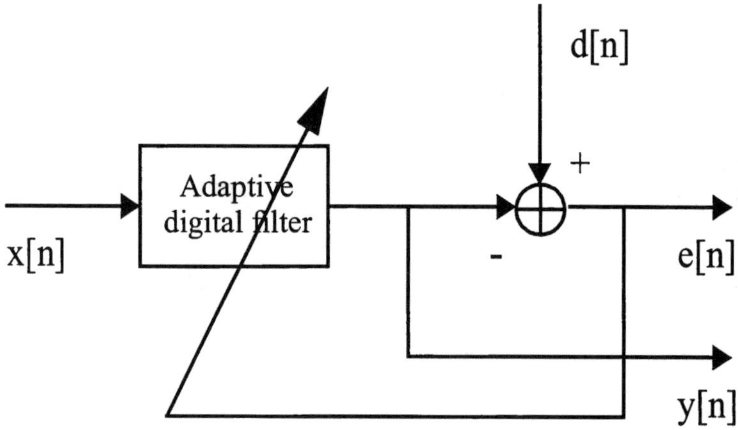

Figure 6.1 Generalized system of adaptive digital signal processing. The output digital signal $y[n] = x[n] * h_a[n]$, where $h_a[n]$ is an adaptive digital filter, and error digital signal $e[n] = d[n] - y[n]$.

time operation, and therefore presents the challenge of producing a real-time implementation with high sample rates. Since the early 1980s, powerful digital signal processors have allowed many real-time adaptive DSP systems to be developed. Since 1990, there has been a large growth in the application of real-time adaptive digital signal processing to solve many problems. We thus find that adaptive digital filters are successfully applied in many fields such as high-fidelity audio, speech, video, wireless and broadband communications, biomedical engineering (see Miao [4-6]), control, radar, sonar, ultrasonics, seismology, and military weapons.

Adaptive digital filters consist of the three basic components: the digital filter, the error, and the adaptation functions as shown in Figure 6.1. The goal of this DSP system is to adapt the digital filter in such a way that the input digital signal $x[n]$ is filtered to produce an output digital signal $y[n]$ that will minimize the error digital signal $e[n]$, when subtracted from the desired digital signal $d[n]$. The arrow through the adaptive digital filter is standard notation to indicate that the digital filter is adaptive. Hence, all of the digital filter coefficients can be updated in terms of some functions of the error digital signal. In addition, the adaptive digital algorithm should be stable with any feedback system.

The adaptive digital filter can be FIR, IIR, or even nonlinear digital filter. To ensure adaptive algorithm stability and mathematical tractability, most adaptive digital filters use the FIR type. However, in the last few years, adaptive digital IIR filters have been increasingly used with stable forms in a number of real-world applications, such as the active noise control and adaptive differential pulse code modulation (ADPCM) techniques. Further, current research has uncovered a number of useful nonlinear adaptive digital filters based on decision feedback and neural networks by Haykin [7], Miao and Clements [8], and Miao [9].

The adaptive digital filter has many applications. Although these applications are indeed quite different in nature, they have a basic common feature in terms of input digital signal $x[n]$, output digital signal $y[n]$, error digital signal $e[n]$, and desired digital signal $d[n]$. The applications of adaptive digital filters can be classified into four fundamental classes based on the implementation architecture as shown in Figure 6.2.

System Identification

In the class of identification architecture, the adaptive digital filter is used to provide a linear model that represents the best fit to an unknown system. The unknown system and the adaptive digital filter are the same input. Then, output of the unknown system provides the desired signal response for the adaptive digital filter as shown in Figure 6.2(a). In some cases, if the unknown system is dynamic, the linear model will be time varying in the system identification. Applications include room acoustic identification, channel estimation, and echo cancellation.

Prediction

In the prediction architecture as shown in Figure 6.2(b), the adaptive digital filter is used to provide the best prediction of the current value of a random input signal. Depending on the application of interest, the system operates as a predictor if the output of

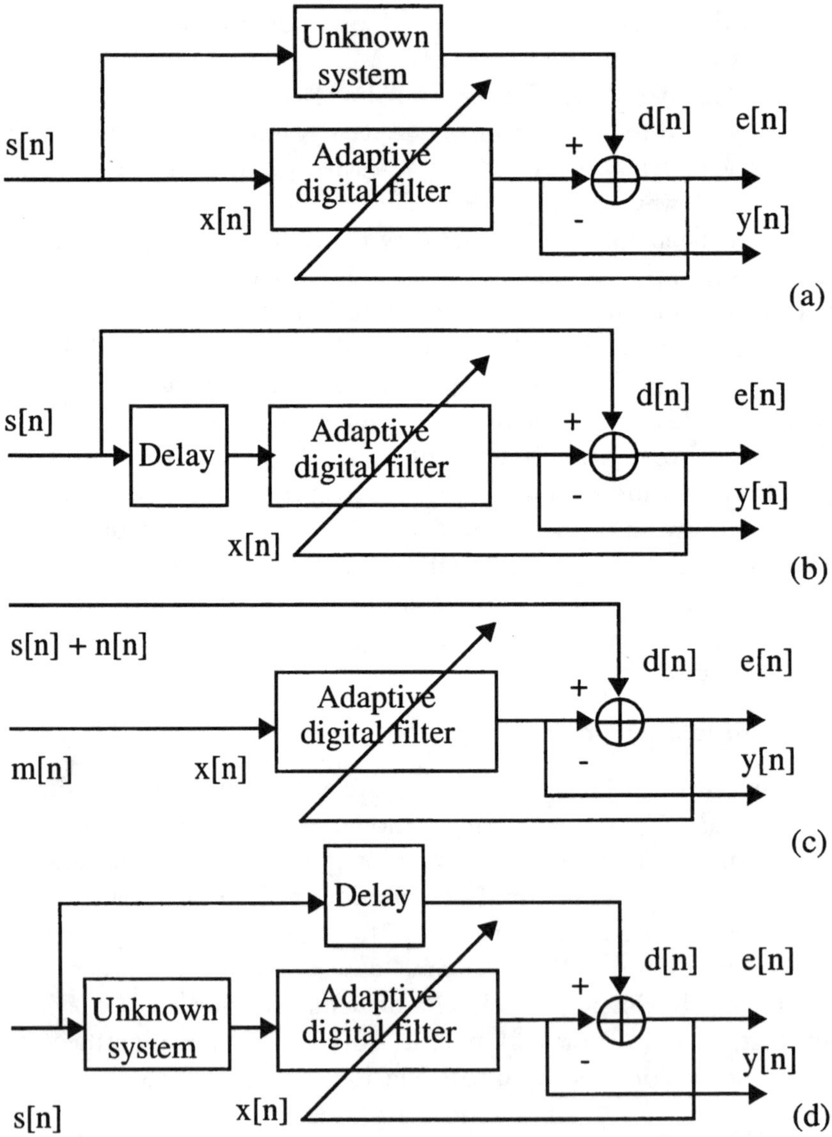

Figure 6.2 Applications based on four basic implementation architectures of an adaptive digital filter. (a) System identification. (b) Prediction. (c) Interference cancellation. (d) Inverse system identification.

the adaptive digital filter serves as the output signal. Otherwise, the system operates as a prediction error filter if the prediction error acts as the output signal. Applications include periodic noise suppression, periodic signal extraction, linear predictive coding, ADPCM, autoregressive spectrum analysis, and signal detection.

Interference Cancellation

In the architecture of interference cancellation, the adaptive digital filter is used to cancel unknown interference contained in an input signal. In Figure 6.2(c), the input digital signal serves as the desired response for the adaptive digital filter. The reference signal is supplied to the input such that the information-bearing signal component is essentially undetectable. Applications consist of adaptive noise cancellation, echo cancellation, adaptive beamforming, and biomedical signal processing.

Inverse System Identification

In the architecture of inverse system identification as shown in Figure 6.2(d), the adaptive digital filter is used to provide the inverse model that is the best fit to an unknown noisy system. The inverse model ideally has a transfer function that is equal to the reciprocal of the unknown transfer function in such a way that the combination of the two constitutes an ideal transmission medium. Applications have adaptive equalization in digital communications, predictive deconvolution, blind equalization, adaptive control system, and high-fidelity low-frequency room equalization.

The key goal of the adaptive digital filter is to minimize the error digital signal $e[n]$. The degree of success of this minimization for the error digital signal, $e[n]$, will clearly depend on the nature of the input signal, the length of the adaptive filter, and the adaptive algorithm used. In the following sections, we will discuss the several adaptive algorithms in detail.

This chapter is organized as follows. Section 6.1 introduced the adaptive filter and its applications in system identification, prediction, interference cancellation, and inverse system identification. Section 6.2 describes a class of linear optimum discrete-time filters known as Wiener filters. The discrete-time Kalman filter based on

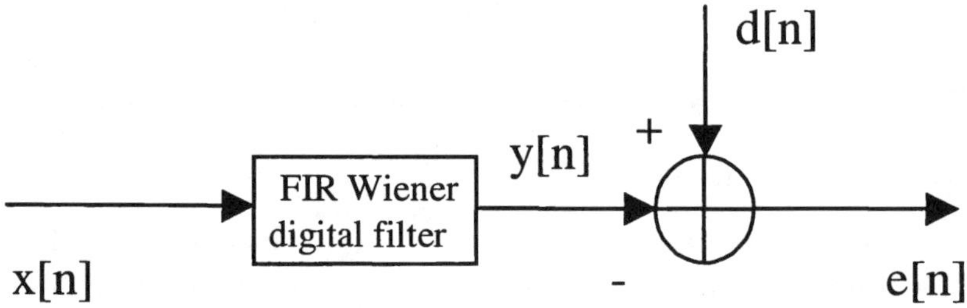

Figure 6.3 Illustration of the general FIR Wiener filter problem based on the Wiener-Hopf optimum solution for the mean-square error minimization.

the recursive least squares solution for nonstationary processes is introduced in Section 6.3, with emphasis on the Kalman filter problem and the Kalman filtering approach. Section 6.4 discusses the LMS algorithms, convergence analysis, and excess mean-squared error. Section 6.5 discusses the recursive least squares (RLS) algorithms, with a focus on the procedure for the RLS algorithms and convergence analysis, and compares the RLS, LMS, and Kalman filters. The chapter is summarized in Section 6.6.

6.2 Wiener Filter Theory

This section describes a class of *linear optimum discrete-time filters* that is known as *Wiener filters* [10]. We consider only the design of FIR Wiener filters, although the theory includes the FIR filter as a special case. The IIR Wiener filter is, in general, noncausal and therefore unrealizable. The main result that we will focus on here is the derivation of the discrete-time Wiener-Hopf equations, which provide the FIR filter coefficients of the optimum filter in the sense of minimum mean-square error (MMSE).

An FIR Wiener filter, as shown in Figure 6.3, is used to produce the MMSE estimate, $y[n]$ for $d[n]$, given two wide-sense stationary processes jointly, $x[n]$ and $d[n]$, that are statistically related to each other. It is also assumed that the autocorrelation functions, $r_x(k)$

and $r_d(k)$, and the cross-correlation function, $r_{dx}(k)$, are known.

Determining a set of coefficients for Wiener filter, $w[n]$, minimizes the mean of the squared error of the filter output as compared to some desired output, $d[n]$. The error signal $e[n]$ is written

$$\begin{aligned} e[n] &= d[n] - y[n] \\ &= d[n] - \sum_{k=0}^{M-1} w[k]x[n-k], \end{aligned} \tag{6.1}$$

or in vector form,

$$e[n] = d[n] - \mathbf{w}^T[n]\mathbf{x}[n]. \tag{6.2}$$

The mean of the squared error is a function of the coefficient weight vector $\mathbf{w}[n]$ chosen, and is written

$$\begin{aligned} \xi &= E\{e[n]e^*[n]\} \\ &= E\{|e[n]|^2\} \\ &= E\{|d[n] - \mathbf{w}^T[n]\mathbf{x}[n]|^2\} \\ &= E\{d[n]d^*[n] - d[n]\mathbf{x}^T[n]\mathbf{w}[n] \\ &\quad - \mathbf{w}^T[n]\mathbf{x}[n]d^*[n] + \mathbf{w}^T[n]\mathbf{x}[n]\mathbf{x}^T[n]\mathbf{w}[n]\} \\ &= var\{d[n]\} - \mathbf{r}_{dx}[n]\mathbf{w}[n] - \mathbf{w}^T[n]\mathbf{r}_{dx}[n] \\ &\qquad + \mathbf{w}^T[n]\mathbf{R}_x[n]\mathbf{w}[n], \end{aligned} \tag{6.3}$$

where $\mathbf{r}_{dx}[n] = E\{\mathbf{x}[n]d^*[n]\}$ is the product of the cross-correlation function, and $\mathbf{R}_x[n] = E\{\mathbf{x}[n]\mathbf{x}^T[n]\}$ is the autocorrelation function.

To minimize ξ, it is necessary and sufficient that the derivative of ξ with respect to w be equal to zero,

$$\mathbf{R}_x[n]\mathbf{w}[n] = \mathbf{r}_{dx}[n]. \tag{6.4}$$

where $\mathbf{R}_x[n]$ is a $M \times M$ Hermitian Toeplitz matrix of autocorrelation, $\mathbf{w}[n]$ is the vector of filter coefficients, and $\mathbf{r}_{dx}[n]$ is the vector of cross-correlation between the desired signed $d[n]$ and the observed signal $x[n]$. The matrix form in (6.4) is sometimes called the *Wiener-Hopf equation.*

If the matrix $\mathbf{R}_x[n]$ is invertible, then $\mathbf{w}[n]$ can be solved by

$$\mathbf{w}[n] = \mathbf{R}_x^{-1}[n]\mathbf{r}_{dx}[n]. \tag{6.5}$$

Equation (6.5) is called the *normal equation*, since the error signal is orthogonal to each element of the input vector as follows:

$$E\{\mathbf{x}[n]e^*[n]\} = 0. \tag{6.6}$$

The minimum mean-square error of the estimate of $d[n]$ can be calculated from (6.3) as follows:

$$\begin{aligned} \xi &= E\{|e[n]|^2\} \\ &= E\{e[n](d[n] - \mathbf{w}^T[n]\mathbf{x}[n])\} \\ &= E\{e[n]d[n]\} - \mathbf{w}[n]E\{\mathbf{x}^T[n]e^*[n]\}. \end{aligned} \tag{6.7}$$

Note that since $\mathbf{w}[n]$ is the solution to the Wiener-Hopf equations, then $E\{\mathbf{x}^T[n]e^*[n]\} = 0$ because the error signal is orthogonal to each element of the input signal vector. Therefore, the second term in (6.7) is equal to zero and

$$\begin{aligned} \xi_{min} &= E\{e[n]d^*[n]\} \\ &= E\{(d[n] - \mathbf{w}^T[n]\mathbf{x}[n])d^*[n]\} \\ &= E\{d[n]d^*[n]\} - \mathbf{w}^T[n]E\{\mathbf{x}[n]d^*[n]\} \\ &= R_d(0) - \mathbf{R}_{dx}^T[n]\mathbf{w}[n]. \end{aligned} \tag{6.8}$$

The minimum error can also be expressed in terms of autocorrelation matrix $\mathbf{R}_x[n]$ and cross-correlation vector $\mathbf{r}_{dx}[n]$ in the following:

$$\xi_{min} = R_d(0) - \mathbf{r}_{dx}^T[n]\mathbf{R}_x^{-1}[n]\mathbf{r}_{dx}[n]. \tag{6.9}$$

Now, it is useful to consider what must be known to solve the Wiener-Hopf equation and to estimate minimum error ξ_{min} as follows:

1. Calculate the autocorrelation matrix $\mathbf{R}_x[n]$ of input vector, $\mathbf{x}[n]$.
2. Determine the cross-correlation vector $\mathbf{r}_{dx}[n]$ between input vector, $\mathbf{x}[n]$, and desired response, $d[n]$.

3. Estimate the variance $\mathbf{R}_d(0)$ of the desired signal, $d[n]$.
4. Compute the optimal coefficients $\mathbf{w}[n]$ and the minimum error ξ_{min}.

From the point of view of DSP implementation, the Wiener-Hopf equation is a single-step algorithm without feedback and can be used to solve different problems including system identification, noise cancellation, equalization, and so on. However, the Wiener-Hopf solution is not a useful algorithm in the real-time DSP operation because of the heavy computation. Note that autocorrelation matrix $\mathbf{R}_x[n]$ and cross-correlation matrix $\mathbf{R}_{dx}[n]$ need approximately $2MN$ multiply and addition (MAC) operations, where N is the number of samples in each sample frame, and M is the adaptive filter length. The inversion matrix $\mathbf{R}_x^{-1}[n]$ requires about M^3 MACs and the matrix vector multiplication needs M^2 MACs. As can be seen, the total number of computations for one-step algorithm is $(2NM+M^3+M^2)$ MACs. For example, with the filter length $M = 50$ and the samples N = 1,000, computing the $\mathbf{R}_x[n]$ and $\mathbf{R}_{dx}[n]$ will require 227,500 MACs to find the $MMSE$ solution. In addition, the optimal coefficient weight vector $\mathbf{w}[n]$ must be computed again if the signals $x[n]$ and $d[n]$ are nonstationary. This leads to a very high computation load for each iteration. Therefore, the direct implementation of the Wiener-Hopf equation is not recommended for real-time DSP operations in practical applications.

In the context of the Wiener filter, we recommend the reference books by Kailath [11], Haykin [7], and Hayes [12].

Example 6.1 Noise Cancellation Using Wiener Filtering

In this example, we introduce noise cancellation using a Wiener filter as shown in Figure 6.4. The goal of this problem is to estimate the noise signal $N_1[n]$. Then, the estimated noise, $y[n] = \hat{N}_2[n]$, is subtracted from the signal $x[n]$ to form an estimated error source $e[n]$ as follows:

$$\begin{aligned} e[n] &= x[n] - y[n] \\ &= d[n] + N_1[n] - \hat{N}_2[n]. \end{aligned} \tag{6.10}$$

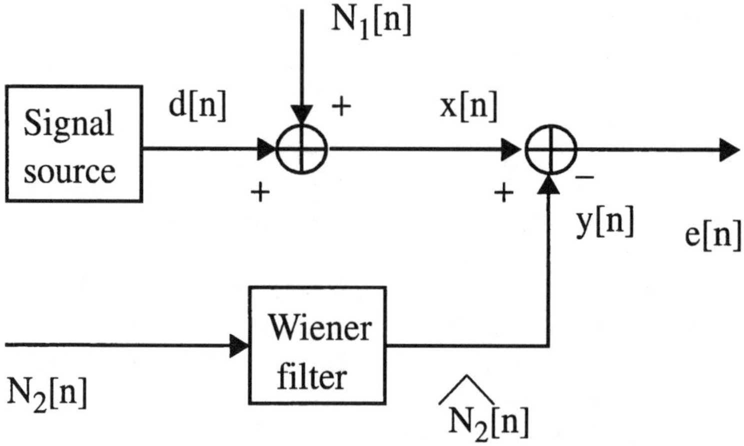

Figure 6.4 Noise cancellation using a Wiener FIR filter.

Assume that the desired signal $d[n]$ is a sinusoidal signal

$$d[n] = cos\left(\frac{1}{400}\pi n + \frac{1}{2}\pi\right), \tag{6.11}$$

and white noise sequences are

$$\begin{aligned} N_1[n] &= 0.85N_1[n-1] + G[n] \\ N_2[n] &= 0.65N_2[n-1] + G[n], \end{aligned}$$

where $G[n]$ is zero-mean and unit variance, and is uncorrelated with signal source $d[n]$. Then, we can show that $\hat{r}_{N1N2} = \hat{r}_{xN2}$. The autocorrelation function $\hat{r}_{N2}$ can be estimated using

$$\hat{r}_{N2}[k] = \frac{1}{N}\sum_{n=0}^{N-1} N_2[n]N_2[n-k], \tag{6.12}$$

and an estimate of the cross-correlation function is

$$\hat{r}_{xN2}[k] = \frac{1}{N}\sum_{n=0}^{N-1} x[n]N_2[n-k]. \tag{6.13}$$

Therefore, the Wiener-Hopf equations are $\hat{\mathbf{R}}_{N2}[n]\mathbf{w}[n] = \hat{\mathbf{r}}_{xN2}[n]$.

Figure 6.5 shows the simulation results of noise cancellation using the Wiener filter. A total of 4,000 signal samples are used and the Wiener filter has a filter order, $N = 22$. The estimated sinusoidal signal is shown in Figure 6.5(e).

6.3 Discrete-Time Kalman Filter

The Wiener filters that we developed in the above section assume that the signals are wide-sense stationary processes. Additionally, Wiener filters can be equally derived to estimate nonstationary processes. However, in a case of nonstationary processes, Wiener filters may be impractical because of a very high computation load for updating the filter coefficients with each iteration. Since most signals in practical applications are nonstationary, the usefulness of the Wiener filter in the real-time DSP applications is limited. In this section, we introduce the *discrete-time Kalman filter* based on the recursive least-squares solution to the linear optimum filtering problem for nonstationary processes.

6.3.1 The Kalman Filtering Problem

Figure 6.6 shows a linear and discrete-time dynamical system illustrated by using the signal flow-graph. The state vector, denoted by $\mathbf{x}[n]$, is defined as a set of quantities that uniquely describe the dynamical behavior of the linear discrete-time system. In order to estimate $\mathbf{x}[n]$, we use the observation vector $\mathbf{z}[n]$, which contains a set of observed data as shown in Figure 6.6. Then we can develop a set of equations based on the signal flow-graph as follows:

First assume that the random process to be estimated can be expressed as a state equation in the following form:

$$\mathbf{x}[n] = \phi[n]\mathbf{x}[n-1] + \mathbf{w}[n-1], \tag{6.14}$$

where $\mathbf{x}[n]$ is an $n \times 1$ process state vector, $\phi[n]$ is an $n \times n$ matrix relating $\mathbf{x}[n]$ to $\mathbf{x}[n-1]$, and $\mathbf{w}[n-1]$ is an $n \times 1$ white noise vector with known covariance structure

$$E\{\mathbf{w}[k]\mathbf{w}^T[i]\} = \begin{cases} \mathbf{Q}[k] & k = i \\ 0 & k \neq i. \end{cases} \tag{6.15}$$

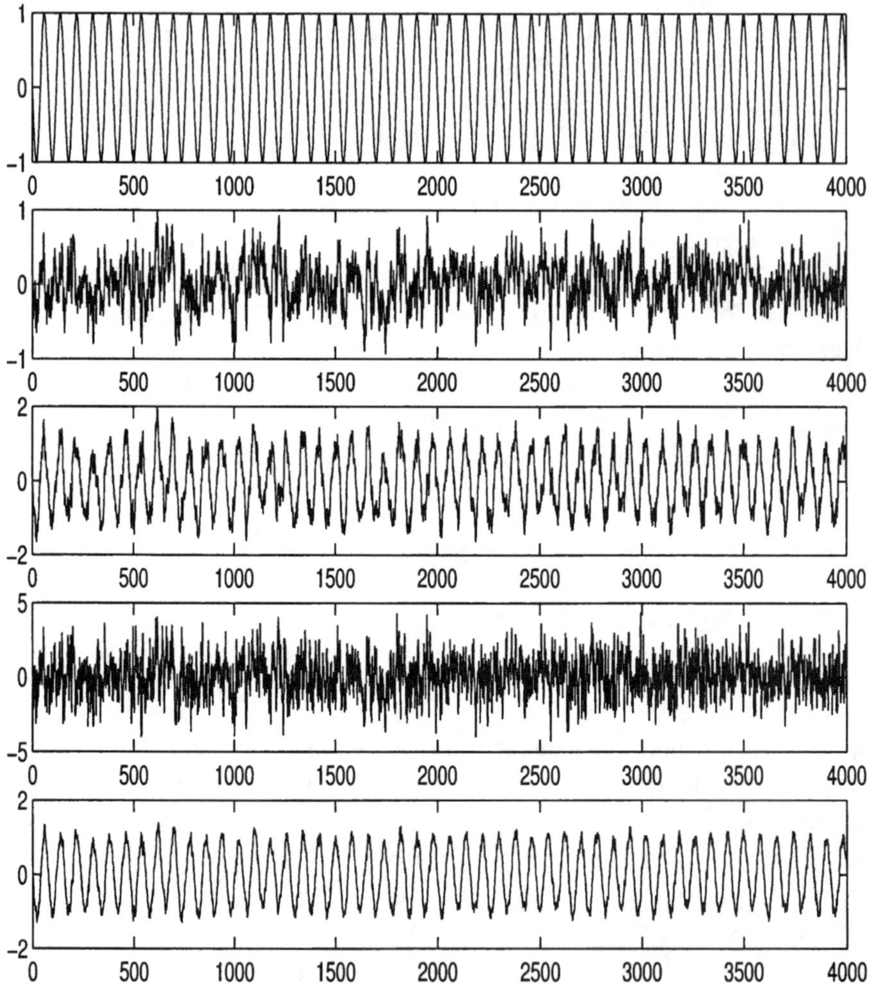

Figure 6.5 The simulation results of noise cancellation. The label art components from (a) to (e) are from top to bottom in the graph. (a) The desired signal. (b) The noise signal, N_1. (c) The desired signal corrupted by the noise signal. (d) The reference noise signal, N_2. (e) The estimated sinusoid signal after using signal $x[n]$ subtracted from output of Wiener filter with 22 filter coefficients.

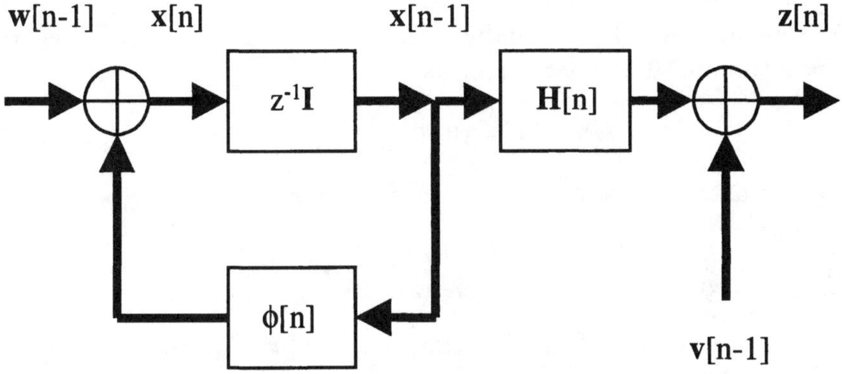

Figure 6.6 Representation of a linear and discrete-time dynamical system by using signal flow-graph.

An observation equation with a linear relationship is assumed to occur at discrete-time n,

$$\mathbf{z}[n] = \mathbf{H}[n]\mathbf{x}[n-1] + \mathbf{v}[n-1], \tag{6.16}$$

where $\mathbf{z}[n]$ is an $m \times 1$ observation vector, $\mathbf{H}[n]$ is an $m \times n$ matrix relating the observation and process state vectors, and $\mathbf{v}[n-1]$ is assumed to be an $m \times 1$ observation error vector, which is a white sequence with known covariance structure

$$E\{\mathbf{v}[k]\mathbf{v}^T[i]\} = \begin{cases} \mathbf{R}[k] & k = i \\ 0 & k \neq i. \end{cases} \tag{6.17}$$

The noise vectors $\mathbf{w}[n]$ and $\mathbf{v}[n]$ are statistically independent, that is, we have zero cross-correlation

$$E\{\mathbf{w}[k]\mathbf{v}^T[i]\} = 0, \tag{6.18}$$

for all k and i.

Thus, the Kalman filtering approach for a linear optimum filtering problem may be stated as follows: The components of the state vector $\mathbf{x}[n]$ are to be estimated using the entire observed data $\mathbf{z}[n]$ such that the mean square error is minimum.

6.3.2 The Kalman Filtering Approach

At this point, we assume that an initial estimate of the process at discrete-time $n-1$ is available and denoted by $\hat{\mathbf{x}}[n-1]$. Thus, we define the initial estimate error as

$$\hat{\mathbf{e}}[n-1] = \mathbf{x}[n-1] - \hat{\mathbf{x}}[n-1], \tag{6.19}$$

and the associated error covariance matrix as

$$\begin{aligned}\hat{\mathbf{P}}[n-1] &= E\{\hat{\mathbf{e}}[n-1]\hat{\mathbf{e}}^T[n-1]\} \\ &= E\{(\mathbf{x}[n-1] - \hat{\mathbf{x}}[n-1]) \\ &\quad \cdot (\mathbf{x}[n-1] - \hat{\mathbf{x}}[n-1])^T\}.\end{aligned} \tag{6.20}$$

Note that, in some cases, we begin the estimation problem without initial measurements. In such cases, if the process mean is zero, then the initial estimate $\hat{\mathbf{x}}[n-1]$ is zero as well. Therefore, the associated error covariance matrix in (6.20) can be rewritten as

$$\hat{\mathbf{P}}[n-1] = E\{\mathbf{x}[n-1]\mathbf{x}^T[n-1]\}. \tag{6.21}$$

The optimum estimate of the state vector based on all past measurements at discrete-time, n, may be expressed in the form

$$\begin{aligned}\hat{\mathbf{x}}[n] &= \phi[n]\hat{\mathbf{x}}[n-1] + \mathbf{K}[n]\hat{\mathbf{v}}[n-1] \\ &= \phi[n]\hat{\mathbf{x}}[n-1] + \mathbf{K}[n](\mathbf{z}[n] - \mathbf{H}[n]\hat{\mathbf{x}}[n-1]),\end{aligned} \tag{6.22}$$

where $\mathbf{K}[n]$ is the *Kalman gain* vector yet to be determined. Hence, the problem is to find the Kalman gain that yields an updated estimate that is optimal in some sense. To solve this problem, we then derive the expression for the error covariance matrix associated with the update estimate as follows:

$$\begin{aligned}\hat{\mathbf{P}}[n] &= E\{\hat{\mathbf{e}}[n]\hat{\mathbf{e}}^T[n]\} \\ &= E\{(\mathbf{x}[n] - \hat{\mathbf{x}}[n])(\mathbf{x}[n] - \hat{\mathbf{x}}[n])^T\} \\ &= (\mathbf{I} - \mathbf{K}[n]\mathbf{H}[n])\hat{\mathbf{P}}[n-1](\mathbf{I} - \mathbf{K}[n]\mathbf{H}[n])^T \\ &\quad + \mathbf{K}[n]\mathbf{R}[n]\mathbf{K}^T[n].\end{aligned} \tag{6.23}$$

For brevity, we omit some of the details in developing the results in (6.23).

To do the optimization, we differentiate $\hat{\mathbf{P}}[n]$ with respect to $\mathbf{K}[n]$,

$$\frac{d\hat{\mathbf{P}}[n]}{d\mathbf{K}[n]} = 2\mathbf{K}[n](\mathbf{H}[n]\hat{\mathbf{P}}[n-1]\mathbf{H}^T[n] + \mathbf{R}[n]) - 2(\mathbf{H}[n]\hat{\mathbf{P}}[n-1])^T, \tag{6.24}$$

and set this derivative to zero, as well as solve for the optimal Kalman gain. The result is

$$\mathbf{K}[n] = \hat{\mathbf{P}}[n-1]\mathbf{H}^T[n](\mathbf{H}[n]\hat{\mathbf{P}}[n-1]\mathbf{H}^T[n] + \mathbf{R}[n])^{-1}. \tag{6.25}$$

Equation (6.25) is called the *optimal Kalman gain*, which minimizes the mean-square estimation error.

Substituting (6.25) into (6.23) with some simplification of the formula, the covariance matrix associated with the optimal state estimate can be expressed as follows:

$$\hat{\mathbf{P}}[n] = (\mathbf{I} - \mathbf{K}[n]\mathbf{H}[n])\hat{\mathbf{P}}[n-1]. \tag{6.26}$$

We now have to update the estimate of state vector $\hat{\mathbf{x}}[n]$ by using (6.14). Note that we can ignore the term of $\mathbf{w}[n]$ because it has zero mean and is not correlated with any of the previous $\mathbf{w}[n-k]$. Thus, the updated estimate of state vector $\hat{\mathbf{x}}[n]$ is easily projected ahead via the transition matrix,

$$\hat{\mathbf{x}}[n+1] = \phi[n+1]\hat{\mathbf{x}}[n]. \tag{6.27}$$

Therefore, the estimate error for the next step is

$$\begin{aligned}\hat{\mathbf{e}}[n+1] &= \mathbf{x}[n+1] - \hat{\mathbf{x}}[n+1] \\ &= \phi[n+1]\mathbf{x}[n] + \mathbf{w}[n] - \phi[n+1]\hat{\mathbf{x}}[n] \\ &= \phi[n+1]\hat{\mathbf{e}}[n] + \mathbf{w}[n].\end{aligned} \tag{6.28}$$

Because of the zero cross-correlation between $\mathbf{w}[n]$ and $\hat{\mathbf{e}}[n]$, the error covariance matrix associated with the next step of state vector estimation $\hat{\mathbf{x}}[n+1]$ is obtained by

$$\begin{aligned}\hat{\mathbf{P}}[n+1] &= E\{\hat{\mathbf{e}}[n+1]\hat{\mathbf{e}}^T[n+1]\} \\ &= E\{(\phi[n+1]\hat{\mathbf{e}}[n] + \mathbf{w}[n]) \\ &\qquad (\phi[n+1]\hat{\mathbf{e}}[n] + \mathbf{w}[n])^T\} \\ &= \phi[n+1]\mathbf{P}[n]\phi^T[n+1] + \mathbf{Q}[n].\end{aligned} \tag{6.29}$$

Equations (6.22), (6.25), (6.26), (6.27), and (6.29) comprise the recursive equations of the Kalman filter. The implementation of the Kalman filter is summarized in the following steps:

1. Initialize the prior estimate $\hat{\mathbf{x}}[n-1]$ and its error covariance $\hat{\mathbf{P}}[n-1]$.
2. Compute the Kalman gain vector using (6.25).
3. Update the estimate $\hat{\mathbf{x}}[n]$ with measurement $\mathbf{z}[n]$ using (6.22).
4. Calculate the error covariance $\hat{\mathbf{P}}[n]$ for the updated estimate using (6.26).
5. Project ahead via the transition matrix $\phi[n+1]$ using (6.27) and (6.29). The process returns to step 2.

It is interesting to reflect on the Kalman filter in perspective before we provide the example. The Kalman filter is a linear, discrete-time, finite-dimensional system that is well suited to a digital computer or a digital signal processor. A distinctive feature of a Kalman filter is its mathematical formulation in terms of state-space descriptions. Another feature of a Kalman filter is that it computes its solution recursively. Note that each updated estimate of the state is computed from the previous estimate and the new input discrete-time data. Thus only the previous estimate requires storage during computation. An additional property of the Kalman filter is that it minimizes the trace of the filtered state error correlation matrix, or the optimal Kalman gain $\mathbf{K}[n]$. This means that the Kalman filter is also the optimum filter in some sense of the minimum mean-square error. In addition, the Kalman filter provides a unifying framework for the derivation of an important family of filters known as recursive least squares (RLS) adaptive filters, which will be discussed in Section 6.5.

There are many technical resources on the Kalman filter. For a detailed treatment of the Kalman filter, the reader is referred to the reference books by Kailath [11], Sorenson [13], Sibul [14], Brown and Hwang [15], and Haykin [16].

Example 6.2 Signal Process Estimation Using a Kalman Filter

Suppose that $x[n]$ has the following signal description:

$$x[n] = 0.95x[n-1] + w[n], \tag{6.30}$$

where $w[n]$ is white noise with zero mean and variance $\sigma_w^2 = 0.5$. The linear observation equation is assumed to have the form

$$z[n] = x[n-1] + v[n-1], \tag{6.31}$$

where $v[n-1]$ is also white noise with zero mean and unit variance, and is uncorrelated with $w[n]$. Therefore, $\phi[n] = 0.95$, $\mathbf{H}[n] = 1$, $\mathbf{Q}[n] = 0.5$, and $\mathbf{R}[n] = 1$. The update Kalman filter state estimation equation is

$$\hat{x}[n] = 0.95\hat{x}[n-1] + K[n](z[n] - \hat{x}[n-1]), \tag{6.32}$$

where $K[n]$ is the Kalman gain to be determined. The Kalman gain is given by

$$K[n] = \hat{P}[n-1](\hat{P}[n-1] + 1)^{-1}, \tag{6.33}$$

and the error covariance matrix is

$$\hat{P}[n] = (1 - K[n])\hat{P}[n-1], \tag{6.34}$$

and

$$\hat{P}[n+1] = (0.95)^2\hat{P}[n] + 0.5. \tag{6.35}$$

Let $\hat{P}[0] = 1$, then the Kalman gain $K[n]$ and error covariance $\hat{P}[n]$ for 200 iterations are plotted in Figure 6.7. As we can see, after 77 iterations, the Kalman filter converges into its steady-state equation as follows:

$$\hat{x}[n] = 0.95\hat{x}[n-1] + 0.8368(z[n] - \hat{x}[n-1]), \tag{6.36}$$

and the Kalman gain obtains 0.8368.

6.4 The LMS Adaptive Filters

In this section, we develop a set of widely used adaptive filter techniques called *least mean square* (LMS) algorithms. The LMS algorithm is the simplest and most often used algorithm in applications today. The LMS algorithm serves as standard against which other adaptive filter algorithms are benchmarked.

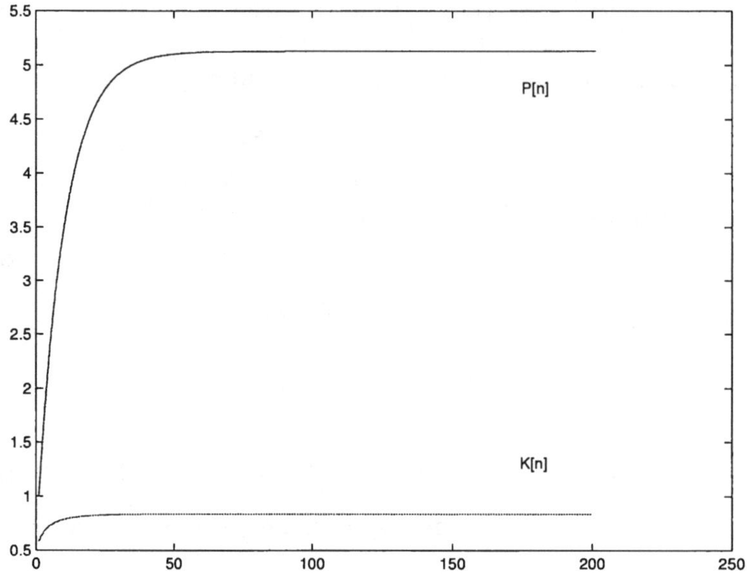

Figure 6.7 The simulation results of signal process estimation using a Kalman filter. The Kalman gain achieves 0.8368 after 77 iterations.

6.4.1 The LMS Algorithms

Consider an adaptive FIR filter used for estimating a desired signal $d[n]$ from an input signal $x[n]$ as shown in Figure 6.8. The transfer function is described as follows:

$$\begin{aligned} y[n] &= \sum_{q=0}^{M-1} b_n[q]x[n-q]. \\ &= \mathbf{B}_n^T \mathbf{x}[n], \end{aligned} \tag{6.37}$$

where $b_n[q]$ is a time-varying set of coefficients for the FIR filter and $\mathbf{B}_n^T = \{b_n[1], b_n[2], ..., b_n[M-1]\}$. The goal of the adaptive process is to adjust the filter coefficients in such a way that they move from their current position toward the MMSE solution. The mean square error $\xi[n]$ is defined by

$$\xi[n] = E\{|e[n]|^2\}, \tag{6.38}$$

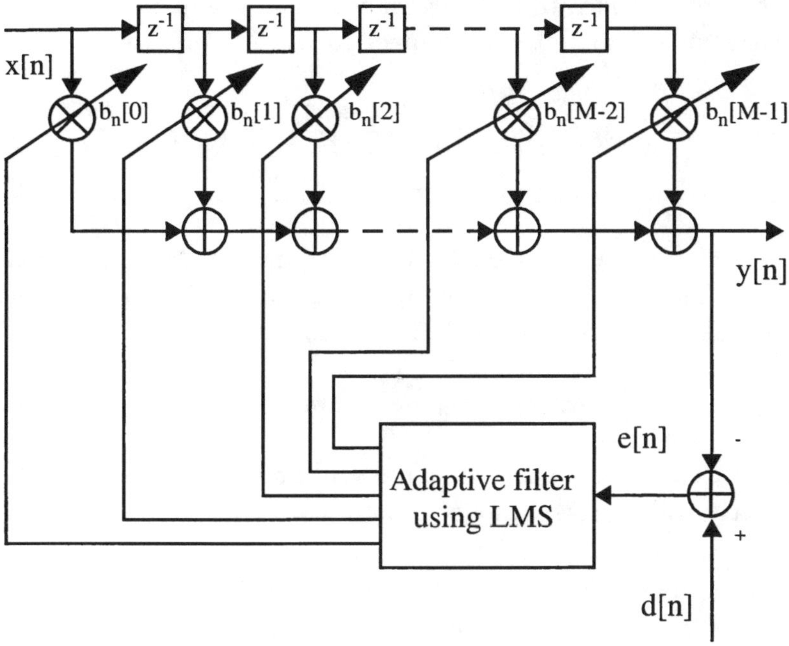

Figure 6.8 Adaptive FIR filter using LMS algorithm.

where the error signal $e[n]$ in Figure 6.8 is

$$\begin{aligned} e[n] &= d[n] - y[n] \\ &= d[n] - \mathbf{B}_n^T \mathbf{x}[n]. \end{aligned} \tag{6.39}$$

The LMS algorithm updates the filter coefficients based on the method of *steepest descent.* [1] This can be expressed in vector notation as follows:

$$\mathbf{B}_{n+1} = \mathbf{B}_n - \frac{1}{2}\mu \bigtriangledown \xi[n], \tag{6.40}$$

1. The steepest descent algorithm is an iterative procedure that minimizes the MMSE at time n using an estimate of the coefficient vector $\mathbf{B}_n$. At time $n+1$, a new estimate of the coefficient vector $\mathbf{B}_{n+1}$ is formed by adding a small correction with step size μ to $\mathbf{B}_n$, which is used to bring the coefficient vector $\mathbf{B}_n$ closer to the desired solution.

where $\mathbf{B}_{n+1}$ is the coefficient vector at time $n+1$, and μ is a *step size* to control the rate of convergence. The gradient, which is the derivative of $E\{|e(n)|^2\}$ with respect to $\mathbf{B}_n$, is equal to

$$\begin{aligned} \nabla\xi[n] &= \frac{\partial E\{|e(n)|^2\}}{\partial \mathbf{B}_n} \\ &= -2E\{e[n]\mathbf{x}[n]\}, \end{aligned} \tag{6.41}$$

where $\mathbf{x}[n]$ is the input signal vector and $e[n]$ is the error signal as shown in Figure 6.8. Thus, the update equation for coefficients $\mathbf{B}_n$ in the steepest descent algorithm in (6.40) becomes

$$\mathbf{B}_{n+1} = \mathbf{B}_n + \mu E\{e[n]\mathbf{x}[n]\}. \tag{6.42}$$

Note that if the measured signal $x[n]$ and the desired signal $d[n]$ are jointly wide-sense stationary, then we obtain

$$\begin{aligned} E\{e[n]\mathbf{x}[n]\} &= E\{(d[n] - \mathbf{B}_n^T\mathbf{x}[n])\mathbf{x}[n]\} \\ &= E\{d[n]\mathbf{x}[n]\} - E\{\mathbf{x}[n]\mathbf{x}[n]\mathbf{B}_n^T\} \\ &= \mathbf{r}_{dx}[n] - \mathbf{R}_x[n]\mathbf{B}_n. \end{aligned} \tag{6.43}$$

Thus, in this case, the steepest descent algorithm in (6.42) becomes

$$\mathbf{B}_{n+1} = \mathbf{B}_n + \mu(\mathbf{r}_{dx}[n] - \mathbf{R}_x[n]\mathbf{B}_n). \tag{6.44}$$

Note that the second term in (6.44) is zero when $\mathbf{B}_n = \mathbf{R}_x^{-1}[n]\mathbf{r}_{dx}[n]$. Therefore, (6.44) indicates that the steepest descent adaptive filter coefficients $\mathbf{B}_n$ converge to the solution of the Wiener-Hopf equations when

$$\lim_{n\to\infty} \mathbf{B}_n = \mathbf{R}_x^{-1}[n]\mathbf{r}_{dx}[n], \tag{6.45}$$

and the step size μ satisfies the condition

$$0 < \mu < \frac{2}{\lambda_{max}}, \tag{6.46}$$

where λ_{max} is the maximum eigenvalue of the autocorrelation matrix $\mathbf{R}_x[n]$.

The steepest descent algorithm has a practical limitation because the expectation $E\{e[n]\mathbf{x}[n]\}$ is generally unknown. Thus, we must estimate it with the sample mean:

$$E\{e[n]\mathbf{x}[n]\} = \frac{1}{N}\sum_{k=0}^{N-1} e[n-k]\mathbf{x}[n-k]. \tag{6.47}$$

Substituting (6.47) into (6.42), the update equation for $\mathbf{B}_n$ becomes

$$\mathbf{B}_{n+1} = \mathbf{B}_n + \frac{\mu}{N}\sum_{k=0}^{N-1} e[n-k]\mathbf{x}[n-k]. \tag{6.48}$$

In a special case of (6.48), if we use only one sample for estimating the sample mean, the update equation for the coefficient vector $\mathbf{B}_n$ has a particularly simple form:

$$\mathbf{B}_{n+1} = \mathbf{B}_n + \mu e[n]\mathbf{x}[n]. \tag{6.49}$$

Equation (6.49) is known as the *LMS algorithm*. Equivalently, we may write the adaptive LMS-based FIR filter in the form of three basic relationships:

1. Design parameters: M is the FIR filter order, and μ is the step size.
2. Initialization filter coefficients: $\mathbf{B}_0 = \mathbf{0}$.
3. LMS computation: For $n = 0, 1, 2, ...N$,

   (a) Calculate the filter output:

   $$y[n] = \mathbf{B}_n^T\mathbf{x}[n] \tag{6.50}$$

   (b) Estimate the signal error:

   $$e[n] = d[n] - y[n] \tag{6.51}$$

   (c) Update the coefficient adaptation:

   $$\mathbf{B}_{n+1} = \mathbf{B}_n + \mu e[n]\mathbf{x}[n]. \tag{6.52}$$

For DSP implementation, the LMS algorithm with M coefficients only needs M multiplications and M additions to update the filter coefficients. Also, one addition is needed to compute the error signal $e[n]$, and one multiplication is needed to form the product $\mu e[n]\mathbf{x}[n]$. Then, M multiplications and $M-1$ additions are needed to calculate the output signal of the adaptive filter. A total of $2(M+1)$ MACs per sample output are therefore required. For example, with the adaptive filter length $M = 50$ and the samples $N = 1{,}000$, computing the LMS algorithm will require 102,000 MACs for computation.

6.4.2 Convergence Analysis of the LMS Algorithm

The selection of the step size parameter μ in (6.49) must be made carefully. If the step size μ is too small, the coefficient vector $\mathbf{B}_n$ will adapt very slowly and may not react to changes in the input signal vector. On the other hand, if the step size μ is too large, the LMS adaptive system will unduly respond to noise in the signal and may not converge to the MMSE solution.

The convergence analysis of the LMS algorithm should be considered by using a statistical framework where coefficient vector $\mathbf{B}_n$ is treated as a vector of random variables. Substituting (6.39) into (6.49), we obtain the update equation of the LMS coefficient as follows:

$$\mathbf{B}_{n+1} = \mathbf{B}_n + \mu\{d[n] - \mathbf{B}_n^T\mathbf{x}[n]\}\mathbf{x}[n]. \tag{6.53}$$

Taking the expectation of both sides of (6.53) results in

$$E\{\mathbf{B}_{n+1}\} = E\{\mathbf{B}_n\} + \mu E\{d[n]\mathbf{x}[n]\} - \mu E\{\mathbf{x}[n]\mathbf{x}^T[n]\mathbf{B}_n\}. \tag{6.54}$$

In order to perform the analysis for (6.54), we assume that the signal $x[n]$ and the desired signal $d[n]$ are jointly wide-sense stationary. In addition, the signal $x[n]$ and the coefficient vector $\mathbf{B}_n$ of the LMS adaptation are statistically independent. Note that this independence assumption for statistical analysis of the LMS algorithm is often approximately true, and its use is attributed to Haykin [16]. With this assumption, (6.54) can be rewritten as

$$\begin{aligned} E\{\mathbf{B}_{n+1}\} &= E\{\mathbf{B}_n\} + \mu\mathbf{r}_{dx}[n] - \mu E\{\mathbf{x}[n]\mathbf{x}^T[n]\}E\{\mathbf{B}_n\} \\ &= (\mathbf{I} - \mu\mathbf{R}_x[n])E\{\mathbf{B}_n\} + \mu\mathbf{r}_{dx}[n]. \end{aligned} \tag{6.55}$$

Note that the LMS adaptive filter also converges to the solution of the Wiener-Hopf equation when $\mathbf{r}_{dx}[n] = \mathbf{R}_x[n]\mathbf{B}$ in (6.44), where $\mathbf{B}$ is the Wiener-Hopf solution. Thus (6.55) becomes

$$E\{\mathbf{B}_{n+1}\} = (\mathbf{I} - \mu\mathbf{R}_x[n])E\{\mathbf{B}_n\} + \mu\mathbf{R}_x[n]\mathbf{B}. \tag{6.56}$$

Subtracting $\mathbf{B}$ from both sides in (6.56) gives

$$E\{\mathbf{B}_{n+1}\} - \mathbf{B} = (\mathbf{I} - \mu\mathbf{R}_x[n])(E\{\mathbf{B}_n\} - \mathbf{B}). \tag{6.57}$$

Since $\mathbf{R}_x[n]$ is an autocorrelation matrix that is symmetric and positive definite, by using the spectral decomposition theorem as presented in the Appendix, $\mathbf{R}_x[n]$ can be decomposed into

$$\mathbf{R}_x[n] = \mathbf{V}[n]\Lambda\mathbf{V}^T[n], \tag{6.58}$$

where Λ is a diagonal matrix consisting of the eigenvalues of the autocorrelation matrix $\mathbf{R}_x[n]$, and $\mathbf{V}[n]\mathbf{V}^T[n] = \mathbf{I}$. Then, substituting (6.58) into (6.57) and using the fact that $\mathbf{V}[n]\mathbf{V}^T[n] = \mathbf{I}$, we obtain

$$\mathbf{V}^T[n](E\{\mathbf{B}_{n+1}\} - \mathbf{B}) = (\mathbf{I} - \mu\Lambda)\mathbf{V}^T[n](E\{\mathbf{B}_n\} - \mathbf{B}). \tag{6.59}$$

Let $\mathbf{u}_n = \mathbf{V}^T[n](E\{\mathbf{B}_n\} - \mathbf{B})$. Equation (6.59) yields

$$\mathbf{u}_{n+1} = (\mathbf{I} - \mu\Lambda)\mathbf{u}_n. \tag{6.60}$$

Now, given an initial vector $\mathbf{u}_0$, it follows that

$$\mathbf{u}_{n+1} = (\mathbf{I} - \mu\Lambda)^{n+1}\mathbf{u}_0. \tag{6.61}$$

Since $(\mathbf{I} - \mu\Lambda)$ is a diagonal matrix, the ith components of the vector $\mathbf{u}_{n+1}$ may be expressed as

$$u_{n+1}(i) = (1 - \mu\lambda_i)^{n+1}u_0(i). \tag{6.62}$$

Note that if we want $\mathbf{B}_{n+1}$ to converge to $\mathbf{B}$, then the vector $\mathbf{u}_{n+1}$ must converge to zero. This requires

$$|1 - \mu\lambda_i| < 1, \qquad i = 0, 1, 2, ..., M, \tag{6.63}$$

where M is the number of filter coefficients. Therefore, to prevent the LMS adaptation from becoming unstable, the value of the step size μ must be chosen such that

$$0 < \mu < \frac{2}{\lambda_{max}}, \tag{6.64}$$

where λ_{max} is the maximum eigenvalue of the autocorrelation matrix $\mathbf{R}_x[n]$.

In practical application of the adaptive LMS filter, knowledge of λ_{max} is not available. The trace of $\mathbf{R}_x[n]$ may be taken as a conservative estimate for λ_{max}. Thus, (6.64) can be reformulated as

$$0 < \mu < \frac{2}{tr\{\mathbf{R}_x[n]\}}, \tag{6.65}$$

where $tr\{\mathbf{R}_x[n]\}$ denotes the trace of autocorrelation matrix $\mathbf{R}_x[n]$. If the signal $x[n]$ is wide-sense stationary, then $\mathbf{R}_x[n]$ is a Toeplitz matrix. Thus the trace $tr\{\mathbf{R}_x[n]\}$ becomes

$$tr\{\mathbf{R}_x[n]\} = (M+1)E\{|x[n]|^2\}, \tag{6.66}$$

where $E\{|x[n]|^2\}$ can be estimated by

$$E\{|x[n]|^2\} = \frac{1}{M}\sum_{k=0}^{M-1} |x(n-k)|^2. \tag{6.67}$$

As a result, (6.64) may be expressed as

$$0 < \mu < \frac{2}{\frac{M+1}{M}\sum_{k=0}^{M-1} |x(n-k)|^2}. \tag{6.68}$$

6.4.3 Mathematical Analysis of Excess Mean-Squared Error

In practical applications, the LMS coefficients $\mathbf{B}_n$ tend to fluctuate about their optimum value when $\mathbf{B}_n$ begins to converge in the mean $\mathbf{B}$. This is because of the noisy gradient vectors that are used to produce the small corrections to $\mathbf{B}_n$. Thus, $\mathbf{B}_n$ does not converge to $\mathbf{B}$ exactly. As a result, the mean square error is larger than the MMSE by an amount referred to as the *excess mean-squared error*.

Using (6.39) and $\Psi = \mathbf{B}_n - \mathbf{B}$, we write the error as follows:

$$\begin{aligned} e[n] &= d[n] - \mathbf{B}_n^T \mathbf{x}[n] \\ &= d[n] - (\mathbf{B} + \Psi)^T \mathbf{x}[n] \\ &= e_{min}[n] + \Psi^T \mathbf{x}[n], \end{aligned} \tag{6.69}$$

where $e_{min}[n] = d[n] - \mathbf{B}^T \mathbf{x}[n]$ is the error if the solution of the Wiener-Hopf equations was used. If the LMS filter is in steady-state such that $E\{\Psi\} = 0$, then the mean square error can be expressed as

$$\begin{aligned} \xi[n] &= E\{|e[n]|^2\} \\ &= \xi_{min} + \xi_{emse}[n], \end{aligned} \tag{6.70}$$

where $\xi_{min} = E\{|e_{min}[n]|^2\}$ is the minimum mean-square error, and $\xi_{emse}[n]$ is the excess mean-squared error.

Haykin [16] showed that the excess mean-squared error is less than the minimum mean-squared error

$$\xi_{emse}[\infty] = \xi_{min} \left[\mu \sum_{i=1}^{M} \frac{\lambda_i}{2 - \mu\lambda_i} \right] \tag{6.71}$$

if the step size μ satisfies (6.64) and the condition

$$\sum_{i=1}^{M} \frac{2\lambda_i}{2 - \mu\lambda_i} < 1, \tag{6.72}$$

where λ_i is the ith eigenvalue of the autocorrelation matrix $\mathbf{R}_x[n]$.

Note that if $\mu \ll \frac{2}{\lambda_{max}}$, then $\mu\lambda_i \ll 2$, and

$$\begin{aligned} \mu \sum_{i=1}^{M} \frac{\lambda_i}{2 - \mu\lambda_i} &\approx \frac{\mu}{2} \sum_{i=1}^{M} \lambda_i \\ &= \frac{\mu}{2} tr\{\mathbf{R}_x[n]\}. \end{aligned} \tag{6.73}$$

Substituting (6.73) into (6.71), we obtain the excess mean-squared error

$$\xi_{emse}[\infty] \approx \xi_{min} \left[\frac{\mu}{2} tr\{\mathbf{R}_x\} \right]. \tag{6.74}$$

Note that the excess mean-squared error $\xi_{emse}[\infty]$ is proportional to the step size μ when $n \to \infty$.

The LMS misadjustment $\mathcal{M}$ is the ratio of the steady-state value $\xi_{emse}[\infty]$ of the excess mean-squared error to the minimum mean-squared error ξ_{min}. Using (6.71), the LMS misadjustment $\mathcal{M}$ is given by

$$\begin{aligned} \mathcal{M} &= \frac{\xi_{emse}[\infty]}{\xi_{min}} \\ &= \mu \sum_{i=1}^{M} \frac{\lambda_i}{2 - \mu\lambda_i}. \end{aligned} \tag{6.75}$$

The LMS misadjustment $\mathcal{M}$ is less than unity if the step-size μ holds the condition in (6.72). If the step size is small such that $\mu \ll \frac{2}{\lambda_{max}}$, then the LMS misadjustment is approximately obtained by

$$\mathcal{M} \approx \frac{\mu}{2} tr\{\mathbf{R}_x[n]\}. \tag{6.76}$$

The trade-off analysis among the rate of convergence, the amount of excess mean-square error, and the ability of the adaptive filter to track the signal is important. Since the filter coefficients $\mathbf{B}_n$ are far from the optimum solution when the LMS adaptation algorithm begins, the step size μ can be large in order to move the coefficients vector $\mathbf{B}_n$ rapidly toward the desired solution. However, the step size μ should be decreased to reduce the excess mean-square error $\xi_{emse}[\infty]$ when the LMS filter begins to converge in the MMSE solution. Therefore, using the LMS algorithm with a time-varying step size $\mu[n]$ is desirable in many applications.

Example 6.3 Adaptive Decision Feedback Equalizer Using LMS Algorithm

In this example, we introduce the adaptive decision feedback equalizer (DFE) using the LMS algorithm. The DFE has been widely used in many communications applications, such as wireless GSM and broadband DSL, for reliable communication of digital data across nonideal channels. The nonideal channel causes channel dispersion that results in a distortion of the pulse shape. This causes neighboring pulses to interfere with each other in a phenomenon

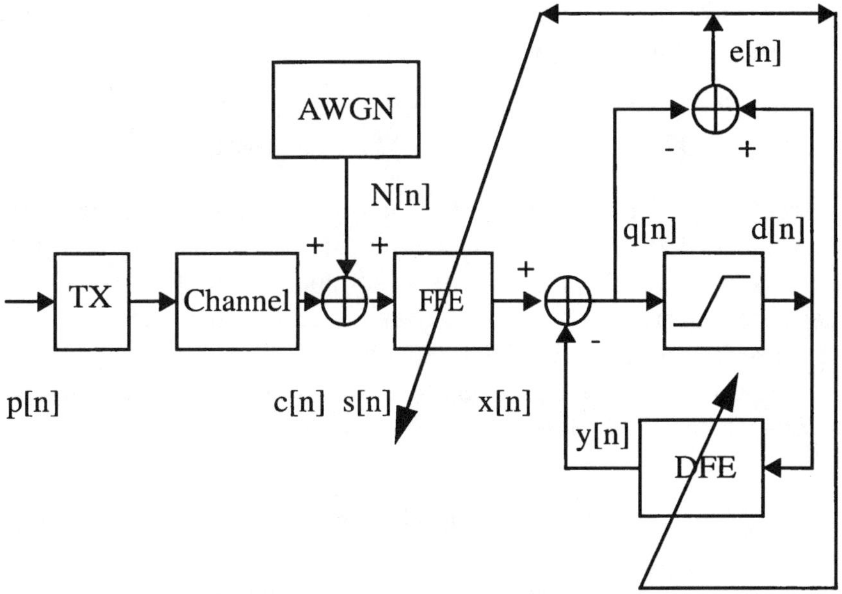

Figure 6.9 Block diagram of communication system with adaptive DFE equalizer.

known as *intersymbol interference* (ISI). The block diagram of the communication system with DFE equalizer used in this example is shown in Figure 6.9. The DFE equalizer consists of a feed-forward equalizer (FFE) and a decision feedback equalizer (DFE). The DFE is driven by decisions from the output of the detector (or slice), and its taps are adjusted to cancel the ISI on the current symbol that results from past detected symbols (or postcursors). The FFE is used to remove the ISI on the future symbol (or precursors). The detector (or slice) is defined by

$$d[n] = \begin{cases} 1 & q[n] \geq 0 \\ -1 & q[n] < 0. \end{cases} \tag{6.77}$$

The pulse shape, $p[n]$, produces the random sequence with 1 or -1. The additive white Gaussian noise (AWGN) is white Gaussian noise with zero mean and variance $\sigma^2_{AWGN} = 0.005$ that is uncorrelated with the pulse shape $p[n]$. The FFE and DFE have 30 taps to be adjusted during training. The impulse response of the channel is assumed to be

$$h[n] = \begin{cases} \frac{1}{2}\{1 + cos[\frac{2\pi}{W}(n-2)]\}, & n = 1, 2, 3, 4, 5, 6 \\ 0 & \text{otherwise}, \end{cases} \tag{6.78}$$

where W is used to control the amount of amplitude distortion. This affects the eigenvalues of the autocorrelation matrix $\mathbf{R}_q[n]$ of the received signal $q[n]$. In this example, we use $W = 2.65$. The channel input pulse $p[n]$ through the channel $h[n]$ provides the desired response for the adaptive DFE equalizer.

The discrete-time equations for the adaptive DFE equalizer are now described. The signal after the channel is given by

$$s[n] = \sum_{k=0}^{6} h[k]p[n-k] + N[n], \tag{6.79}$$

where $N[n]$ is AWGN. The discrete-time signal $x[n]$ from the output of FFE is obtained by

$$x[n] = \mathbf{F}_n \mathbf{s}[n], \tag{6.80}$$

where $\mathbf{F}_n$ is the FFE filter coefficients. The discrete-time signal $y[n]$ from the output of DFE is given by

$$y[n] = \mathbf{D}_n \mathbf{d}[n], \tag{6.81}$$

where $\mathbf{D}_n$ is the DFE filter coefficients, and the detector is $d[n] = Q\{q[n]\}$ and has output of 1 or -1. Then, the discrete-time error signal is

$$e[n] = d[n] - q[n]. \tag{6.82}$$

Therefore, the updated FFE coefficients are

$$\mathbf{F}_{n+1} = \mathbf{F}_n + \alpha_f e[n]\mathbf{s}[n], \tag{6.83}$$

and the updated DFE coefficients are given by

$$\mathbf{D}_{n+1} = \mathbf{D}_n + \alpha_d e[n]\mathbf{d}[n], \tag{6.84}$$

where α_f and α_d are the step sizes for FFE and DFE, respectively.

For initialization of the simulation, we set the step size $\alpha_f = 0.0003$, and $\alpha_d = -0.0003$. We also define the center of $\mathbf{F}_0$ coefficients as one, other coefficients as zero, and all the $\mathbf{D}_0$ values also as zero. The decision direct method (or blind training method) is used for the training. Figure 6.10 shows the simulation results after 6,000 iterations. The impulse response of channel $h[n]$ is plotted in Figure 6.10(a). The coefficients of $\mathbf{F}_n$ and $\mathbf{D}_n$ are shown at 6,000 iterations in Figure 6.10(b) and (c), respectively. The learning curve for an adaptive decision feedback equalizer using step sizes of $\alpha_f = 0.0003$ and $\alpha_d = -0.0003$ is plotted in Figure 6.10(d).

The adaptive equalizer is a very important topic in telecommunications. We recommend the references by Proakis [17, 18], Stojanovic, Proakis, and Catipovic [19], Rappaport [20], and Gibson [21].

6.5 Recursive Least Squares Algorithms

The recursive least squares (RLS) algorithm is derived from the least squares solution of the general adaptive signal processor. The RLS algorithm utilizes information contained in the particular processed input sample data. In other words, for different input sample data, we get different adaptive filters. As a result, the adaptive filter coefficients, which minimize the least squares error, will be optimal for the given sample data rather than statistically optimal over a particular class of processes.

To derive the RLS adaptive filter, we consider minimizing the total sum of squared errors for all input sample data at discrete-time n. The weighted least squares error is as follows:

$$\xi[n] = \sum_{i=0}^{n} \lambda^{n-i} |e[i]|^2, \tag{6.85}$$

where $0 < \lambda \leq 1$ is an exponential weighting factor and $e[i]$ is defined by

$$\begin{aligned} e[i] &= d[i] - y[i] \\ &= d[i] - \mathbf{w}^T[n]\mathbf{x}[i], \end{aligned} \tag{6.86}$$

$\mathbf{w}[n]$ is the *coefficient-weight* vector at discrete-time n, and $\mathbf{x}[i]$ is the *coefficient-input* vector at discrete-time i.

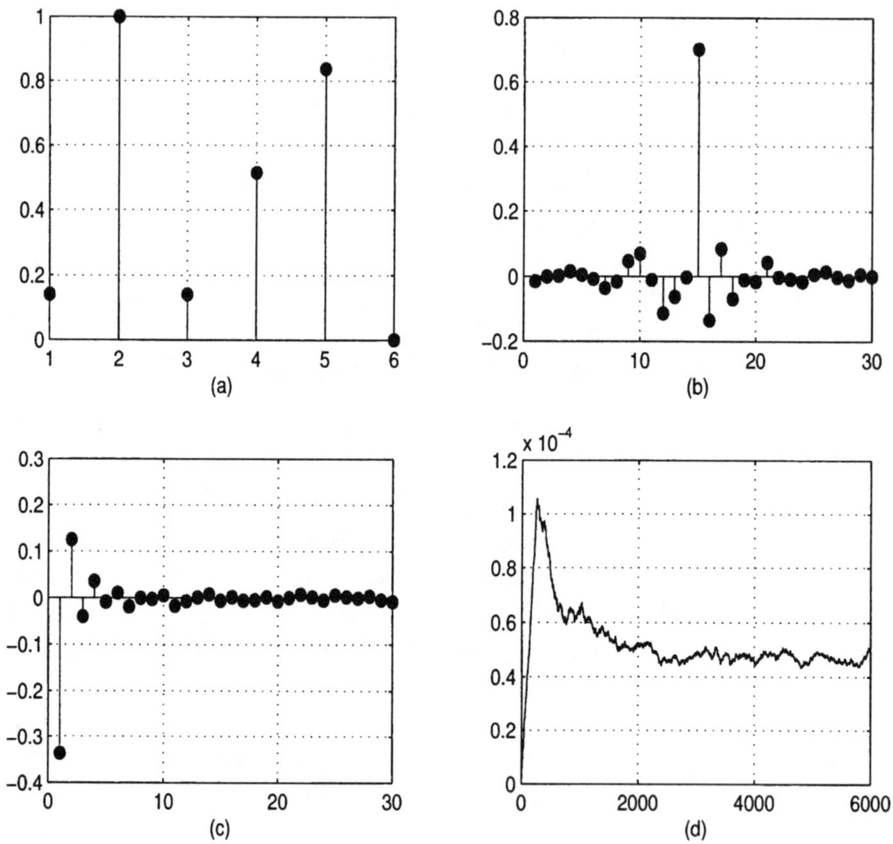

Figure 6.10 The simulation results of adaptive DFE equalizer. (a) Impulse response of channel model. (b) Coefficients of $\mathbf{F}_n$ at 6,000 iterations. (c) Coefficients of $\mathbf{D}_n$ at 6,000 iterations. (d) Learning curve with average mean-squared error for LMS algorithm after 6,000 iterations.

The optimum value of the coefficient-weight vector $\mathbf{w}[n]$, which is generated by setting the derivative of $\xi[n]$ with respect to $w_k[n]$ equal to zero for $k = 0, 1, 2, ..., M$, is defined by the normal equations written in matrix form:

$$\mathbf{R}_x[n]\mathbf{w}[n] = \mathbf{r}_{dx}[n], \tag{6.87}$$

where $\mathbf{R}_x[n]$ is an $(M + 1) \times (M + 1)$ exponentially weighted deterministic autocorrelation matrix for $x[n]$ as follows,

$$\mathbf{R}_x[n] = \sum_{i=0}^{n} \lambda^{n-i}\mathbf{x}^*[i]\mathbf{x}^T[i], \tag{6.88}$$

and where $\mathbf{r}_{dx}[n]$, the deterministic cross-correlation between $d[n]$ and $x[n]$ is expressed by

$$\mathbf{r}_{dx}[n] = \sum_{i=0}^{n} \lambda^{n-i} d[i]\mathbf{x}^*[i]. \tag{6.89}$$

In order to compute the least squares estimate of the coefficient-weight vector $\mathbf{w}[n]$ in (6.87), we need to determine the inverse of the autocorrelation matrix $\mathbf{R}_x[n]$,

$$\mathbf{w}[n] = \mathbf{R}_x^{-1}[n]\mathbf{r}_{dx}[n]. \tag{6.90}$$

Equation (6.90) is known as the least squares estimate $\mathbf{w}[n]$ for the tap-weight vector. However, it is too time consuming to perform such an implementation directly. In practice, we should avoid such performance if the number of tap weights, M, is relatively high. Therefore, we compute the least squares estimate of the coefficient-weight vector $\mathbf{w}[n]$ recursively for $n = 1, 2, 3, ..., M$.

Since $\mathbf{R}_x[n]$ and $\mathbf{r}_{dx}[n]$ both depend on n, we have the following recursive equation for updating the value of the autocorrelation matrix:

$$\mathbf{R}_x[n] = \lambda\mathbf{R}_x[n-1] + \mathbf{x}^*[n]\mathbf{x}^T[n], \tag{6.91}$$

where $\mathbf{R}_x[n-1]$ is the value of the autocorrelation matrix at discrete-time $n - 1$, and the matrix product $\mathbf{x}^*[n]\mathbf{x}^T[n]$ can be considered as the "correction" term in the updating operation.

Similarly, the cross-correlation matrix may be updated recursively as follows:

$$\mathbf{r}_{dx}[n] = \lambda \mathbf{r}_{dx}[n-1] + d[n]\mathbf{x}^*[n]. \quad (6.92)$$

Inverting the autocorrelation matrix $\mathbf{R}_x[n]$ can be accomplished using the matrix inversion lemma as follows:

$$\begin{aligned} \mathbf{A}^{-1} &= (\mathbf{B}^{-1} + \mathbf{C}\mathbf{D}^{-1}\mathbf{C}^T)^{-1} \\ &= \mathbf{B} - \mathbf{B}\mathbf{C}(\mathbf{D} + \mathbf{C}^T\mathbf{B}\mathbf{C}^{-1})\mathbf{C}^T\mathbf{B}. \end{aligned} \quad (6.93)$$

Let $\mathbf{A} = \mathbf{R}_x[n]$, $\mathbf{B}^{-1} = \lambda \mathbf{R}_x[n-1]$, $\mathbf{C} = \mathbf{x}[n]$, and $\mathbf{D} = 1$. Then, substituting these definitions into (6.93), we obtain the recursive equation of the inverse autocorrelation matrix as follows:

$$\begin{aligned} \mathbf{R}_x^{-1}[n] &= \lambda^{-1}\mathbf{R}_x^{-1}[n-1] \\ &\quad - \frac{\lambda^{-2}\mathbf{R}_x^{-1}[n-1]\mathbf{x}[n]\mathbf{x}^T\mathbf{R}_x^{-1}[n-1]}{1 + \lambda^{-1}\mathbf{x}[n]\mathbf{R}_x^{-1}[n-1]\mathbf{x}[n]}. \end{aligned} \quad (6.94)$$

To simplify the formula, we let

$$\mathbf{P}[n] = \mathbf{R}_x^{-1}[n], \quad (6.95)$$

where $\mathbf{P}[n]$ is referred to as the *inverse autocorrelation matrix*, and

$$\mathbf{k}[n] = \frac{\mathbf{P}[n-1]\mathbf{x}[n]}{\lambda + \mathbf{x}^T[n]\mathbf{P}_x[n-1]\mathbf{x}[n]}, \quad (6.96)$$

where $\mathbf{k}[n]$ is called the *gain vector*. Using the above definitions, we rewrite (6.94) as follows:

$$\mathbf{P}[n] = \frac{1}{\lambda}\{\mathbf{I} - \mathbf{k}[n]\mathbf{x}^T[n]\}\mathbf{P}[n-1]. \quad (6.97)$$

Equation (6.97) is sometimes referred to as the *Riccati* formula for the RLS algorithm. Finally, the desired recursive equation for updating the coefficient-weight vector is,

$$\begin{aligned} \mathbf{w}[n] &= \mathbf{w}[n-1] + \Delta\mathbf{w}[n-1] \\ &= \mathbf{w}[n-1] + \mathbf{k}[n]\xi[n], \end{aligned} \quad (6.98)$$

where $\xi[n]$ is the *priori estimation error*, which is defined by

$$\xi[n] = d[n] - \mathbf{w}^T[n-1]\mathbf{x}[n]. \tag{6.99}$$

Note that the implementation of the RLS algorithm requires the initialization of the recursion formula of (6.97) by choosing a starting value $\mathbf{P}[0]$, which is the nonsingularity of the autocorrelation matrix $\mathbf{R}_x[n]$. In order to obtain the initialization value, a simpler procedure is to change the expression of (6.91) slightly for the autocorrelation matrix $\mathbf{R}_x[n]$ by writing

$$\mathbf{R}_x[n] = \lambda\mathbf{R}_x[n-1] + \delta\lambda^n\mathbf{I}, \tag{6.100}$$

where $\mathbf{R}_x[n-1]$ is given by (6.88), $\mathbf{I}$ is the $(M+1) \times (M+1)$ identity matrix, and δ is a small positive constant. When $n = 0$, (6.100) becomes

$$\mathbf{R}_x[0] = \delta\mathbf{I}. \tag{6.101}$$

Thus, the initial value of $\mathbf{P}[n]$ is equal to

$$\begin{aligned} \mathbf{P}[0] &= \mathbf{R}_x^{-1}[0] \\ &= \frac{1}{\delta}\mathbf{I}. \end{aligned} \tag{6.102}$$

For convenience, we summarize all of the steps of the RLS algorithm as follows:

1. Select the filter order M, exponential weighting factor, λ, and a small positive constant, δ, to initialize function matrix $\mathbf{P}[0]$ by using (6.102).
2. Initialize the RLS algorithm by setting $\mathbf{P}[0] = \frac{1}{\delta}\mathbf{I}$, where δ is a small positive constant, and $\mathbf{w}[0] = \mathbf{0}$.
3. For each instant of time, $n = 1, 2, ..., N$, we calculate:
   (a) The priori estimation error using (6.99).
   (b) The gain vector using (6.96).
   (c) The coefficient weight update vector using (6.98).
   (d) The *Riccati* formula using (6.97).

For DSP implementation, the RLS algorithm requires $[3(N+1)(N+2)]$ MACs. We leave the detailed calculation of this number for the reader. Note that the computational complexity of the RLS algorithm is much higher than for the LMS algorithm. However, the RLS algorithm generally converges faster than the LMS algorithm.

For a fundamental treatment of the least squares method, we cite a paper by Cadzow [22]. For the RLS methods based on the slide window, we suggest a paper by Zhao et al. [23]. For a method of state-space in the RLS approach, we refer the reader to a paper by Sayed and Kailath [24].

6.5.1 Convergence Analysis of the RLS Algorithm

To analyze the convergence of the RLS algorithm, we first set $\lambda = 1$ for mathematical simplification. Then, the autocorrelation matrix $\mathbf{R}_x[n]$ in (6.88) can be rewritten as

$$\mathbf{R}_x[n] = \sum_{i=0}^{n} \mathbf{x}^*[i]\mathbf{x}^T[i], \tag{6.103}$$

and the cross-correlation matrix (6.89) becomes

$$\mathbf{r}_{dx}[n] = \sum_{i=0}^{n} d[i]\mathbf{x}^*[i]. \tag{6.104}$$

In order to perform the analysis, (6.86) can be rewritten as

$$d[i] = \mathbf{w}_c^T \mathbf{x}[i] + e_w[i], \tag{6.105}$$

where $e_w[i]$ is white noise with zero mean and variance σ^2, and the weight vector $\mathbf{w}_c$ is constant. Thus, substituting (6.105) into (6.104), we obtain

$$\begin{aligned} \mathbf{r}_{dx}[n] &= \sum_{i=0}^{n} (\mathbf{w}_c^T \mathbf{x}[i] + e_w[i])\mathbf{x}^*[i] \\ &= \sum_{i=0}^{n} \mathbf{x}^*[i]\mathbf{x}^T[i]\mathbf{w}_c + \sum_{i=0}^{n} e_w[i]\mathbf{x}^*[i] \\ &= \mathbf{R}_x[n]\mathbf{w}_c + \sum_{i=0}^{n} e_w[i]\mathbf{x}^*[i]. \end{aligned} \tag{6.106}$$

So, by substituting (6.106) into (6.90), we can rewrite (6.90) as follows:

$$\begin{aligned} \mathbf{w}[n] &= \mathbf{R}_x^{-1}[n](\mathbf{R}_x[n]\mathbf{w}_c + \sum_{i=0}^{n} e_w[i]\mathbf{x}^*[i]) \\ &= \mathbf{w}_c + \mathbf{R}_x^{-1}[n]\sum_{i=0}^{n} e_w[i]\mathbf{x}^*[i]. \qquad (6.107) \end{aligned}$$

Now, we need to make two assumptions as follows:

1. The input signal vectors, $\mathbf{x}[1]$, $\mathbf{x}[2]$, ..., $\mathbf{x}[n]$, are statistically independent.
2. The measurement error signal $e_w[i]$ is also independent of the input signal vector $\mathbf{x}[i]$ for $n = 1, 2, ..., n$.

With the above two assumptions, we take the expectation value for both sides of (6.107) with $n \geq M$ given by

$$\begin{aligned} E\{\mathbf{w}[n]\} &= E\{\mathbf{w}_c\} + E\left\{\mathbf{R}_x^{-1}[n]\sum_{i=0}^{n} e_w[i]\mathbf{x}^*[i]\right\} \\ &= \mathbf{w}_c, \qquad (6.108) \end{aligned}$$

where M is the number of coefficients in the adaptive filter and the measurement error signal $e_w[i]$ has zero mean. As we can see, (6.108) indicates that the RLS algorithm is convergent in the mean value, $\mathbf{w}_c$, for $n \geq M$. This, in turn, means that the RLS algorithm does not wait for convergence for $n \to \infty$.

6.5.2 A Comparison Discussion

The rate of convergence of the RLS algorithm is usually an order of magnitude faster than the convergence of the LMS algorithm. This is because there are fundamental differences between the RLS algorithm and the LMS algorithm. Note that the step size μ in the LMS algorithm is used, while the inverse of the autocorrelation matrix $\mathbf{R}_x^{-1}[n]$ of the input signal vector $\mathbf{x}[n]$ is used in the RLS algorithm. This results in a profound impact on the convergence behavior of the RLS algorithm. Therefore, the RLS algorithm does not wait for convergence for $n \to \infty$.

It can be shown [16] that the mean square error of the RLS algorithm with setting $\lambda = 1$ is given by

$$\xi[n] = \sigma^2 + \frac{M\sigma^2}{n - M - 1}, \qquad n > M + 1 \tag{6.109}$$

where σ^2 is the variance of the measurement error $e_w[n]$, and M is the number of filter coefficients in the adaptive filter. Note that the second term in (6.109) is zero approximately when $n \to \infty$. This means that the RLS algorithm produces zero excess mean-squared error when $n \to \infty$ during operating in a stationary environment. Recall that the excess mean-squared error in the LMS algorithm is

$$\xi_{emse}[\infty] \approx \xi_{min} \left[\frac{\mu}{2} tr[\mathbf{R}_x]\right], \tag{6.110}$$

which is proportional to the step size μ when $n \to \infty$.

The RLS algorithm can be viewed as a special case of the Kalman filter. Note that the RLS algorithm is deterministic, while the Kalman filter is stochastic. However, there exists a one-to-one correspondence between their individual variables that has been verified by Haykin [16]. In addition, the Kalman filter problem sets the stage for deriving the corresponding variants of the RLS algorithm, such as square-root (QR) adaptive filtering algorithms for RLS estimation, known as the QR-RLS algorithm, extended QR-RLS algorithm, and inverse QR-RLS algorithm [24].

Example 6.4. Adaptive Linear Prediction Using the RLS Algorithm

Adaptive linear prediction using the exponentially weighted RLS algorithm is shown in Figure 6.11. Assume that $s[n]$ is a second-order autoregressive process,

$$s[n] = 0.95s[n-1] - 0.55s[n-2], \tag{6.111}$$

and $g[n]$ is white noise with zero mean and unit variance. Then, signal $x[n]$ is generated according to the difference equation as follows:

$$x[n] = 0.95x[n-1] - 0.55x[n-2] + g[n]. \tag{6.112}$$

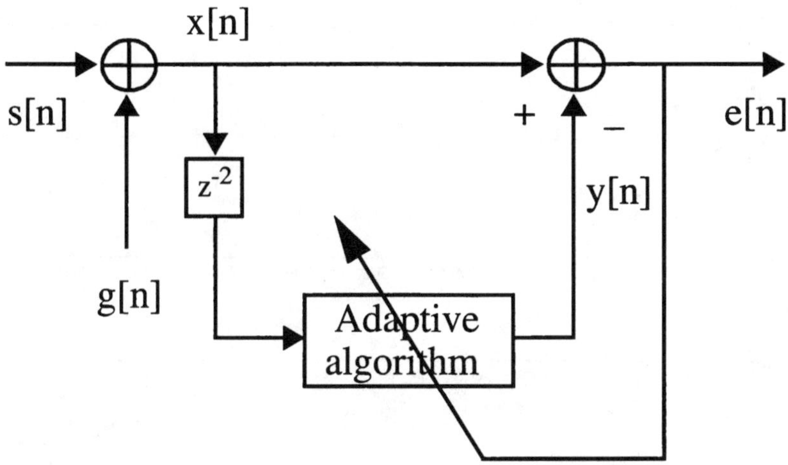

Figure 6.11 Adaptive linear predication using the exponentially weighted RLS algorithm.

Therefore, the estimated second-order adaptive linear predictor as shown in Figure 6.11 has a form

$$y[n] = w_n[1]x[n-1] + w_n[2]x[n-2], \tag{6.113}$$

where $w_n[1]$ and $w_n[2]$ need to be updated using the exponentially weighted RLS algorithm. Then, the error signal is $e[n] = x[n] - y[n]$ and the estimated average mean square error is given by:

$$\xi_{min}[n] = \frac{1}{N} \sum_{k=0}^{N-1} |e[k]|^2. \tag{6.114}$$

For initialization of the RLS algorithm, we set the exponential weight $\lambda = 0.999$, and $\delta = 0.0001$. The simulation results are shown in Figure 6.12(a) and (b). As can be seen, the filter coefficients $w_n[1]$ and $w_n[2]$ are converged to 0.95 and -0.55, respectively, after 1,000 iterations.

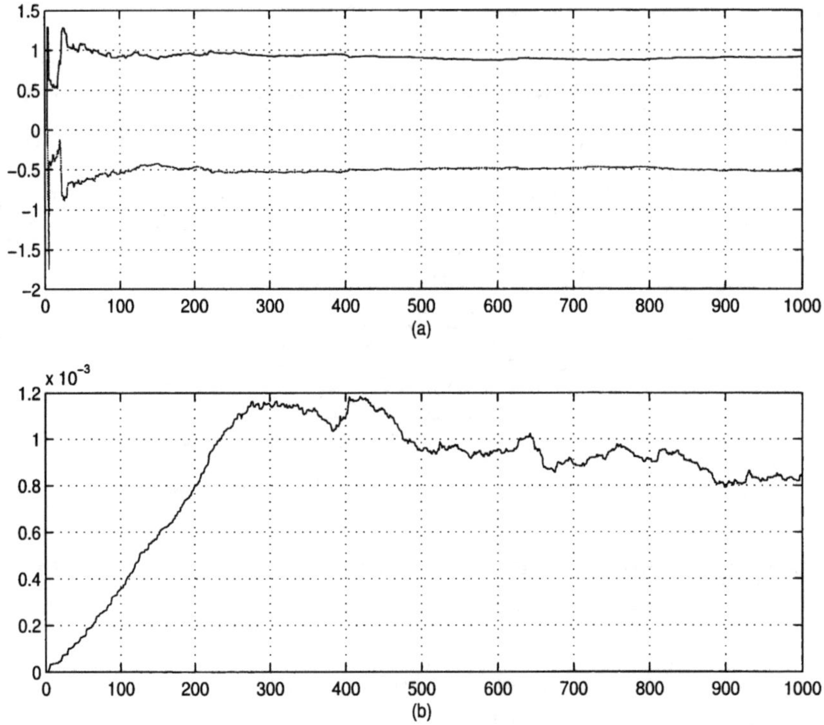

Figure 6.12 Adaptive FIR filter using the exponentially weighted RLS algorithm. (a) Adaptive FIR filter coefficients $w_n[1]$ and $w_n[2]$ during 1,000 iterations. (b) Estimated average mean square errors.

6.6 Summary

In this chapter, we introduced adaptive filters, an attractive option when signals result from operations in an environment of unknown statistics. First, we presented a class of linear optimum discrete-time filters known as Wiener filters and Kalman filters. Then we introduced adaptive algorithms including the LMS algorithms and the RLS algorithms.

In the Wiener filters, the Wiener-Hopf equation is a single-step algorithm, which specifies the coefficients of the optimum filter based on MMSE. It can be used to solve problems with stationary properties. However, the Wiener-Hopf method is not a useful algorithm in nonstationary environments for real-time DSP operations because of the high computational complexity. Alternatively, the minimization of the mean square error may be accomplished recursively by using the LMS algorithm. The LMS is a simple adaptive algorithm that often performs well enough to be used successfully in many applications. However, the convergence rate of the LMS algorithm is slow due to the fact that there is only a single parameter of step size to control the rate of adaptation. Faster converging algorithms are obtained if we use the Kalman approach or RLS algorithm for adaptive filtering. In general, the Kalman and RLS algorithms provide faster convergence and better tracking of time-varying communications channel characteristics than does the LMS algorithm. However, the price paid for faster convergence is an increase in computational complexity per iteration. In addition, these algorithms are considerably more sensitive to round-off noise, especially in fixed-point DSP implementation. Therefore, understanding the numerical stability of this class of algorithms is necessary before using them for adaptive filtering in practical applications.

Another type of adaptive filter is established by using neural networks techniques. Neural networks have a built-in capability to adapt their synaptic weights to changes in the nearby environment. Since the neural network is a nonlinear device, we may refer to the neural network as a nonlinear adaptive filter. For a technical treatment of nonlinear adaptive filters, we recommend the reference book by Haykin [7], and cite the application references by Miao and Clements [8], and Miao [9].

In practical applications, the concerns are usually measured in terms of hardware complexity, chip area, processing consumption, computation speed, and cost. These factors are directly related to the adaptive filter order and which implementation architecture is used. The computational efficiency of the implement architecture is very important for adaptive filter operations. In practical applications, when the adaptive filters are operated in multirate signal processing systems, efficient implementation architectures of adaptive filter banks are of special importance in minimizing the computational complexities. We will introduce the topic of multirate adaptive filter banks in discrete-time multirate signal processing in the next chapter.

References

[1] Widrow, B., and M. E. Hoff, "Adaptive Switch Circuits," *IRE WESCON Conv. Rec.*, Part IV, pp. 96-104, 1960.

[2] Widrow, B., et al., "Adaptive Noise Cancelling: Principles and Applications," *Proc. of IEEE*, Vol. 63, pp. 1692-1716, 1975.

[3] Widrow, B., et al., "Stationary and Nonstationary Learning Characteristics of the LMS Adaptive Filter," *Proc. of IEEE*, Vol. 64, pp. 1151-1162, 1976.

[4] Miao, J., "Discrete-Time Adaptive Self-Tuning Control," *Proc. of Int. Conf. Digital Signal Processing Applications and Technology*, pp. 900-911, Berlin, Germany, October 1991.

[5] Miao, J., "Microcomputer-Based Adaptive Control System of Blood Pressure," *Proc. of ISMM: Int. Symp. Microcomputer Applications*, pp. 28-31, ACTC Press, Long Beach, California, December 1991.

[6] Miao, J., "Automatic Drug Infusion System Using Self-Tuning Adaptive Controller," *Proc. of 13th Ann. IEEE Int. Conf. of Engineering in Medical and Biology*, Nagel, J., and W. Smith (eds.), Vol. 13, pp. 2145-2146, Orlando, Florida, October 1991.

[7] Haykin, S., *Neural Networks: A Comprehensive Foundation*, IEEE Press, New York, 1994.

[8] Miao, J., and M. A. Clements, "Optimal Discriminant Feature-Based Waveform Classification with Neural Networks," *Proc. of IEEE Int. Conf. on Acoustics, Speech and Signal Processing*, Vol. 4, pp. 3462-3465, Atlanta, Georgia, May 1996.

[9] Miao, J., "Component Feature-Based Digital Waveform Analysis and Classification," Ph.D. Dissertation, Georgia Institute of Technology, Atlanta, Georgia, April 1996.

[10] Wiener, N., *Extrapolation, Interpolation, and Smoothing of Stationary Time Series with Engineering Applications*, MIT Press, Cambridge, MA, 1949.

[11] Kailath, T, *Lectures on Wiener and Kalman Filtering*, 2d printing, Springer-Verlag, Wien, New York, 1981.

[12] Hayes, M. H., *Statistical Digital Signal Processing and Modeling*, John Wiley & Sons, Inc., New York, 1996.

[13] Sorenson, H. W. (eds.), *Kalman Filtering: Theory and Application*, IEEE Press, New York, 1985.

[14] Sibul, L. H. (eds.), *Adaptive Signal Processing*, IEEE Press, New York, 1987.

[15] Brown, R. G., and P. Y. C. Hwang, *Introduction to Random Signals and Applied Kalman Filtering*, 2d ed., John Wiley & Sons, Inc., New York, 1992.

[16] Haykin, S., *Adaptive Filter Theory*, 3d ed., Prentice Hall, Upper Saddle River, New Jersey, 1996.

[17] Proakis, J. G., "Adaptive Equalization Techniques for Acoustic Telemetry Channels," *IEEE Trans. on Oceanic Engineering*, Vol. 16, No. 1, pp. 21-31, January 1991.

[18] Proakis, J. G., "Adaptive Equalization for TDMA Digital Mobile Radio," *IEEE Trans. on Vehicular Technology*, Vol. 40, No. 2, pp. 333-341, May 1991.

[19] Stojanovic, M., J. G. Proakis, and J. A. Catipovic, "Analysis of the Impact of Channel Estimation Errors on the Performance of a Decision-Feedback Equalizer in Fading Multipath Channels," *IEEE Trans. on Communications*, Vol. 43, No. 2, pp. 877-885, April 1995.

[20] Rappaport, T. S., *Wireless Communications: Principles & Practice*, IEEE Press, New York, and Prentice Hall, Upper Saddle River, New Jersey, 1996.

[21] Gibson, J. D. (eds.), *The Mobile Communications Handbook*, 2d ed., IEEE Press and CRC Press, 1999.

[22] Cadzow, J. A., "Signal Processing via Least Squares Error Modeling," *IEEE Signal Processing Magazine*, pp. 12-31, October 1990.

[23] Zhao, K., F. Ling, H. Lev-Ari, J. G. Proakis, "Slide Window Order-Recursive Least-Squares Algorithms," *IEEE Trans. on Signal Processing*, Vol. 42, No. 8, pp. 1961-1972, August 1994.

[24] Sayed, A. H., and T. Kailath, "A State-Space Approach to Adaptive RLS Filtering," *IEEE Signal Processing Magazine*, Vol. 11, pp. 18-60, 1994.

7

Discrete-Time Multirate Signal Processing

7.1 Introduction

In many discrete-time signal processing systems, it is often necessary that the output-sampling rate be different from the input-sampling rate. In other words, one discrete-time signal must be converted into another with the same spectral characteristics, but at a different sampling rate. Unlike in a fixed-sampling-rate discrete-time signal processing system, the sample spacing can vary depending on the bandwidth of the signal under consideration. This process allows efficient processing of signals by keeping the sampling rate as small as possible within various processing steps. This strategy is referred to as *discrete-time multirate signal processing*. In these systems, the basic building blocks are *decimators*, *interpolators*, *upsamplers*, and *downsamplers*.

Multirate signal processing systems have many practical applications including:

- Wireless communications such as cellular phones, handheld devices, and wireless Internet appliances;
- Broadband communications including cable modem and digital subscriber line (DSL);

- Speech processing;
- Video and image compression;
- Digital audio processing;
- Spectrum analysis and radar systems;
- Adaptive signal processing such as adaptive equalizers;
- Multichannel analysis and synthesis filter banks;
- Discrete-time multidimensional signal processings;
- Subband coding of digital waveforms;
- Rational sampling rate conversions;
- Multirate analog-to-digital or digital-to-analog converters.

However, in this chapter, our emphasis will be on the theory of multirate signal processing systems and their applications. The focus will be on efficient architectures for reducing computational complexity and system delay in both hardware and software DSP implementations. This often results in efficient use of processing power, chip size area, and cost for high-tech products.

This chapter is organized as follows. In Section 7.1, we introduced discrete-time multirate signal processing. Section 7.2 introduces the basic building blocks of decimation and interpolation. An efficient polyphase architecture for multirate signal implementation is given in Section 7.3. We also present some fundamental building blocks, and address the efficient polyphase structure for decimation and interpolation filters, as well as for fractional rate conversions. Section 7.4 discusses efficient design techniques for multiband filters including halfband, Mth-band, narrowband, and wideband filters. Section 7.5 provides the multistage design and implementation of discrete-time multirate signal processing including decimation and interpolation systems. Multirate filter banks are presented in Section 7.6, with an emphasis on the QMF bank and its polyphase representation, as well as the perfect reconstruction system. In Section 7.7, we introduce the theory of uniform DFT filter banks including polyphase decomposition, and maximally decimated and perfect reconstruction approaches. The multirate adaptive filter bank based on subband and polyphase matrix schemes is presented in Section 7.8. Section 7.9 concludes with a summary.

7.2 Decimation and Interpolation Systems

In this section, we introduce the basic building blocks of discrete-time multirate signal processing including upsamplers, downsamplers, decimation and interpolation filters, and interconnection methods, along with their frequency domain characterizations. We also discuss the computational complexities of implementation for these multirate signal processing systems.

7.2.1 Decimation System

The operation of reducing the sampling rate as shown in Figure 7.1(a) is called *downsampling*. Downsampling takes a discrete-time input sequence $x[n]$ and produces a new discrete-time output sequence $y_d[n]$ as follows:

$$y_d[n] = x[Mn], \tag{7.1}$$

where M is an integer. For example, Figure 7.1(b) depicts the idea of downsampling for $M = 2$. As we can see, the downsampling function keeps only those discrete-time input samples $x[n]$ that occur at integer multiples of M.

For downsampling by M, we can develop the frequency z-domain relationship between the output $Y_d(z)$ and input $X(z)$ as follows:

The z-transform $Y_d(z)$ of a discrete-time sequence $y_d[n]$ is defined as

$$\begin{aligned} Y_d(z) &= \sum_{n=-\infty}^{\infty} y_d[n]z^{-n} \\ &= \sum_{n=-\infty}^{\infty} x[Mn]z^{-n}. \end{aligned} \tag{7.2}$$

Let us define an intermediate sequence $x_M[n]$,

$$x_M[n] = p[n]x[n], \tag{7.3}$$

where $p[n]$ is the sequence given by

$$p[n] = \begin{cases} 1 & n \text{ an integer multiple of } M \\ 0 & \text{otherwise.} \end{cases} \tag{7.4}$$

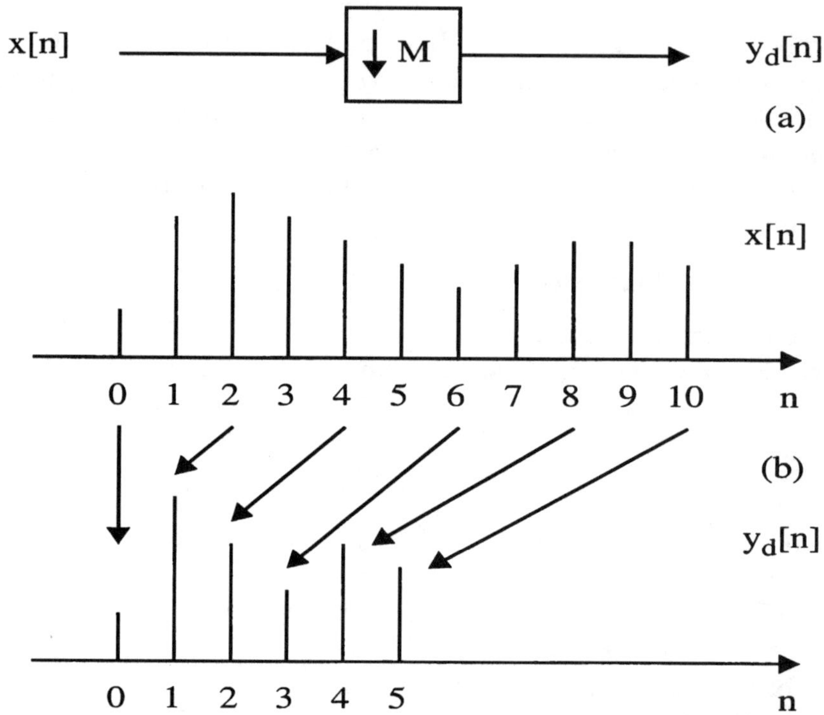

Figure 7.1 Downsampling function in a discrete-time multirate signal processing system. (a) Downsampling by a factor of M. (b) An example of downsampling for $M = 2$.

The sequence $p[n]$ can be expressed by its Fourier series

$$p[n] = \frac{1}{M} \sum_{k=0}^{M-1} e^{jk2\pi n/M}. \tag{7.5}$$

Since $y_d[n] = x[Mn]$, (7.2) can be rewritten

$$\begin{aligned} Y_d(z) &= \sum_{n=-\infty}^{\infty} x[Mn]z^{-n} \\ &= \sum_{n=-\infty}^{\infty} x_M[Mn]z^{-n} \end{aligned}$$

$$= \sum_{m=-\infty}^{\infty} x_M[m] z^{-m/M}$$
$$= X_M(z^{1/M}). \tag{7.6}$$

Notice that (7.6) is only valid when m is a multiple of M. Otherwise, (7.6) is equal to zero.

On the other hand, the z-transform $X_M(z)$ based on (7.3), (7.4), and (7.5) can be obtained by

$$\begin{aligned} X_M(z) &= \frac{1}{M}\sum_{k=0}^{M-1}\sum_{n=-\infty}^{\infty} x[n]e^{jk2\pi n/M} z^{-n} \\ &= \frac{1}{M}\sum_{k=0}^{M-1}\sum_{m=-\infty}^{\infty} x[n](ze^{-jk2\pi/M})^{-n} \\ &= \frac{1}{M}\sum_{k=0}^{M-1} X(ze^{-jk2\pi/M}). \end{aligned} \tag{7.7}$$

Therefore, using (7.6) and (7.7), the z-transform of $Y_d(z)$ can be expressed in terms of $X(z)$ by

$$Y_d(z) = \frac{1}{M}\sum_{k=0}^{M-1} X(z^{1/M} e^{-jk2\pi/M}). \tag{7.8}$$

In general, the frequency components expressed by $X(z^{\frac{1}{M}})|_{z=e^{j\omega}}$ of $y_d[n]$ will overlap with its shifted replicas. In this case, we cannot recover the input signal $x[n]$ from $y_d[n]$. This distortion is referred to as *aliasing*. Aliasing can be avoided if the original sampling rate is at least M times the *Nyquist* rate, or if the bandwidth of the input sequence $x[n]$ is first bandlimited by filtering with cutoff no higher than π/M.

Lowpass digital filtering followed by downsampling is called *decimation* as shown in Figure 7.2. Thus, the output signal $x_h[n]$ can be downsampled without aliasing if the input signal $x[n]$ is first filtered by an ideal lowpass filter with a cutoff frequency π/M as illustrated in Figure 7.3.

For direct implementation, the decimation system needs N multiplications and $N-1$ additions per output point if the lowpass

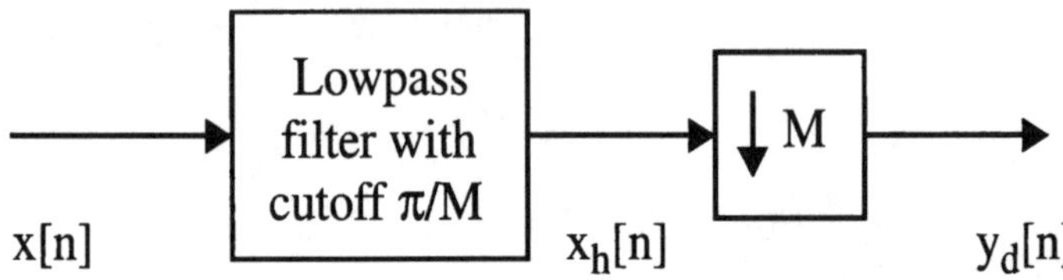

Figure 7.2 General decimation system for sampling rate reduction by M.

filter is $(N-1)$th order FIR. For example, if the total number of samples of input signal $x[n]$ is L, direct implementation of the decimation system requires LN multiplications and additions (MACs) to complete this process. This result amounts to inefficient implementation with respect to special hardware or DSPs. Therefore, we will address this problem and provide an efficient architecture for the implementation based on polyphase decomposition in Section 7.3.3.

7.2.2 Interpolation System

Upsampling by an integer factor of L involves the operation depicted in Figure 7.4(a). The relationship between input and output is given by

$$y_u[n] = \begin{cases} x[\frac{n}{L}] & n = 0, \pm L, \pm 2L, \ldots \\ 0 & \text{otherwise.} \end{cases} \tag{7.9}$$

This is equivalent to

$$y_u[n] = \sum_{k=-\infty}^{\infty} x[k]\delta[n-kL]. \tag{7.10}$$

In other words, the discrete-time output sequence $y_u[n]$ is obtained by inserting $L-1$ zero-valued samples between adjacent samples of input sequence $x[n]$ as shown in Figure 7.4(b).

The operation of the upsampling system in Figure 7.4(a) can also be represented in the frequency domain. The z-transform of the

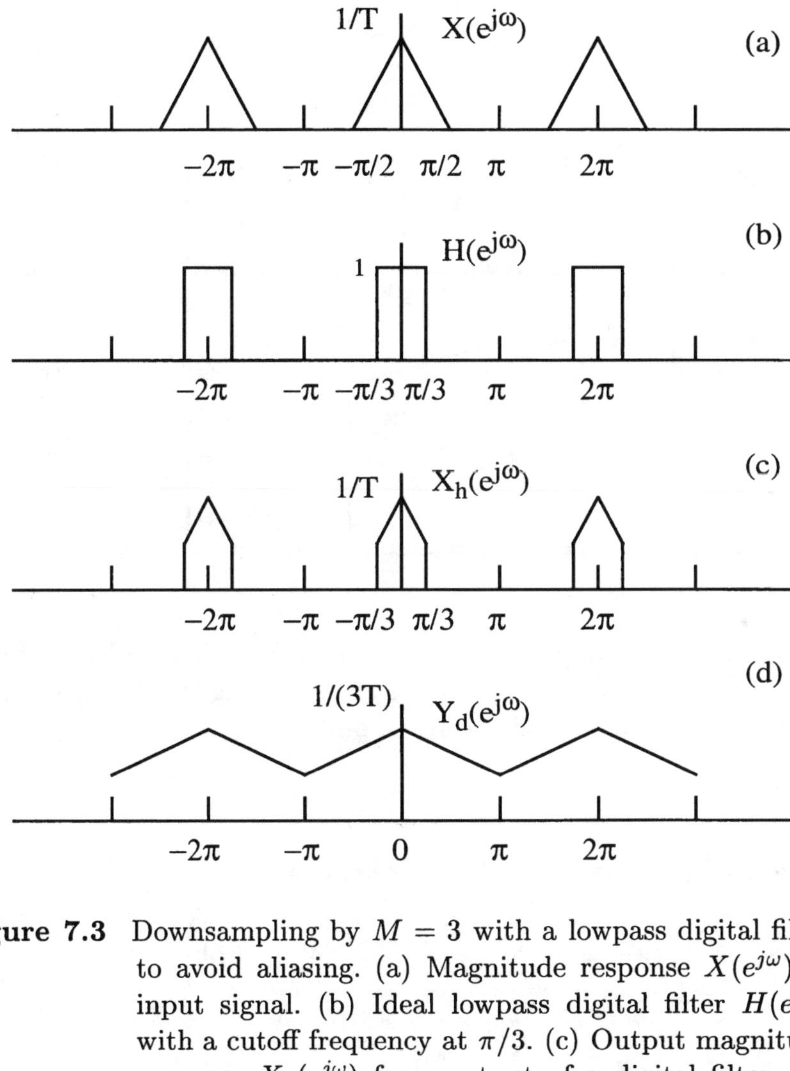

Figure 7.3 Downsampling by $M = 3$ with a lowpass digital filter to avoid aliasing. (a) Magnitude response $X(e^{j\omega})$ of input signal. (b) Ideal lowpass digital filter $H(e^{j\omega})$ with a cutoff frequency at $\pi/3$. (c) Output magnitude response $X_h(e^{j\omega})$ from output of a digital filter. (d) Magnitude response $Y_d(e^{j\omega})$ after downsampling by $M = 3$.

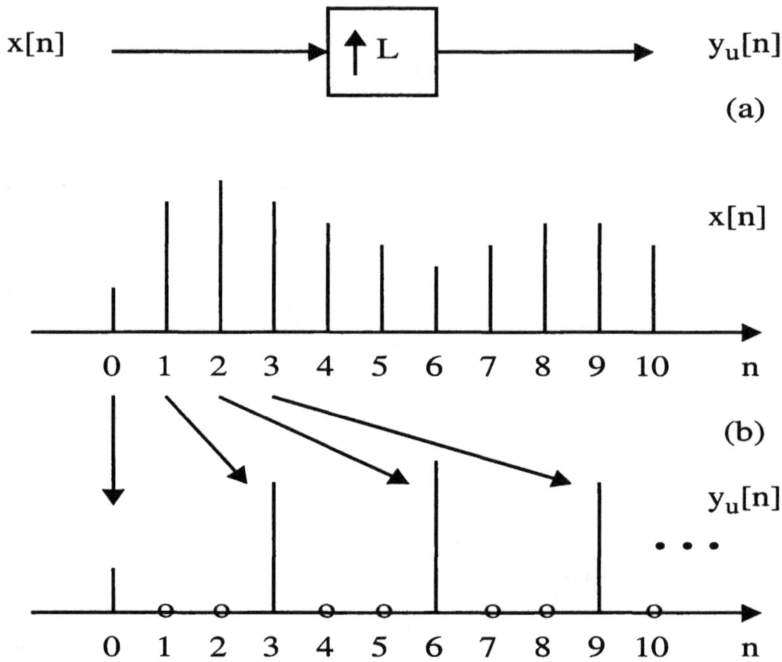

Figure 7.4 Upsampling function in a discrete-time multirate signal processing system. (a) Upsampling by a factor of L. (b) An example of upsampling with $L = 3$.

output sequence $y_u[n]$ can be expressed as

$$\begin{aligned} Y_u(z) &= \sum_{n=-\infty}^{\infty} \left(\sum_{k=-\infty}^{\infty} x[k]\delta[n-kL] \right) z^{-n} \\ &= \sum_{k=-\infty}^{\infty} x[k] z^{-Lk} \\ &= X(z^L). \end{aligned} \tag{7.11}$$

This means that $Y_u(z)$ is an L-fold compressed version of $X(z)$. In other words, upsampling creates multiple images of $X(e^{j\omega})$ in the frequency domain.

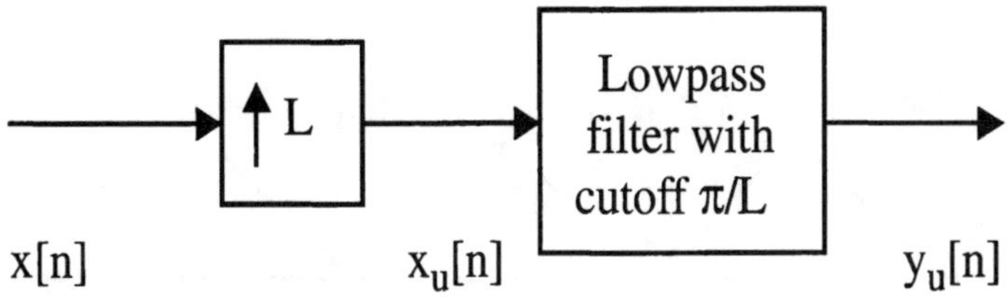

Figure 7.5 General interpolation system for increasing the sampling rate by L.

Figure 7.5 shows an interpolation system including upsampling followed by a lowpass digital filter. The digital filter typically uses the cutoff frequency of π/L along with a filter gain L. The purpose of the digital filter is to suppress all of the images as illustrated in Figure 7.6. The effect of interpolation is that the zero-valued samples introduced by upsampling are filled in with interpolated values.

For direct implementation, the interpolation system needs N multiplications and $N-1$ additions if the lowpass digital filter is $(N-1)$th order FIR. For example, if a total number of samples of input signal $x[n]$ is M, the output size of upsampling with L is LM. This requires (LMN) MACs to complete this process. This result amounts to inefficient implementation for special hardware or DSPs. The efficient implementation of the interpolation system will be discussed when we present polyphase decomposition and structure in Section 7.3.4.

7.2.3 Interconnection System

It has been shown how to decrease or increase the sampling rate of sequences by an integer factor based on the decimation and interpolation. In decimation, downsampling is preceded by a lowpass filtering to avoid aliasing. In interpolation, a filter that follows upsampling eliminates spurious images. Note that aliasing can cause the loss of information because of the overlap of shifted spectrum. Extra images do not lead to any loss of information. However, by

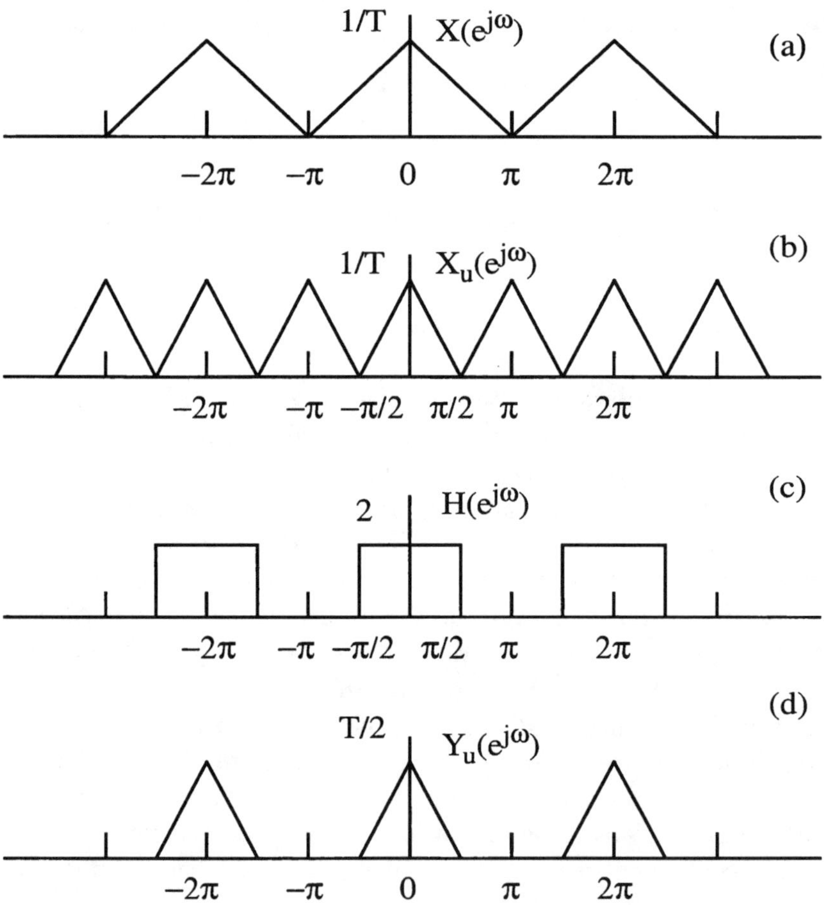

Figure 7.6 Upsampling by $L = 2$ with lowpass digital filter to suppress all the images. (a) Magnitude response $X(e^{j\omega})$ of input signal $x[n]$. (b) Magnitude response $X_u(e^{j\omega})$ of upsampling by $L = 2$. (c) Ideal lowpass digital filter $H(e^{j\omega})$ with cutoff frequency at $\pi/2$ and filter gain 2. (d) Output magnitude response $Y_u(e^{j\omega})$ from output of the digital filter.

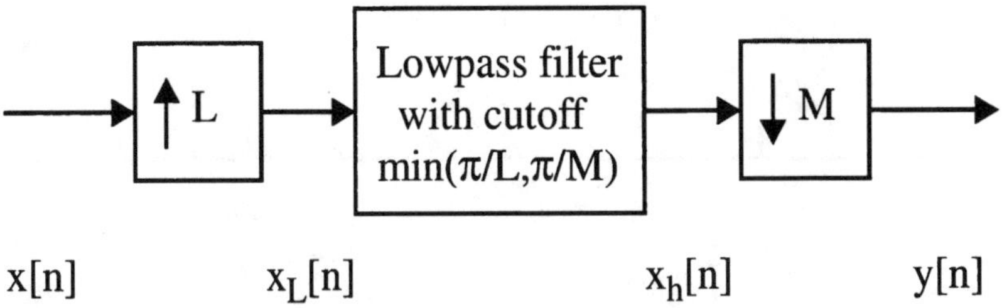

Figure 7.7 General block diagram of interconnection system with a noninteger factor of M/L.

cascading decimation and interpolation systems, it is possible to change the sampling rate by a noninteger factor, M/L.

Figure 7.7 shows a generalized block diagram of integrating decimation and interpolation for changing the sampling rate by rational factor M/L. Note that the interpolation and decimation filters have been combined into one lowpass digital filter $H(z)$ with gain L and cutoff frequency $min\{\pi/L, \pi/M\}$.

For information on the theory of decimation and interpolation systems, we suggest the reference papers by Schafer and Rabiner [1] and Crochiere and Rabiner [2], and the books by Crochiere and Rabiner [3] and Oppenheim and Schafer [4].

Example 7.1 Rational Sampling Rate Using an Interconnection System

As an example, we wish to change the rate by $M/L = 3/2$ as shown in Figure 7.7. Assume that the frequency response $X(e^{j\omega})$ of input sequence $x[n]$ is bandlimited as shown in Figure 7.8(a). An ideal lowpass filter $H(e^{j\omega})$ is used with the cutoff frequency $\pi/3$. The output frequency response $X_h(e^{j\omega})$, with the shadow, is shown in Figure 7.8(b) after the input signal $x[n]$ is first upsampled by 2 and then passed through the ideal lowpass filter $H(e^{j\omega})$. Thus, the signal $x_h[n]$ is downsampled by 3 to produce the desired output signal $y[n]$. As a result, the output frequency represented by $Y(e^{j\omega})$ is also shown in Figure 7.8(b) after the rate conversion.

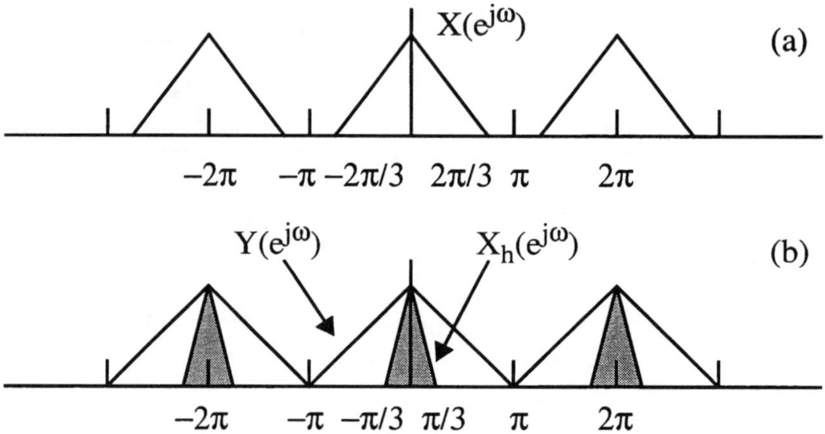

Figure 7.8 Example of an interconnection system with the rational rate using setting $M = 3$ and $L = 2$. (a) Magnitude response $X(e^{j\omega})$ of bandlimited input signal $x[n]$. (b) Magnitude response $X_h(e^{j\omega})$ with the shadow after the decimator $H(e^{j\omega})$, and magnitude response $Y(e^{j\omega})$ of fractional decimation of the input signal $x[n]$.

7.3 Efficient Polyphase Architecture for Implementing a Multirate Signal Processing System

So far, we have shown the basic concepts of discrete-time multirate signal processing systems, including decimation, interpolation, and interconnection systems. We have pointed out that the direct implementation of those systems involves computational complexity and that they should not be recommended for practical applications. Therefore, in this section we introduce efficient architectures for implementation. These methods of efficient implementation are sometimes called *multirate implementations*, *polyphase representation*, or *polyphase decomposition* in the literature.

7.3.1 Basic Operation of Multirate Identity Blocks

A number of downsampling identity blocks of interconnections and equivalencies are shown in Figure 7.9. These basic interconnections

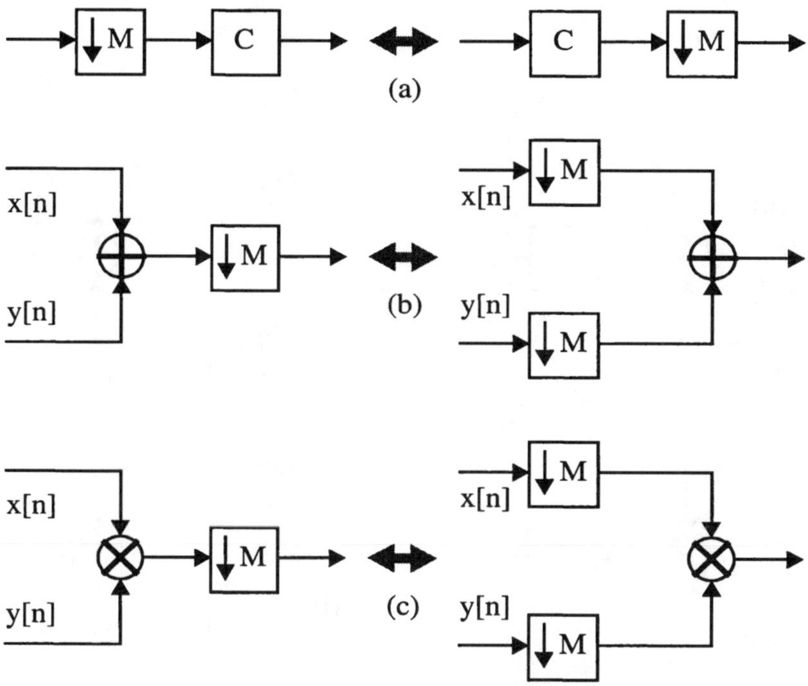

Figure 7.9 Basic identity blocks of downsampling interconnection system. (a) Downsampling with the constant. (b) Superposition of $x[n]$ and $y[n]$ with downsampling. (c) Two signals $x[n]$ and $y[n]$ are multiplied with downsampling.

of identity blocks are often used in the multirate system of signal processing and play an important role in efficient implementation architectures. Figure 7.10 illustrates upsampling identity blocks of interconnections and equivalencies that can easily be verified. We leave the verification of those functions as shown in Figures 7.9 and 7.10 for the reader as an exercise.

Figure 7.11 shows the multirate or noble identities for a discrete-time multirate signal processing system including the decimation and interpolation identity systems. These multirate or noble identities are very important in the theory and implementation of discrete-time multirate signal processing. In fact, these identities, as shown in

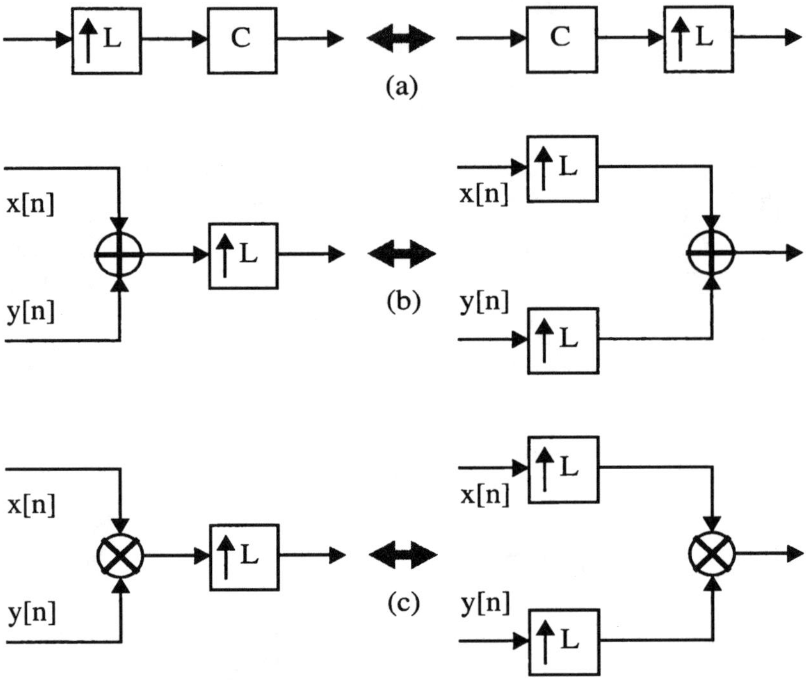

Figure 7.10 Basic identity blocks of upsampling interconnection system. (a) Upsampling with the constant. (b) Superposition of $x[n]$ and $y[n]$ with upsampling. (c) Two signals $x[n]$ and $y[n]$ are multiplied with upsampling.

Figure 7.11, are at the heart of efficient implementation architecture when we use the polyphase decomposition, as well as multirate filter banks. This permits greatly simplified results and leads to computational efficiency in practical applications.

It is not difficult to prove the noble identities as shown in Figure 7.11. For the decimation identity system, based on (7.8), the z-transform $Y_d(z)$ of the output signal sequence $y_d[n]$ is given by

$$Y_d(z) = \frac{1}{M}\sum_{k=0}^{M-1} X_h(z^{1/M}e^{-j2\pi k/M})$$

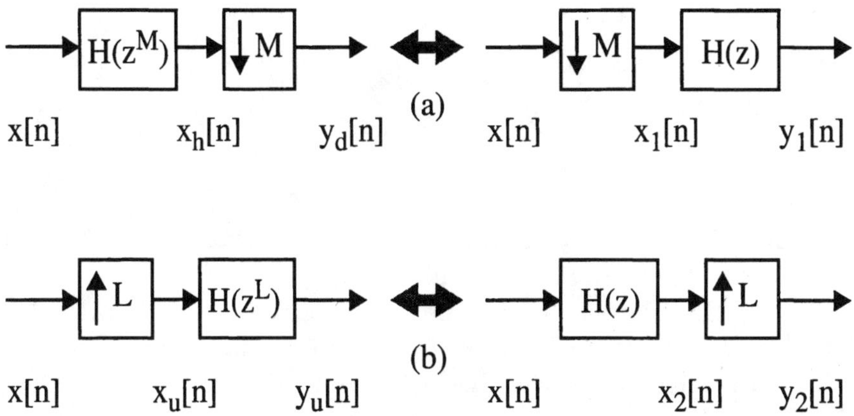

Figure 7.11 Multirate or noble identities for discrete-time multirate signal processing system. (a) Decimation identity system. (b) Interpolation identity system.

$$= \frac{1}{M}\sum_{k=0}^{M-1} X(z^{1/M}e^{-j2\pi k/M})H[(z^{1/M}e^{-j2\pi k/M})^M]$$

$$= \frac{1}{M}\sum_{k=0}^{M-1} X(z^{1/M}e^{-j2\pi k/M})H(z). \quad (7.12)$$

Notice that the z-transform $Y_1(z)$ of output signal sequence $y[n]$ is expressed

$$\begin{aligned} Y_1(z) &= X_1(z)H(z) \\ &= \frac{1}{M}\sum_{k=0}^{M-1} X(z^{1/M}e^{-j2\pi k/M})H(z). \end{aligned} \quad (7.13)$$

As we can see, (7.12) is the same as (7.13).

For an interpolation identity system, based on (7.11), the z-transform $Y_2(z)$ of output signal sequence $y_2[n]$ can be obtained by

$$\begin{aligned} Y_2(z) &= X_2(z^L) \\ &= X(z^L)H(z^L), \end{aligned} \quad (7.14)$$

while the z-transform $Y_u(z)$ of output signal sequence $y_u[n]$ is

$$\begin{aligned} Y_u(z) &= X_u(z)H(z^L) \\ &= X(z^L)H(z^L). \end{aligned} \tag{7.15}$$

Therefore, (7.14) is equal to (7.15). The verifications of the noble identities in Figure 7.11 are completed.

Notice that the theory of noble identities works if and only if filter function $H(z)$ is rational, for example, a ratio of polynomials in terms of z^{-1}. In other words, the theory of noble identities holds whether the FIR or IIR filter function $H(z)$ is causal or noncausal. However, the theory of noble identities may not work when filter function $H(z)$ is nonrational. In practical applications, incorrect use of the noble identities can be avoided in the multirate signal processing system by ensuring that $H(z)$ is rational.

7.3.2 Basic Operation of Multirate Switch Models

An important method to help understand the operation of the polyphase implementation architectures, which we will discuss in Sections 7.3.3 and 7.3.4, is the multirate switch model. Figure 7.12 demonstrates such a model, which is useful for analyzing the polyphase structure of decimation and interpolation implementations. The switch position rotates and only assumes one of the three positions at uniform speed. In Figure 7.12(a), when the switch rotates in counterclockwise direction, it is used for decimation, while in Figure 7.12(b), when the switch rotates in a clockwise direction it is used for interpolation.

An alternative use of the multirate switch model, illustrating the polyphase implementation architecture, is the *serial-to-parallel* or *parallel-to-serial* function, as shown in Figure 7.13; the former is used for decimation and the latter is used for interpolation.

In the next two sections, we will show that these switch models are very useful architectures in polyphase decomposition and how they are applied to the multirate implementation in an efficient way.

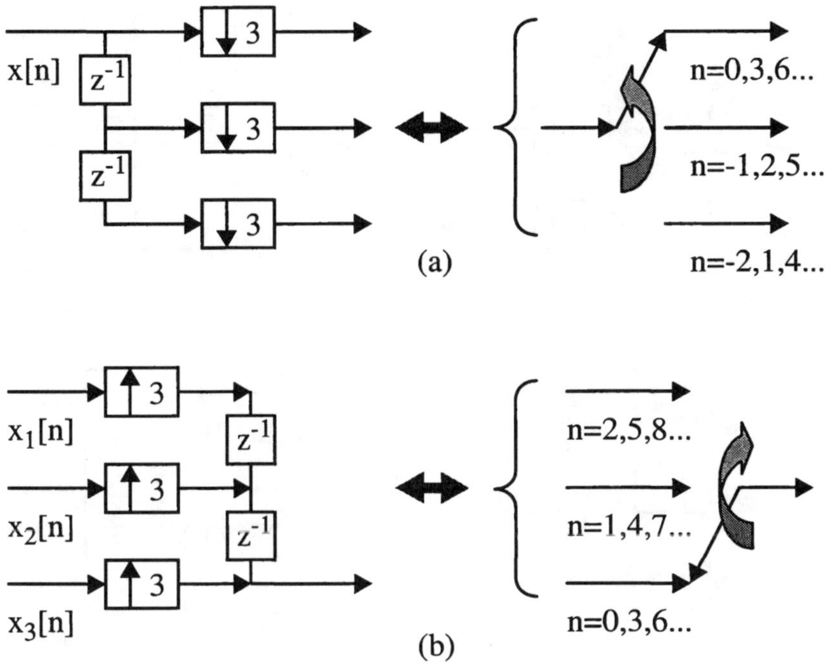

Figure 7.12 The operation of multirate switch models. (a) The switch rotates in a counterclockwise direction for the decimation system. (b) The switch rotates in a clockwise direction for the interpolation system.

7.3.3 Polyphase Structure of the Decimation Filter

In order to illustrate the polyphase structure of the decimation filter, consider the discrete-time lowpass filter $H(z)$ with downsampling, $M = 2$, as shown in Figure 7.2. By separating the even-indexed filter coefficients from the odd-indexed filter coefficients, we can rewrite the filter function $H(z)$ as follows:

$$\begin{aligned} H(z) &= \sum_{n=-\infty}^{\infty} h[n] z^{-n} \\ &= \sum_{n=-\infty}^{\infty} h[2n] z^{-2n} \end{aligned}$$

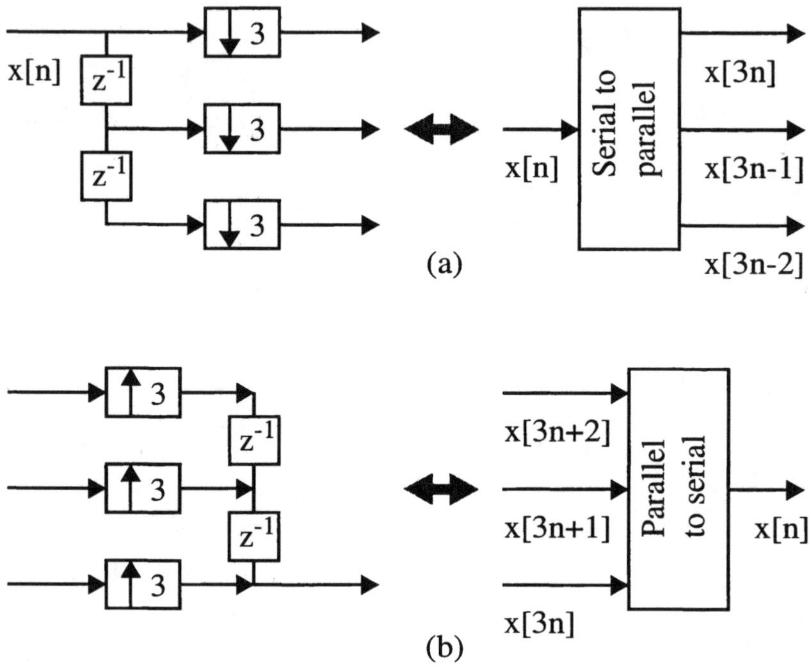

Figure 7.13 The operation of an alternative method for multirate switch models. (a) The serial-to-parallel function for decimation implementation. (b) The parallel-to-serial function for interpolation implementation.

$$+ z^{-1} \sum_{n=-\infty}^{\infty} h[2n+1]z^{-2n}. \tag{7.16}$$

If we define

$$D_0(z) = \sum_{n=-\infty}^{\infty} h[2n]z^{-n} \tag{7.17}$$

and

$$D_1(z) = \sum_{n=-\infty}^{\infty} h[2n+1]z^{-n}, \tag{7.18}$$

then we can reexpress (7.16) as

$$H(z) = D_0(z^2) + z^{-1}D_1(z^2). \tag{7.19}$$

As we can see from (7.19), we have grouped the impulse response of discrete-time filter $h[n]$ into even-indexed samples $h[2n]$ and odd-indexed samples $h[2n+1]$. Therefore, the efficient polyphase implementation of the decimation system can be developed as shown in Figure 7.14.

Given any integer M, it is possible to decompose filter $H(z)$ into the M-channel filter bank with polyphase form as follows:

$$H(z) = \sum_{k=0}^{M-1} z^{-k} D_k(z^M), \tag{7.20}$$

where the impulse response $d_k[n]$ of $D_k(z)$ is obtained by

$$d_k[n] = h[nM + k], \quad 0 \leq k \leq M - 1. \tag{7.21}$$

This representation of (7.20) is referred to as the polyphase decimation decomposition of an M-channel filter bank and is often used for implementation of decimation systems.

The concept of *polyphase decomposition*, which has been described by Bellanger, Bonnerot, and Coudreuse [5], Crochiere and Rabiner [2], and Vaidyanathan [6], is fundamental in multirate signal processing for many practical applications. These include decimation and interpolation, rational sampling rate conversion (see Hsiao [7]), digital transmultiplexer filters (see Bellanger [8]), uniform DFT or DCT filter banks, analysis/synthesis filter banks, fractionally spaced decision-feedback equalizers (see Treichler et al. [9]), and multistandard/multimode wireless devices (see Miao et al. [10]), and multirate adaptive filter banks.

For implementation of the decimation system as shown in Figure 7.14(c), let us consider how efficient the polyphase structure is by comparing it to its direct implementation. Given a decimation filter $H(z)$ of order $N+1$ and P input samples, its direct implementation will require (PN) MACs. For the polyphase structure, $D_0(z)$ requires $N_0 + 1$ multiplications and N_0 additions, and $D_1(z)$ also needs $N_1 + 1$ multiplications and N_1 additions. Note

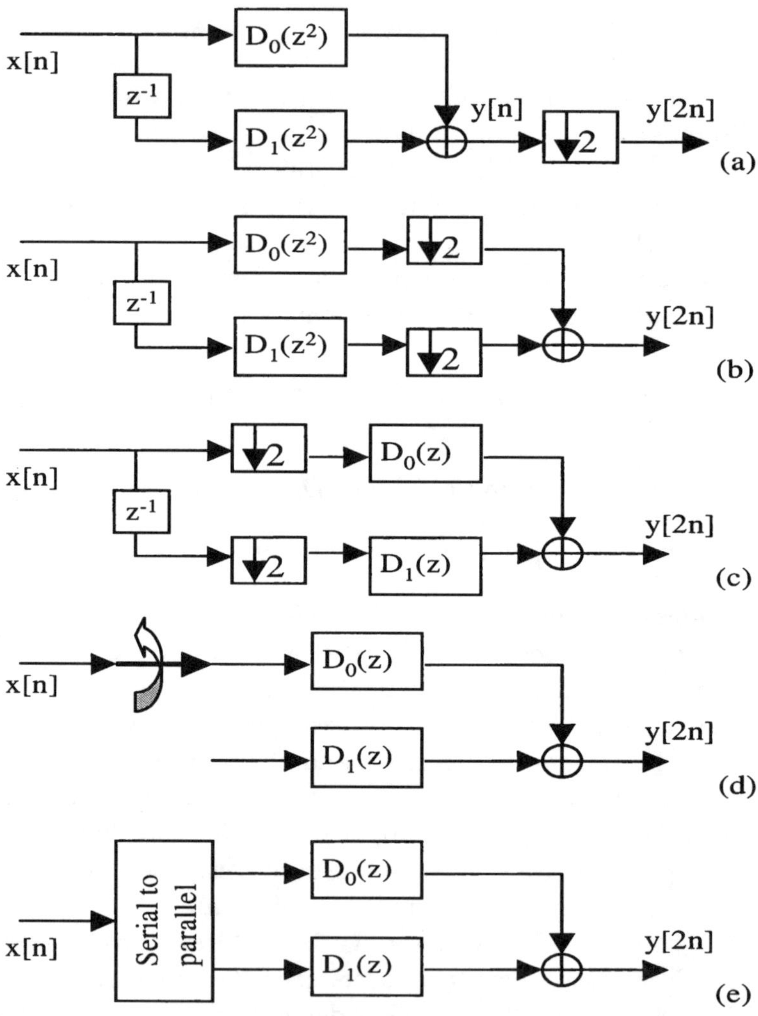

Figure 7.14 (a) Implementation structure of decimation system using (7.19). (b) Downsampling moves into the parallel structure based on the concept of basic identity block for downsampling system. (c) Exchanging the positions between downsampling and filters using noble identities. (d) Multirate switch model with polyphase structure. (e) Serial-to-parallel function with polyphase structure.

that $N+1 = N_0+N_1+2$. The total computational complexity is still $N+1$ multiplications and N adds. However, both filters, $D_0(z)$ and $D_1(z)$, operate at half the data original rate, so approximately 50% less MACs are used in the implementation. This also applies when a decimation filter $H(z)$ is an IIR case.

Example 7.2 Polyphase Decomposition of the Decimation Filters

In this example, we provide the polyphase decomposition of FIR and IIR decimation filters with downsampling by 2. For an FIR decimation filter, assume that the z-transform function is

$$H(z) = a + bz^{-1} + cz^{-2} + dz^{-3} + ez^{-4} + fz^{-5}. \tag{7.22}$$

Using (7.20), we can rewrite $H(z)$ as follows:

$$H(z) = (a + cz^{-2} + ez^{-4}) + z^{-1}(b + dz^{-2} + fz^{-4}). \tag{7.23}$$

Then, the polyphase decomposition of the filter banks is

$$D_0(z) = a + cz^{-1} + ez^{-2}, \quad D_1(z) = b + dz^{-1} + fz^{-2}. \tag{7.24}$$

For an example of an IIR decimation filter, we let

$$H(z) = \frac{\alpha}{1 - \alpha z^{-1}}, \tag{7.25}$$

where $|\alpha| < 1$. We can rewrite $H(z)$ as

$$H(z) = \frac{\alpha}{1 - \alpha^2 z^{-2}} + z^{-1}\left(\frac{\alpha^2}{1 - \alpha^2 z^{-2}}\right). \tag{7.26}$$

Therefore, in this case, the polyphase decomposition of the filter banks is

$$D_0(z) = \frac{\alpha}{1 - \alpha^2 z^{-1}}, \quad D_1(z) = \frac{\alpha^2}{1 - \alpha^2 z^{-1}}. \tag{7.27}$$

Note that the system order of $H(z)$ in (7.26) has been increased after rewriting $H(z)$ in the IIR case. However, the polyphase decomposition of the systems $D_0(z)$ and $D_1(z)$ still keeps the same system order as the original one in the system function $H(z)$. In other words, the polyphase decomposition for an IIR case does not increase the system order.

For a general IIR transfer function, $H(z) = \frac{N(z)}{D(z)}$, the polyphase decomposition is obtained by multiplying the numerator and the denominator with an appropriate factor $P(z)$. Then the modified transfer function can be reexpressed as

$$\begin{aligned} H(z) &= \frac{N(z)P(z)}{D(z)P(z)} \\ &= \frac{N_p(z)}{D_p(z^M)} \\ &= \sum_{k=0}^{M-1} z^{-k} \frac{N_{pk}(z^M)}{D_p(z^M)}. \end{aligned} \tag{7.28}$$

Notice that (7.28) may not work if $N_p(z)$ is nonrational.

7.3.4 Polyphase Structure of the Interpolation Filter

For an efficient polyphase-structured interpolation filter, reconsider discrete-time filter $H(z)$ with upsampling, $L = 2$, as shown in Figure 7.5. By separating the odd-indexed filter coefficients from the even-indexed filter coefficients, we can rewrite filter function $H(z)$ as follows:

$$\begin{aligned} H(z) &= \sum_{n=-\infty}^{\infty} h[n]z^{-n} \\ &= z^{-1} \sum_{n=-\infty}^{\infty} h[2n+1]z^{-2n} + \sum_{n=-\infty}^{\infty} h[2n]z^{-2n}. \end{aligned} \tag{7.29}$$

Now, let

$$I_0(z) = \sum_{n=-\infty}^{\infty} h[2n+1]z^{-n} \tag{7.30}$$

and

$$I_1(z) = \sum_{n=-\infty}^{\infty} h[2n]z^{-n}. \tag{7.31}$$

We then can reexpress (7.29) as

$$H(z) = z^{-1}I_0(z^2) + I_1(z^2). \tag{7.32}$$

Notice that $I_i(z^2) = D_{M-1-i}(z^2)$ from (7.32) and (7.19). Therefore, the efficient polyphase implementation of the interpolation system can be developed as shown in Figure 7.15.

For a general case, given any integer L, it is possible to decompose the filter $H(z)$ in an L-band filter bank with polyphase form as follows:

$$H(z) = \sum_{k=0}^{L-1} z^{-(L-1-k)} I_k(z^L), \tag{7.33}$$

Equation (7.33) is the *polyphase interpolation decomposition* of the L-channel filter bank and is often used for implementation of the interpolation system.

To understand the implementation of the interpolation system as shown in Figure 7.15, consider an interpolation filter $H(z)$ of the order $N+1$ along with upsampling by L and P input samples. The direct implementation of this system would have (PLN) MACs as shown in Figure 7.5. On the other hand, for the polyphase structure, $I_0(z)$ requires N_0+1 multiplications and N_0 additions, and $I_1(z)$ also needs N_1+1 multiplications and N_1 additions. Note that $N+1 = N_0 + N_1 + 2$ still holds. In other words, the total computational complexity is still $N+1$ multiplies and N adds. However, both of the filters $I_0(z)$ and $I_1(z)$ operate at the normal data rate so that approximately (PN) MACs are needed for implementation. This can save up to $\left(\frac{L-1}{L}\right)$ MACs for each operation. Similarly, this is also valid when an interpolation filter is an IIR case.

7.3.5 Polyphase Structure of Fractional Sampling Rates

A general scheme for a fractional sampling rate conversion by a noninteger factor M/L is shown in Figure 7.7. However, we already indicated that it is an inefficient operation for any practical application if we implement the interconnection system directly. In order to obtain a more efficient structure of implementation, we must rearrange the structure of the *interconnection system*; that is, we need to restructure the filter $H(z)$ into a parallel form using polyphase decomposition. Additionally, we move the downsampling block up front and the upsampling block after the filter bank $H(z)$.

To illustrate the most efficient structure of fractional sampling rate conversion, we offer an example with $M = 4$ and $L = 3$ by

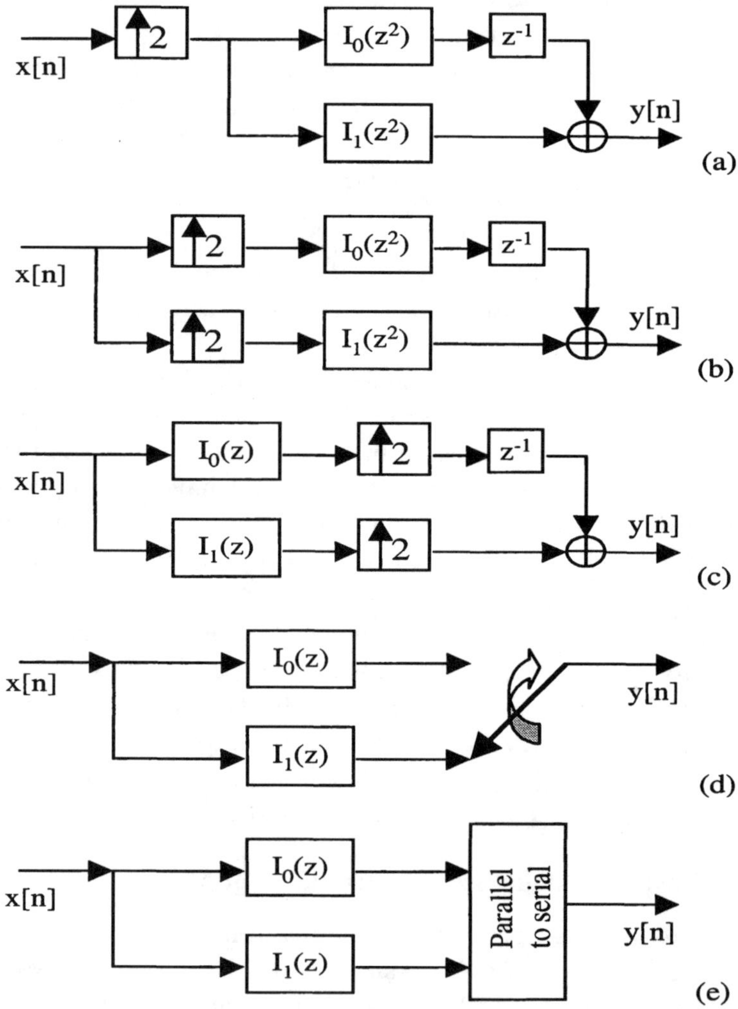

Figure 7.15 (a) Implementation structure of interpolation system using (7.32). (b) Upsampling moves into the parallel structure based on the concept of a basic identity block for upsampling system. (c) Exchanging the positions between upsampling and filters using noble identities. (d) Multirate switch model with a polyphase structure. (e) Serial-to-parallel function with a polyphase structure.

using the theory of polyphase decomposition and noble identities. Figure 7.16 shows a detailed procedure of step-by-step restructuring for maximizing computational efficiency.

By using polyphase interpolation decomposition, we obtain the implementation form of Figure 7.16(a). Now, we can rewrite $z^{-2} = z^6 z^{-8}$ and $z^{-1} = z^3 z^{-4}$ in such a way that Figure 7.16(b) can be redrawn as in Figure 7.16(c) using noble identities. Exchanging the position of downsampling M and upsampling L results in Figure 7.16(d). Notice that these are equivalent if and only if downsampling M and upsampling L are relatively prime. Finally, Figure 7.17 shows the maximum computationally efficient implementation using polyphase decimation decomposition. Then, the filter banks $F_0(z)$, $F_1(z)$, and $F_2(z)$ have the following forms:

$$\begin{aligned} F_0(z) &= F_{01}(z^4) + z^{-1}F_{02}(z^4) + z^{-2}F_{03}(z^4) + z^{-3}F_{04}(z^4), \\ F_1(z) &= F_{11}(z^4) + z^{-1}F_{12}(z^4) + z^{-2}F_{13}(z^4) + z^{-3}F_{14}(z^4), \\ F_2(z) &= F_{21}(z^4) + z^{-1}F_{22}(z^4) + z^{-2}F_{23}(z^4) + z^{-3}F_{24}(z^4). \end{aligned}$$

Therefore, Figure 7.17 is the same as Figure 7.16(a). Alternatively, the switch implementation structure and serial-to-parallel/parallel-to-serial forms of fractional sampling rate conversion are shown in Figures 7.18 and Figure 7.19, respectively.

For implementation, we still need $N+1$ multiplications and N additions as shown in Figures 7.17, 7.18, and 7.19 if the filter $H(z)$ is the N order. However, each multiplication and addition operates at a lower data rate, which is one-fourth the input data rate. So, in general, the polyphase implementation of the fractional sampling rate requires only $(N+1)/M$ multiplications and N/M additions. As we can see, the implementation architecture is most efficient because we have moved the decimation system to the left of the filter banks and the interpolation system to the right of these filter banks.

Polyphase decomposition for the decimation and interpolation systems has been well developed so far. In the context of multirate signal processing systems with polyphase decomposition, we cite the reference papers by Crochiere and Rabiner [2], Neuvo, Dong, and Mitra [11], Vaidyanathan [6], Kovačevia and Vetterli [12], Kalker [13], Miao et al. [14], and the books by Crochiere and Rabiner [3] and Vaidyanathan [15].

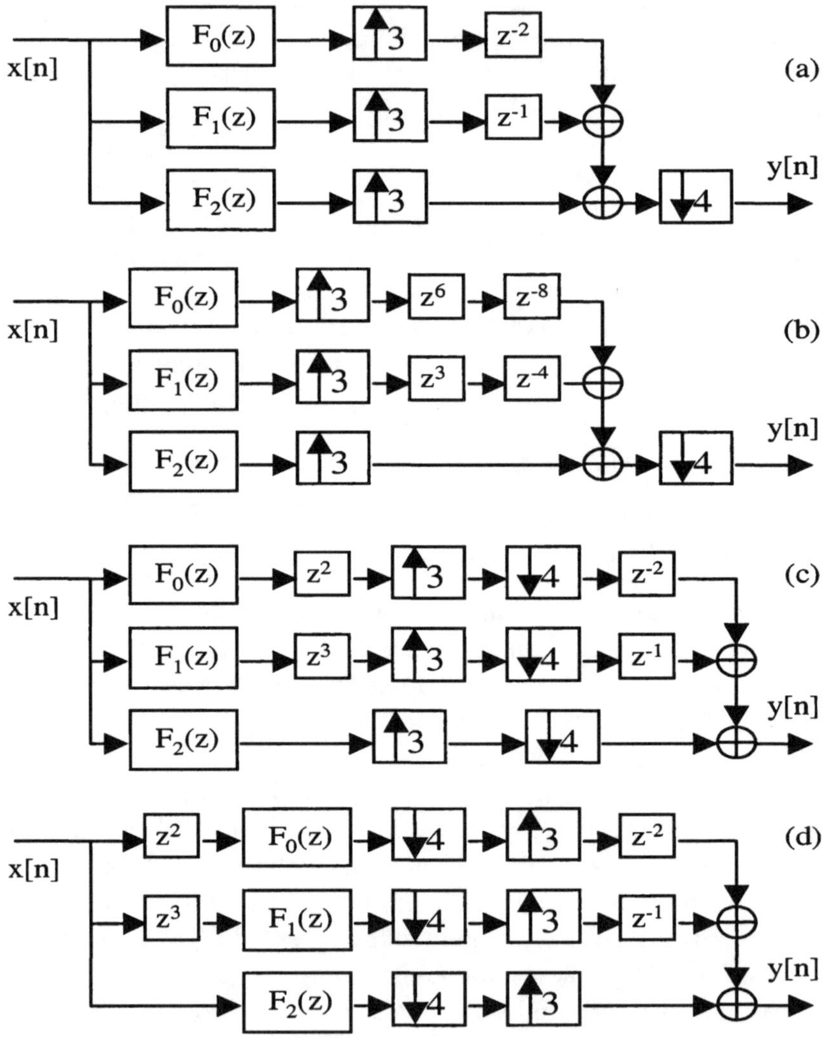

Figure 7.16 The step-by-step procedure of redrawing fractional rate to achieve the maximum computational efficiency. (a) Moving upsampling into the left of computational units. (b) Redrawing the system using $z^{-2} = z^6 z^{-8}$ and $z^{-1} = z^3 z^{-4}$. (c) Modifying the system using noble identities. (d) Exchanging the position of downsampling and upsampling.

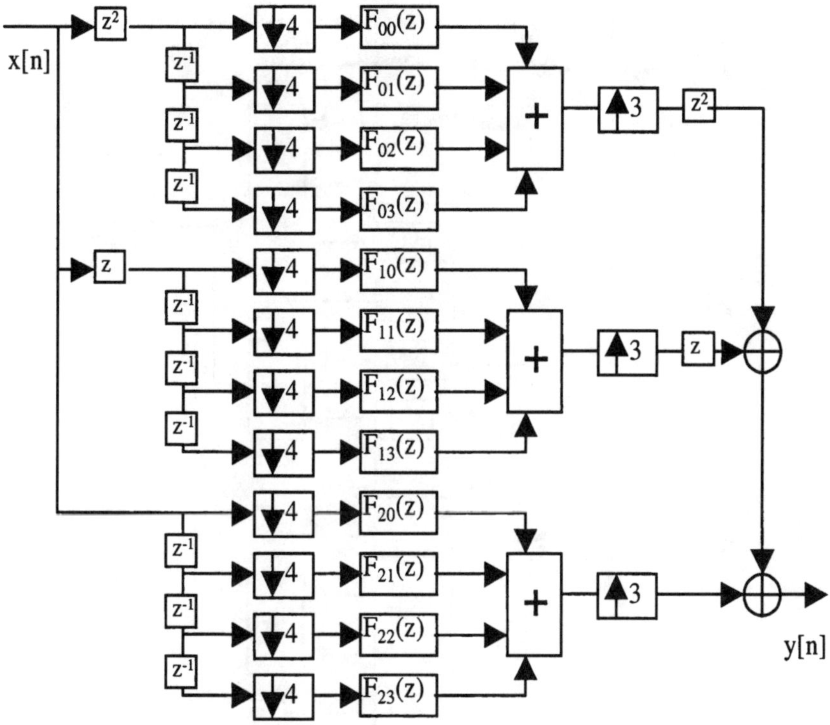

Figure 7.17 Redrawing the fractional rate system using polyphase decimation decomposition for the efficiency architecture.

7.4 Efficient Design Techniques for Multiband Filters

In this section, we introduce a special case of efficient design techniques for multiband filters. Multiband filters are particularly useful for implementations of multirate signal processing. In other words, multiband filters require fewer computations than equivalent decimation or interpolation filters by using the techniques of polyphase decomposition and noble identities.

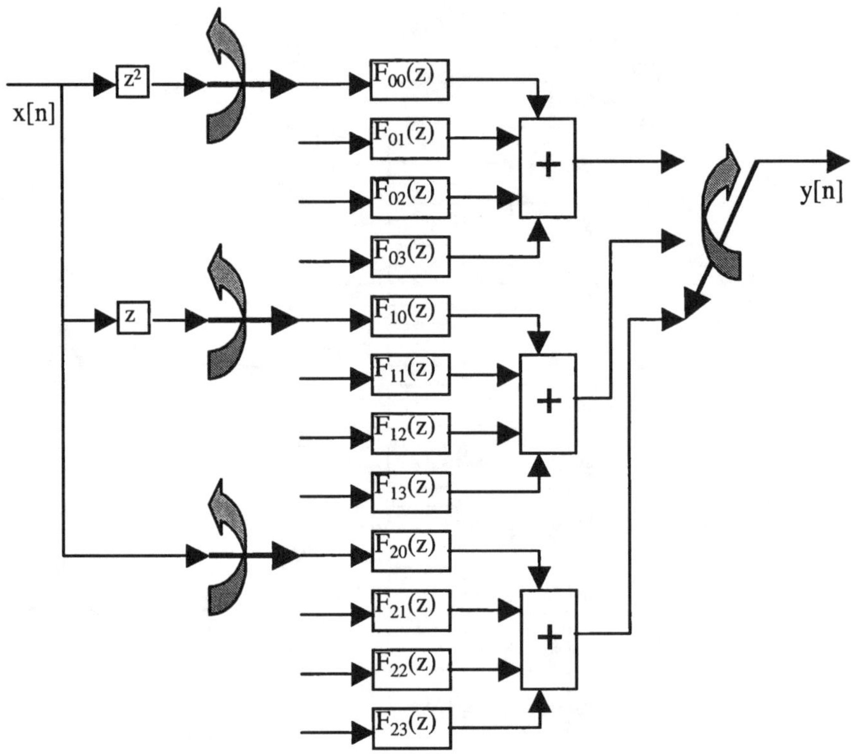

Figure 7.18 The efficient architecture of fractional sampling rate using the switch implementation method.

7.4.1 Halfband Filter

Reconsider the polyphase decimation decomposition in (7.19), assuming that $D_0(z^2) = \alpha$. Thus, we can rewrite the transfer function $H(z)$ as

$$H(z) = \alpha + z^{-1}D_1(z^2). \tag{7.34}$$

In this case, the impulse response of transfer function $H(z)$ satisfies

$$h[2n] = \begin{cases} \alpha & n = 0 \\ 0 & \text{otherwise,} \end{cases} \tag{7.35}$$

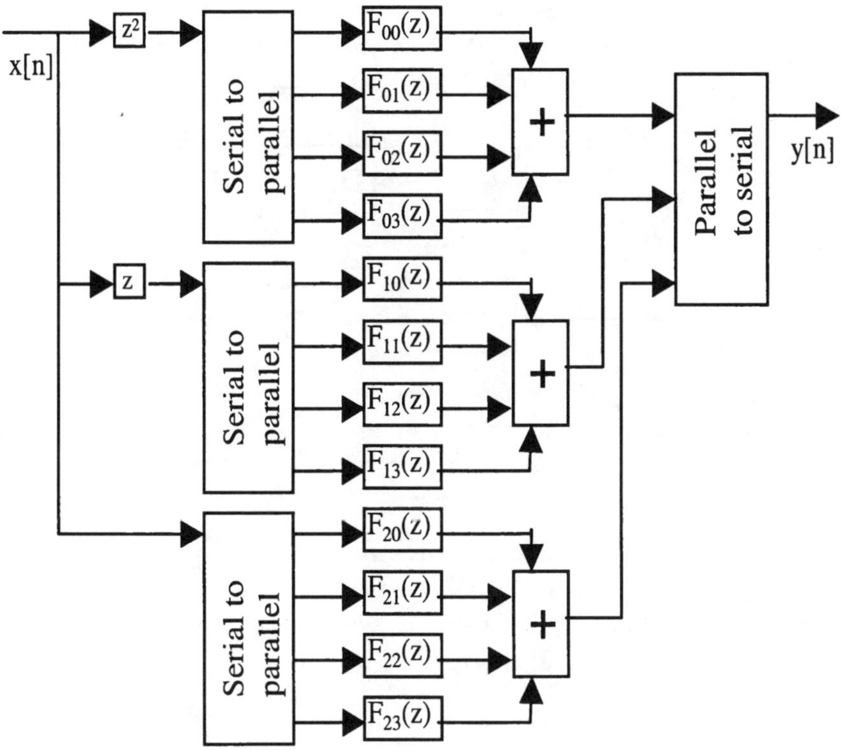

Figure 7.19 The efficient architecture of fractional sampling rate using serial-to-parallel and parallel-to-serial methods.

where α is usually 0.5. As we can see, the even-indexed samples of impulse response $h[n]$ are equal to zero except for $h[0]$. Then, for a real coefficient FIR filter with zero-phase, we obtain

$$H(e^{j\omega}) + H(e^{j(\pi-\omega)}) = 1. \tag{7.36}$$

Therefore, we have a symmetric FIR filter with respect to the halfband frequency at $\pi/2$. We call this filter a *halfband filter*. It has fewer computations than the regular one for implementation of multirate signal processing as shown in Figure 7.20(a).

Notice that the transfer function $I_1(z^2)$ equals the transfer function $D_0(z^2)$ from polyphase interpolation decomposition in (7.32).

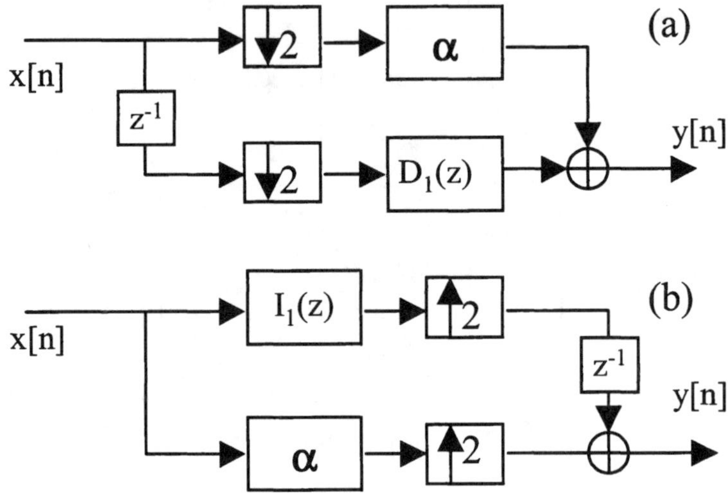

Figure 7.20 An efficient architecture for implementing the multirate system. (a) Decimation system with a halfband filter. (b) Interpolation system with a halfband filter.

So, we can rewrite the halfband filter $H(z)$ for the interpolation system as

$$H(z) = z^{-1}I_0(z^2) + \alpha. \tag{7.37}$$

Equation (7.37) can be expressed using the block diagram in Figure 7.20(b). As we can see, the number of computations for the interpolation system is reduced further. Alternatively, we can assume that the constant α equals zero or all the odd-indexed coefficients in the transfer function $H(z)$ are zeros. Then, one branch without the filters $D_0(z^2)$ and $I_0(z^2)$ in Figure 7.20(a) and (b), respectively, can simply be deleted. In other words, we can directly switch the position between filter $H(z)$ and downsampling/upsampling by using the noble identities. In addition, if we set $\alpha = 1$ for the interpolation system as shown in Figure 7.20(b), then output $y[2n] = x[n]$. This indicates that, although the interpolation filter creates new samples, the filter passes the existing samples to the output without distortion.

Example 7.3 Designing the Lowpass Halfband FIR Filter

Filter design techniques including window design methods, least squares design methods, and optimum equiripple methods can all be used for designing the lowpass halfband FIR filter. One important consideration in the design of halfband filters is to make sure that $\omega_p + \omega_s = \pi$; that is, the frequency band edges ω_p and ω_s are equally distant from $\pi/2$.

Figure 7.21 shows the result with frequency response and impulse responses for a lowpass halfband filter using the optimum equiripple method. As we can see, the even-indexed samples of the impulse response $h[n]$ as shown in Figure 7.21(b) are all zeros except $h[12] = 0.5$.

7.4.2 Mth-band Filter

Reconsider (7.20). A general case of the Mth-band filter using polyphase decomposition is obtained by

$$H(z) = \alpha + \sum_{i=1}^{M-1} z^{-i} D_i(z^M), \tag{7.38}$$

where the constant α is usually set to $1/M$. The impulse response $h[n]$ of such a transfer function $H(z)$ is given by

$$h[Mn] = \begin{cases} \alpha & n = 0 \\ 0 & \text{otherwise.} \end{cases} \tag{7.39}$$

Equation (7.39) is usually called an *Mth-band filter*. It can be shown that (7.38) satisfies

$$\sum_{k=0}^{M-1} H(ze^{-j2k\pi/M}) = M\alpha. \tag{7.40}$$

Therefore, we conclude that all M uniformly shifted versions of transfer function $H(z)$ add up to a constant. Clearly, the halfband filter in (7.34) is a special case of (7.38) when $M = 2$ and $\alpha = 1/2$.

In the context of designing for the halfband and Mth-band FIR filters, we cite the reference papers by Mintzer [16] and Vaidyanathan and Nguyen [17].

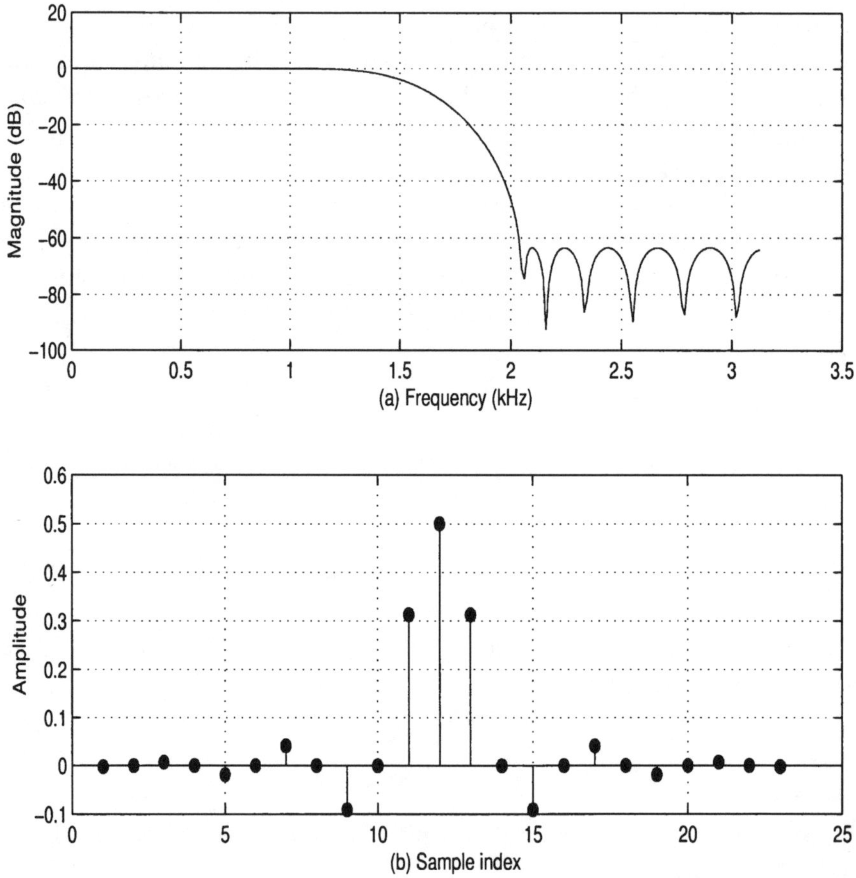

Figure 7.21 A design result of lowpass halfband FIR filters. (a) The magnitude response of lowpass halfband FIR filters. (b) The impulse response of lowpass halfband FIR filters.

Example 7.4 Designing the Lowpass Mth-band FIR Filter

The design of the Mth-band lowpass filter is required to be linear phase with cutoff of frequency at π/M. To guarantee this property, we use the filter design technique of windowing as illustrated in Section 5.4.1. The coefficients of the Mth-band lowpass filter are obtained by

$$h[n] = \left[\frac{sin(n\pi/M)}{n\pi}\right] w[n], \tag{7.41}$$

where, in this example, $w[n]$ is the symmetric Kaiser window. Figure 7.22 shows a design result of an Mth-band lowpass filter with $M = 5$.

As can be seen, the total number of the filter length is equal to 68. The center coefficients of the Mth-band lowpass filter are located at discrete-time index $n = 35$, while others $h[5m+1] = 0$ and $h[5m+34] = 0$, for $n = 1, 2, ..., 7$.

7.4.3 Narrowband and Wideband Filters

In many practical applications of multirate signal processing, it is common to design the decimation and interpolation system with a large integer factor. This leads to a very narrowband filter for decimation and interpolation. The direct design of such a filter is inefficient because the filter has very-high-order coefficients that can result in numerical inaccuracies during implementation. Therefore, in this section we introduce an efficient way of designing two cascade filters to achieve one narrowband filter. In fact, the technique presented here has nothing to do with decimation and interpolation systems and can be applied to any narrowband filter design.

To illustrate the basic idea, we first assume an Nth-order narrowband filter, $H(z)$, as shown in Figure 7.23(a). Then we present an efficient method of designing cascade filters $R(z)$ and $G(z^M)$ in Figure 7.23(b) that is equivalent to the narrowband filter $H(z)$. Figure 7.24(a) shows a frequency specification of a lowpass narrowband filter $H(z)$ with passband ω_p and stopband ω_s. Instead of meeting the frequency specification of the narrowband filter $H(z)$, we design an M-fold stretched frequency specification of filter $G(z)$ as shown in Figure 7.24(b). Notice that filter $G(z)$ has transition bandwidth $\triangle\omega = M(\omega_s - \omega_p)$ so that the filter order of $G(z)$ is

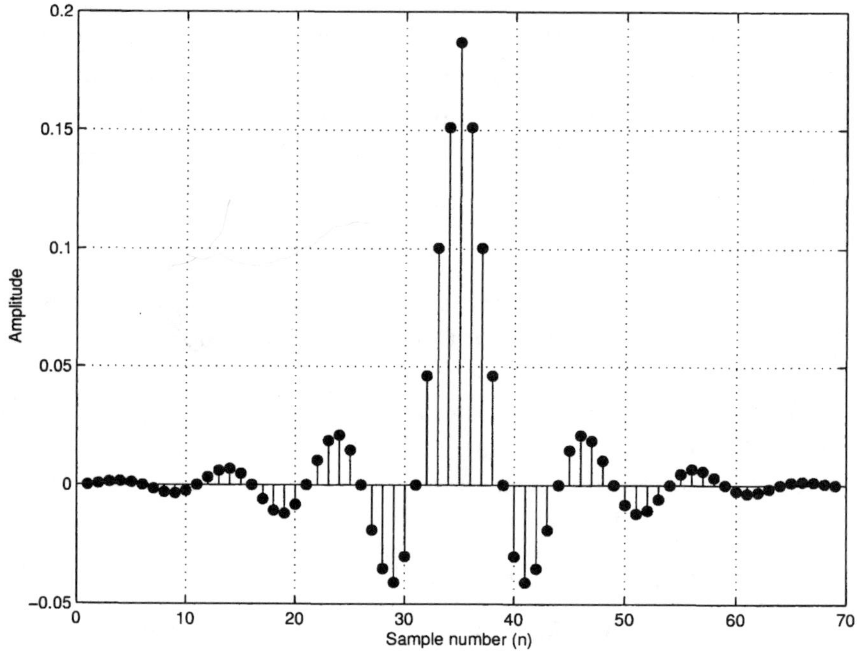

Figure 7.22 The impulse response of an Mth-band lowpass filter with $M = 5$.

(N/M). In other words, the MACs of the filter can be reduced by an integer factor of M.

Figure 7.24(c) shows the magnitude response of Mth-band filter $G(z^M)$. This Mth-band filter $G(z^M)$ has $M - 1$ passband images in addition to the desired passband at $\omega = 0$. Therefore, a wideband filter $R(z)$ has to be designed to eliminate these images as shown in Figure 7.24(d). This wideband filter $R(z)$ is referred to as an *image-suppressor*.

In general, M can be selected as large as the integer result of π/ω_s for the Mth-band filter $G(z^M)$. On the other hand, the transition bandwidth of image-suppressor filter $R(z)$ depends on the desired passband ω_p, stopband ω_s, and an integer factor of M. In other words, the filter $R(z)$ has the passband at ω_p and the stopband at

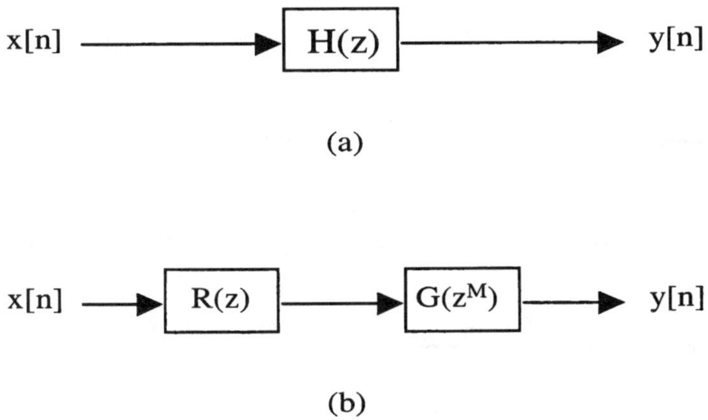

Figure 7.23 Block diagram of design of the narrowband filter. (a) A desired narrowband filter. (b) The efficient design of the narrowband filter.

$\frac{2\pi}{M} - \omega_s$. Finally, two cascade filters $G(z^M)R(z)$ still have the linear phase because both filters $G(z)$ and $R(z)$ have this property.

Example 7.5 An Efficient Design for the Narrowband Filter

Assume that the desired narrowband filter $H(z)$ has the specifications $\omega_p = 0.05\pi$, $\omega_s = 0.10\pi$, $\delta_1 = 0.01$, and $\delta_2 = 0.001$ with a sampling rate F_s of 2,000 kHz. Figure 7.25 shows a direct design result of narrowband filter $H(z)$ that meets the filter requirement of frequency specification using the optimum equiripple technique. This narrowband filter $H(z)$ requires $N = 138$ for the filter order. These filter coefficients are symmetric. So, the computational complexity of the direct design narrowband filter is 69 multiplications and 69 additions for implementation.

For the efficient design of the narrowband filter, we let $M = 2$ for simplicity. Then, the filter $G(z)$ has the frequency specifications $\omega_p = 0.1\pi$, $\omega_s = 0.20\pi$, $\delta_1 = 0.01$, and $\delta_2 = 0.001$. The wideband filter $R(z)$

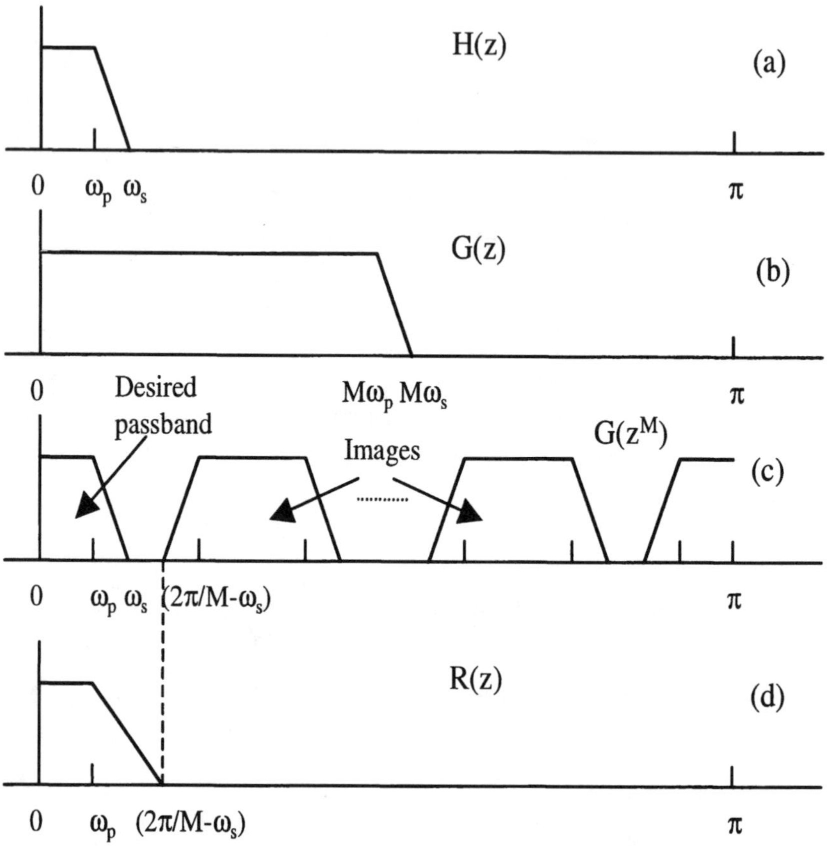

Figure 7.24 Efficient techniques for designing narrowband filters. (a) The desired frequency specification of filter $H(z)$. (b) The filter $G(z)$ with passband at $M\omega_p$ and stopband at $M\omega_s$. (c) The magnitude response of Mth-band $G(z^M)$. (d) The magnitude response of wideband filter $R(z)$.

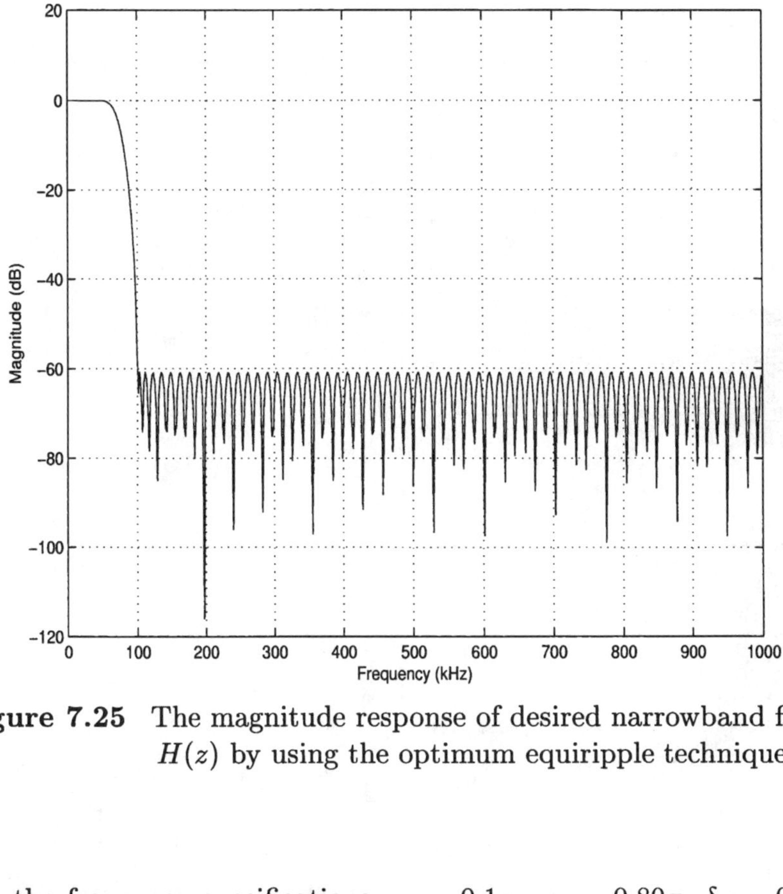

Figure 7.25 The magnitude response of desired narrowband filter $H(z)$ by using the optimum equiripple technique.

has the frequency specifications $\omega_p = 0.1\pi$, $\omega_s = 0.80\pi$, $\delta_1 = 0.01$, and $\delta_2 = 0.001$. The design results of filter frequency response $G(z)$, $G(z^2)$, $R(z)$, and $G(z^2)R(z)$ are shown in Figure 7.26(a), (b), (c), and (d).

The designed filter $G(z^2)$ has 68 coefficients with symmetry. Then, this filter $G(z^2)$ needs 38 multiplications and 38 additions. On the other hand, the designed filter $R(z)$ has 7 filter coefficients with symmetry. Thus, this filter $R(z)$ requires 4 multiplications and 4 additions. Therefore, the total computational complexity of combining two cascade filters $G(z^2)$ and $R(z)$ is 42 multiplications and 42 additions. As we can see, in this example the efficient design of the narrowband filter can save up to 39.13% in terms of MAC

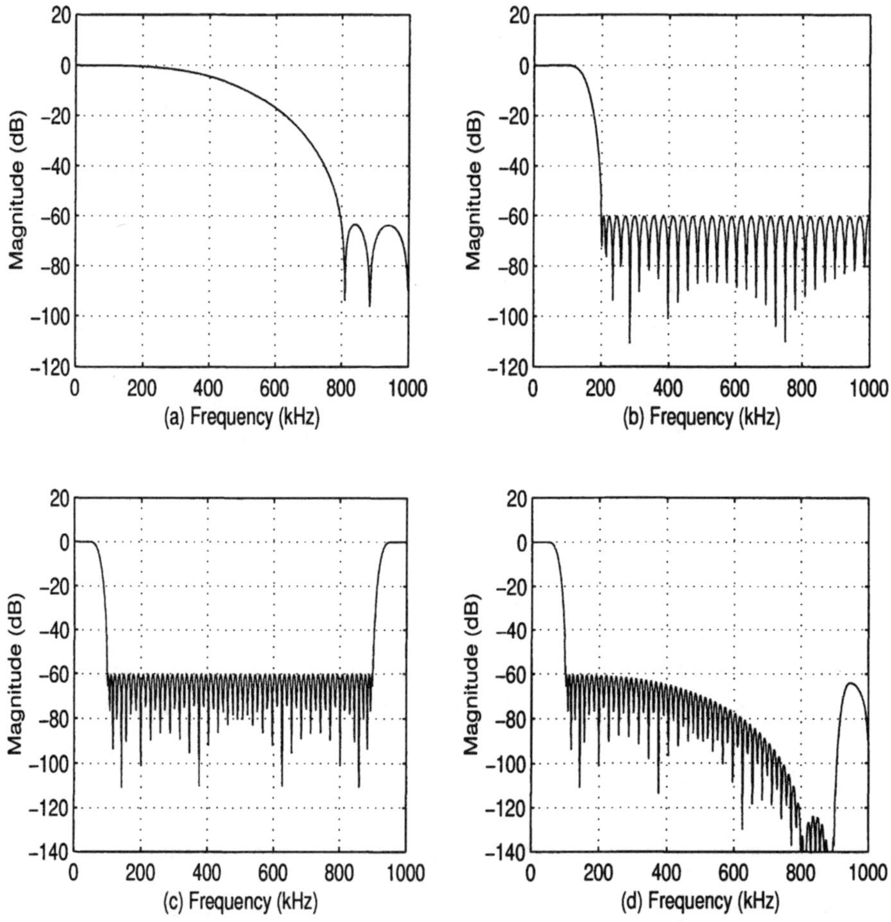

Figure 7.26 The efficient design of narrowband filter $H(z)$ using two cascade filters $R(z)$ and $G(z^2)$. (a) The magnitude response of filter $R(z)$. (b) The magnitude response of filter $G(z)$. (c) The magnitude response of the halfband filter $G(z^2)$. (d) The magnitude response of combining filters $G(z^2)$ and $R(z)$.

operations.

7.5 Multistage Design of Multirate Signal Processing

In practical applications, from the point of view of implementation it is necessary to design the decimation and interpolation systems in multiple stages when downsampling or upsampling with a large integer factor M. The reason behind this is that multistage implementations result in more efficient systems for multirate signal processing. So, in this section we discuss the multistage design and implementation for decimation and interpolation systems.

7.5.1 Multistage Decimation System

Figure 7.27(a) shows the lowpass filter $H(z)$ with frequency cutoff at π/M in the decimation system. The filter $H(z)$ is a narrowband filter when M is a large integer factor. To reduce the computational complexity, we can use the efficient technique to design the narrowband filter by taking $M = M_1 M_2$. This system is shown in Figure 7.27(b). Then, using noble identities, this decimation system can be redrawn into a two-stage system as shown in Figure 7.27(c). Thus, the entire decimation system has been decomposed into two stages: the first stage with downsampling M_1 and the second stage with downsampling M_2. Eventually, we are able to repeat this process by factoring integer M_1 and M_2 further. Finally, this system can be rearranged using polyphase decimation decomposition as shown in Figure 7.27(d). Notice that the polyphase structure of Figure 7.27(d) may be further simplified by moving the downsampling by M_2 block before the filter $R(z)$ to achieve the maximum computational efficiency.

Example 7.6 Two-Stage Mth-Band Decimation System

Reconsidering the decimation system as shown in Figure 7.27(a), assume that the input signal has a sampling frequency of 2,000 kHz and the desired downsampling factor is $M = 40$. The decimation filter $H(z)$ has a passband ripple $\delta_1 = 0.01$ and a stopband ripple $\delta_2 = 0.001$. The cutoff frequency of the decimation filter is $\pi/40$. Then, the decimation filter has passband edge at 0.02π and stopband

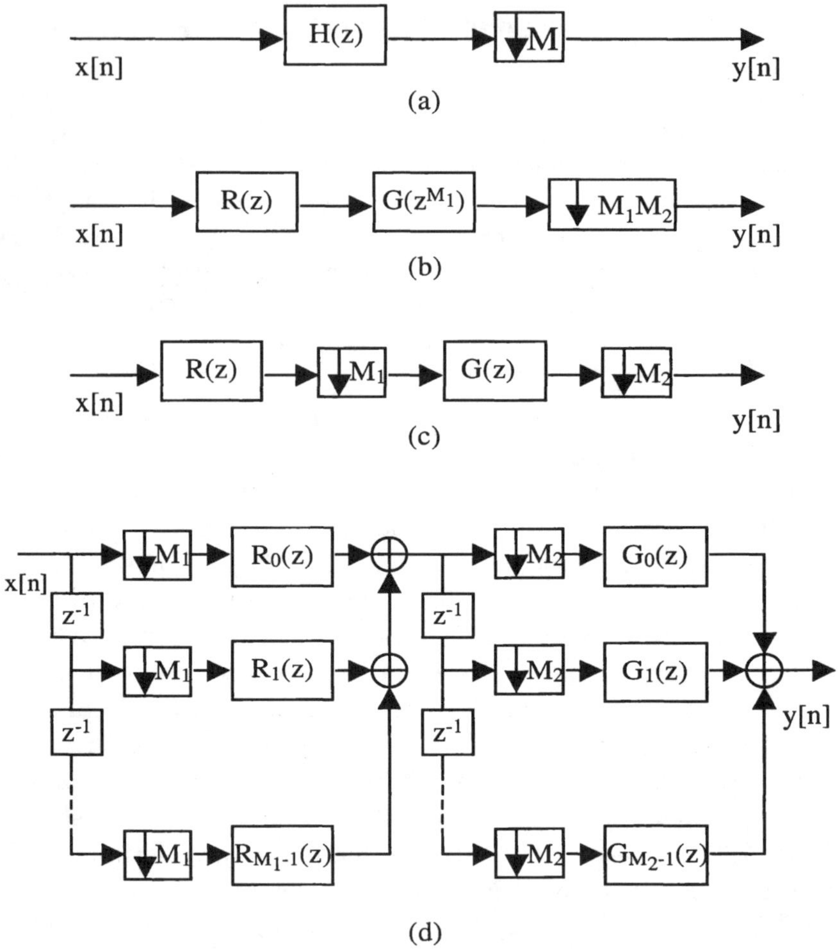

Figure 7.27 The multistage implementation of the decimation system when M is a large integer factor. (a) The decimation system including the filter $H(z)$ and downsampling by M. (b) The representation of decimation system efficiency. (c) Interchange positions between filters $R(z)$, $G(z^{M_1})$, and downsampling blocks M_1 and M_2. (d) The polyphase structure of implementation in a two-stage decimation system.

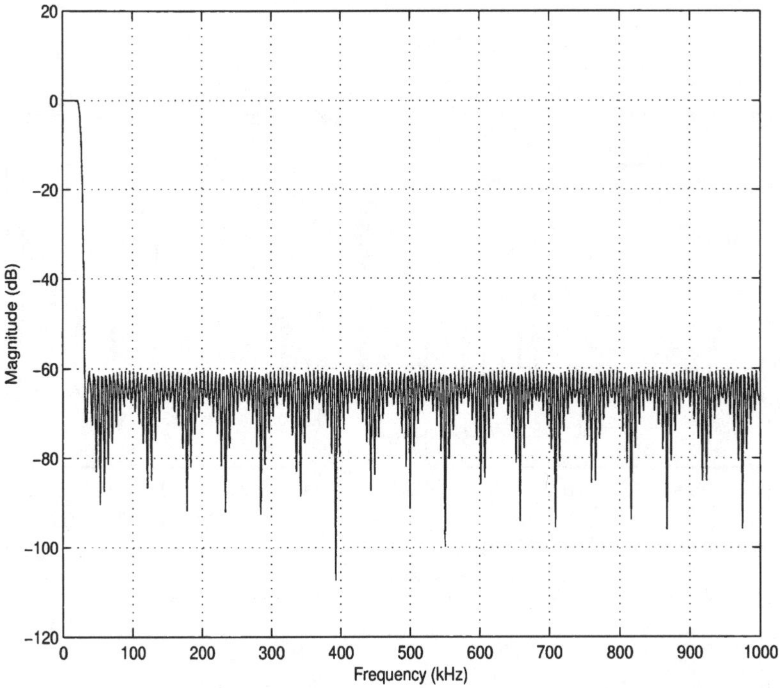

Figure 7.28 The desired frequency specification of decimation filter $H(z)$ using optimum equiripple technique. This filter has the order of 671 with symmetric coefficients.

edge at 0.03π. As we can see, this is a narrowband filter. A direct design of this decimation filter needs a filter order $N = 671$ using the optimum equiripple technique as shown in Figure 7.28.

The narrowband lowpass filter uses an efficient design procedure with multistage implementation as illustrated in Figure 7.27(b), (c), and (d). We split $M = 40$ into $M_1 = 20$ and $M_2 = 2$. The magnitude response of filter $G(z)$ has passband edge at $\omega_p = 0.4\pi$, stopband edge at $\omega_s = 0.6\pi$, $\delta_1 = 0.01$, and $\delta_2 = 0.001$. The design result of this filter using the optimum equiripple technique is shown in Figure 7.29(a). This filter $G(z)$ has a filter order $N_g = 34$. The corresponding filter $G(z^{20})$ with 11 passbands is shown in Figure 7.29(b). On the other hand, the frequency specification of filter $R(z)$ has a passband edge at $\omega_p = 0.02\pi$, a stopband edge at $\omega_s = 0.07\pi$, $\delta_1 = 0.01$, and

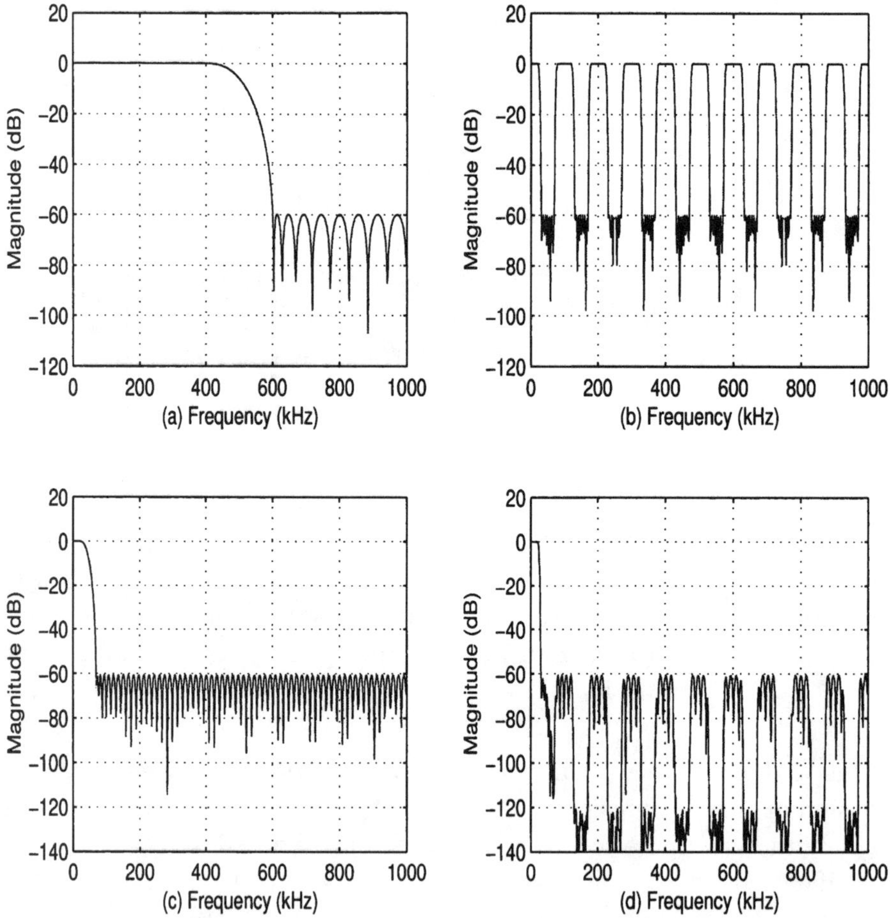

Figure 7.29 Efficient design of the narrowband decimation filter with multistage implementation. (a) The magnitude response of filter $G(z)$. (b) The 11 passbands of filter $G(z^{20})$. (c) The magnitude response of filter $R(z)$. (d) The two cascaded filters of $G(z^{20})$ and $R(z)$.

$\delta_2 = 0.001$. The designed frequency response of filter $R(z)$ using optimum equiripple technique is plotted in Figure 7.29(c). This filter $R(z)$ has a filter order $N_R = 126$. The result of the cascaded filters, $G^{(20)}$ and $R(z)$, is shown in Figure 7.29(d). Therefore, the filter $G(z^{20})R(z)$ does meet the desired frequency specification of the decimation filter $H(z)$.

Notice that the designed filters including $H(z)$, $G(z)$, $G(z^{20})$, $R(z)$, and $G(z^{20})R(z)$ have the property of linear phase. Thus, the filter coefficients are all symmetric. So, the multiplications can be reduced by 50% for all of the filters if the symmetric implementation is used. If the polyphase structure is used, the implementation of filter $H(z)$ would require 8.4 multiplications and 16.8 additions. For the efficient design with multistage implementation, however, the filter $G(z^2)R(z)$ needs only 2 multiplications and 4 additions. Therefore, in this example, we can eliminate roughly 76.2% of the multiplications and additions by using the efficient design along with a multistage implementation of the decimation system.

7.5.2 Multistage Interpolation System

We illustrated the efficient design of a narrowband decimation filter along with multistage implementation in Section 7.5.1. For the interpolation system, multistage implementation for the efficiently designed narrowband interpolation filter works just as well.

Figure 7.30(a) shows the lowpass filter $H(z)$ with a frequency cutoff at π/L in the interpolation system. If L is a large integer factor, then the interpolation filter $H(z)$ is a narrowband filter. To reduce the computational complexity, we use an efficient technique to design the narrowband filter by taking $L = L_1L_2$. This system is shown in Figure 7.30(b). Using the theory of noble identities, this interpolation system can be redrawn into a two-stage system as shown in Figure 7.30(c). Thus, the entire interpolation system has been designed into two stages: the first stage with upsampling L_1 and the second stage with upsampling L_2. For the implementation, this interpolation system can be rearranged into the polyphase structure in Figure 7.30(d). Notice that we may be able to further simplify the polyphase structure of Figure 7.30(d) by moving the upsampling by L_2 block after the filter $R(z)$ to achieve the maximum computational

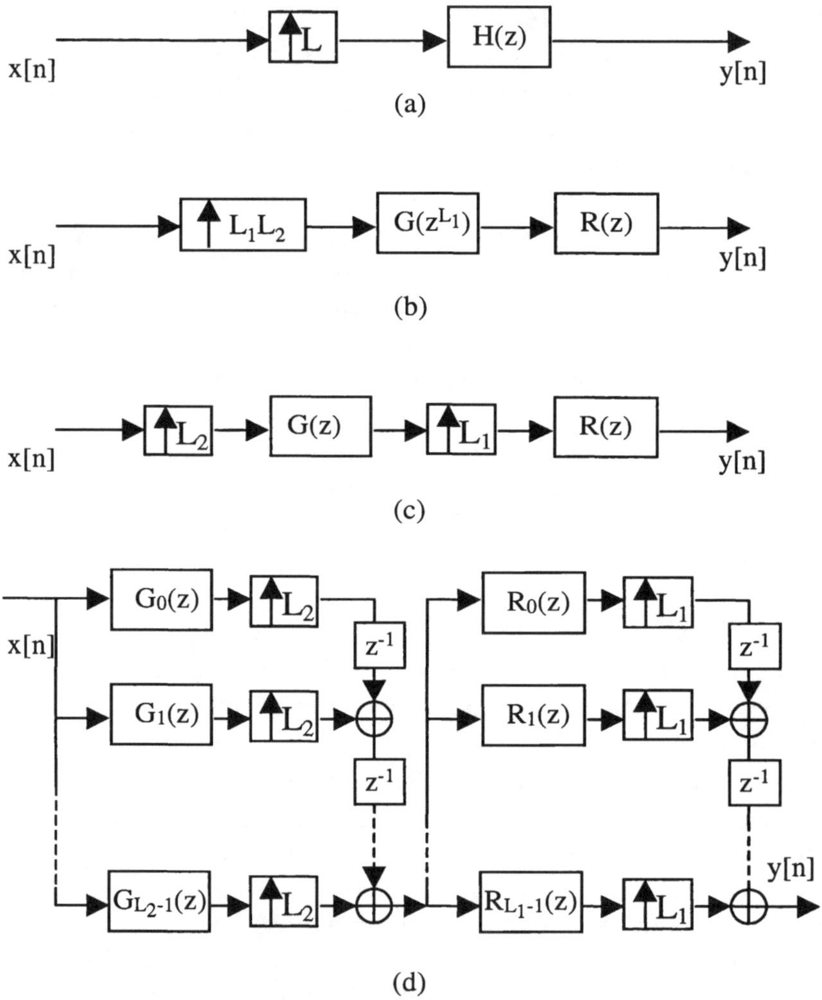

Figure 7.30 The multistage implementation of the interpolation system when L is a large integer factor. (a) The interpolation system including the filter $H(z)$ and upsampling by L. (b) The representation of interpolation system efficiency. (c) Interchange positions between filters $R(z)$, $G(z^{L_1})$, and upsampling blocks L_1 and L_2. (d) The polyphase structure of implementation in a two-stage interpolation system.

efficiency.

Notice that the efficient design technique along with multistage implementation can also work for the multirate signal processing system of fractional sampling rate conversion.

7.6 Multirate Filter Banks

In this section, we discuss multirate filter banks including the analysis filter and synthesis filter banks, with emphasis on the quadrature mirror filter (QMF) bank. We focus on the detailed theory and operation, as well as polyphase decomposition of a QMF bank.

7.6.1 QMF Banks

To illustrate the basic operation of a QMF bank, we first discuss the two-channel QMF as shown in Figure 7.31. The reconstructed signal $\hat{x}[n]$ differs from $x[n]$ because of aliasing, amplitude distortion, and phase distortion. However, it will be shown that the filters, $R_k(z)$ ($k = 0, 1$), can be designed in such a way that all the distortions can be eliminated. Then these results will be extended to the M-channel QMF bank as well.

The output of the analysis filter bank in the z-transform domain as shown in Figure 7.31 can be expressed as follows:

$$X_k(z) = H_k(z)X(z), \quad k = 0, 1. \tag{7.42}$$

Then, the output $C_k(z)$ of the decimated signal $c_k[n]$ in the z-transform domain follows:

$$C_k(z) = \frac{1}{2}\left[X_k(z^{1/2}) + X_k(z^{-1/2})\right]. \tag{7.43}$$

Note that the second term in (7.43) indicates the possibility of aliasing. In fact, actual aliasing depends not only on the frequency spectrum of the input signal $x[n]$ but also on the frequency response of the filters, $H_k(z)$, since $X_k(z) = H_k(z)X(z)$.

The z-transform domain $Y_k(z)$ of the upsampled output is

$$\begin{aligned} Y_k(z) &= C_k(z^2) \\ &= \frac{1}{2}\left[X_k(z) + X_k(-z)\right] \\ &= \frac{1}{2}\left[H_k(z)X(z) + H_k(-z)X(-z)\right]. \end{aligned} \tag{7.44}$$

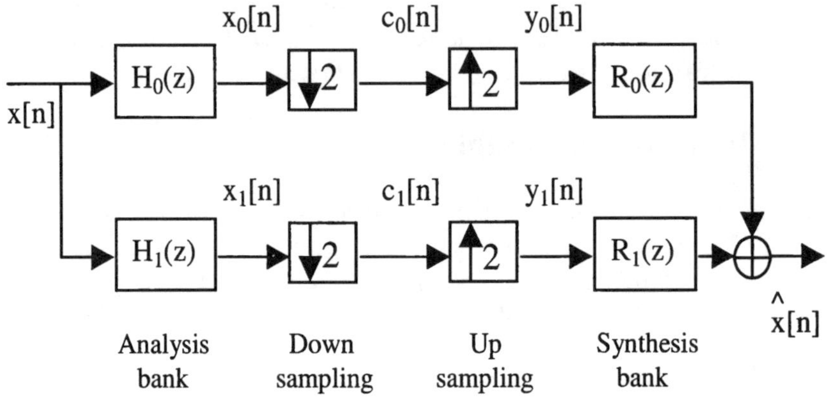

Figure 7.31 The two-channel quadrature mirror filter bank including analysis filter bank and synthesis filter bank.

Therefore, the reconstructed signal $\hat{x}[n]$ in the z-transform domain is given by

$$\hat{X}(z) = R_0(z)Y_0(z) + R_1(z)Y_1(z). \tag{7.45}$$

Substituting from (7.44) into (7.45), (7.45) can be rewritten as

$$\begin{aligned} \hat{X}(z) &= \frac{1}{2}\left[H_0(z)R_0(z) + H_1(z)R_1(z)\right]X(z) \\ &\quad + \frac{1}{2}\left[H_0(-z)R_0(z) + H_1(-z)R_1(z)\right]X(-z). \end{aligned} \tag{7.46}$$

The second term in (7.46) contains aliasing because of the downsampling, and imaging because of the upsampling. We call this term the aliasing component.

To cancel aliasing, we choose the synthesis filters such that the second term in (7.46) is zero. Thus, the synthesis filter bank $R_k(z)$ $(k = 0, 1)$ is as follows:

$$R_0(z) = H_1(-z), \ \ and \ \ R_1(z) = -H_0(-z). \tag{7.47}$$

Therefore, given the frequency specification of analysis filter bank $H_0(z)$ and $H_1(z)$, we can design the synthesis filter bank $R_k(z)$ to cancel aliasing completely using (7.47).

Now, assume that the aliasing term in (7.46) is entirely canceled. We then have the reconstructed signal,

$$\begin{aligned}\hat{X}(z) &= \frac{1}{2}\left[H_0(z)R_0(z) + H_1(z)R_1(z)\right]X(z) \\ &= \frac{1}{2}\left[H_0(z)H_1(-z) - H_1(z)H_0(-z)\right]X(z), \qquad (7.48)\end{aligned}$$

Let $A(z) = \frac{1}{2}\left[H_0(z)H_1(-z) - H_1(z)H_0(-z)\right]$, and $A(e^{j\omega}) = |A(e^{j\omega})|e^{j\phi(\omega)}$, we then express the reconstructed signal in the frequency domain as

$$\hat{X}(e^{j\omega}) = |A(e^{j\omega})|e^{j\phi(\omega)}X(e^{j\omega}). \qquad (7.49)$$

Equation (7.49) indicates that $|A(e^{j\omega})|$ is the amplitude distortion and $e^{j\phi(\omega)}$ is the phase distortion. As we can see, the reconstructed signal $\hat{X}(e^{j\omega})$ suffers from amplitude distortion unless $A(e^{j\omega})$ is allpass, such as $|A(e^{j\omega})| = c \neq 0$, for all ω. It has phase distortion unless $A(e^{j\omega})$ is a generalized linear phase, that is, $\phi(\omega) = a + b\omega$ for any constant a, b.

In order to eliminate aliasing, amplitude distortion, and phase distortion in the QMF bank, we must use the perfect reconstruction property (PRP). This is equivalent to having the condition $A(z) = cz^{-n_0}$, so that a PRP QMF bank has

$$\hat{X}(z) = cz^{-n_0}X(z), \qquad (7.50)$$

where n_0 is a positive integer. In the discrete-time domain, a PRP QMF bank is represented as follows:

$$\hat{x}[n] = cx[n - n_0]. \qquad (7.51)$$

In other words, the reconstructed signal $\hat{x}[n]$ of a PRP QMF bank is a scaled and delayed version of the input signal $x[n]$.

For the M-channel QMF bank as shown in Figure 7.32, the input signal $x[n]$ is passed through an analysis filter bank consisting of M bandpass filters, $H_k(z)$, decomposing the input signal $x[n]$

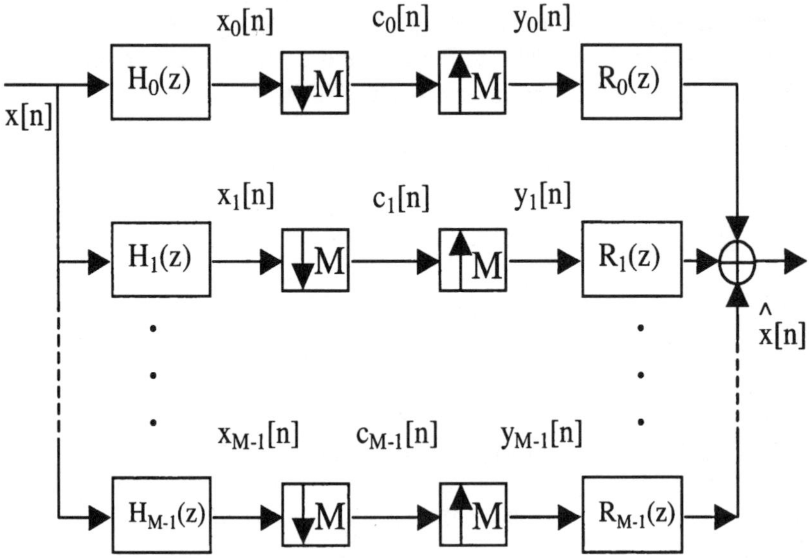

Figure 7.32 The M-channel quadrature mirror filter bank.

into M subband signals $x_k[n]$. The M subband signals $x_k[n]$ occupy contiguous nonoverlapping frequency ranges of bandwidth $2\pi/M$. Then, the subband signals $x_k[n]$ are downsampled by an integer factor M. Sequentially, these downsampled signals $c_k[n]$ are upsampled by M, and passed through a synthesis filter bank composed of M bandpass filters, $R_k(z)$, to remove the images. As a result, the output signal $\hat{x}[n]$ can be perfectly reconstructed as the input signal $x[n]$, with the same sampling rate.

The generalized relation between the input signal $x[n]$ and the output reconstructed signal $\hat{x}[n]$ in the z-transform domain is obtained by

$$\hat{X}(z) = \frac{1}{M} \sum_{k=0}^{M-1} R_k(z) \sum_{i=0}^{M-1} H_k(zW_M^i) X(zW_M^i), \tag{7.52}$$

where $W_M = e^{-j2\pi/M}$. To achieve PRP, the analysis filter bank and synthesis filter bank must be designed such that the output of the reconstructed signal $\hat{x}[n]$ is a delayed and scaled replica of the input signal $x[n]$ with complete cancellation of aliasing, and amplitude and phase distortion. In this case, (7.52) can be rewritten as

$$\hat{X}(z) = Kz^{-n_0}X(z), \tag{7.53}$$

where n_0 is a positive integer and K is a constant.

7.6.2 Polyphase Decomposition of QMF Banks

The polyphase decomposition has been found to be very useful both in the theory of multirate signal processing and in engineering practice. It is often beneficial, both conceptually and computationally, for us to represent the analysis filter and synthesis filter banks in an M-channel QMF bank by using polyphase decomposition.

From (7.20), we know that the analysis filter $H_l(z)$ in Figure 7.32 can be decomposed into the M-channel filter bank in terms of polyphase representation as follows:

$$H_l(z) = \sum_{k=0}^{M-1} z^{-k} D_{lk}(z^M), \quad l = 0, 1, ..., M-1. \tag{7.54}$$

Equation (7.54) can be rewritten in the matrix notation as

$$\mathbf{H}(z) = \mathbf{D}(z^M)\mathbf{Z}_D(z), \tag{7.55}$$

where

$$\mathbf{H}(z) = [H_0(z), H_1(z), ..., H_{M-1}(z)]^T, \tag{7.56}$$

$$\mathbf{Z}_D(z) = [1, z^{-1}, ..., z^{-(M-1)}]^T, \tag{7.57}$$

and

$$\mathbf{D}(z) = \begin{bmatrix} D_{00}(z) & D_{01}(z) & \cdots & D_{0,M-1}(z) \\ D_{10}(z) & D_{11}(z) & \cdots & D_{1,M-1}(z) \\ \vdots & \vdots & \ddots & \vdots \\ D_{M-1,0}(z) & D_{M-1,1}(z) & \cdots & D_{M-1,M-1}(z) \end{bmatrix} \tag{7.58}$$

are referred to as the analysis filter bank, the delay chain, and the polyphase matrix for the analysis filter bank. Figure 7.33 shows the polyphase representation of an analysis filter bank.

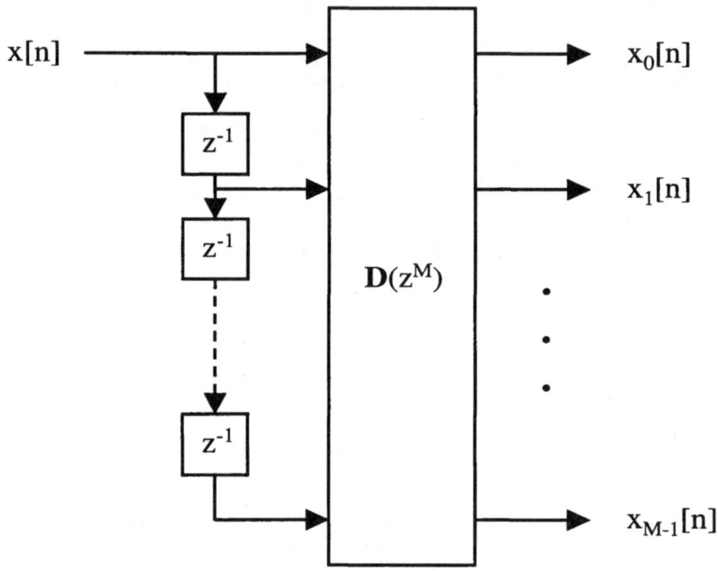

Figure 7.33 The polyphase representation of an analysis filter bank.

In a similar way, based on (7.33), we can express a set of synthesis filters $R_l(z)$ of the M-channel filter bank in terms of polyphase representation as follows:

$$R_l(z) = \sum_{k=0}^{M-1} z^{-(M-1-k)} I_{lk}(z^M), \quad l = 0, 1, ..., M-1. \tag{7.59}$$

Using matrix notation, (7.59) can be rewritten as

$$\mathbf{R}^T(z) = \mathbf{Z}_I^T(z)\mathbf{R}(z^M), \tag{7.60}$$

where

$$\mathbf{R}^T(z) = [R_0(z), R_1(z), ..., R_{M-1}(z)], \tag{7.61}$$

$$\mathbf{Z}_I(z) = [z^{-(M-1)}, z^{-(M-2)}, ..., z^{-1}, 1], \tag{7.62}$$

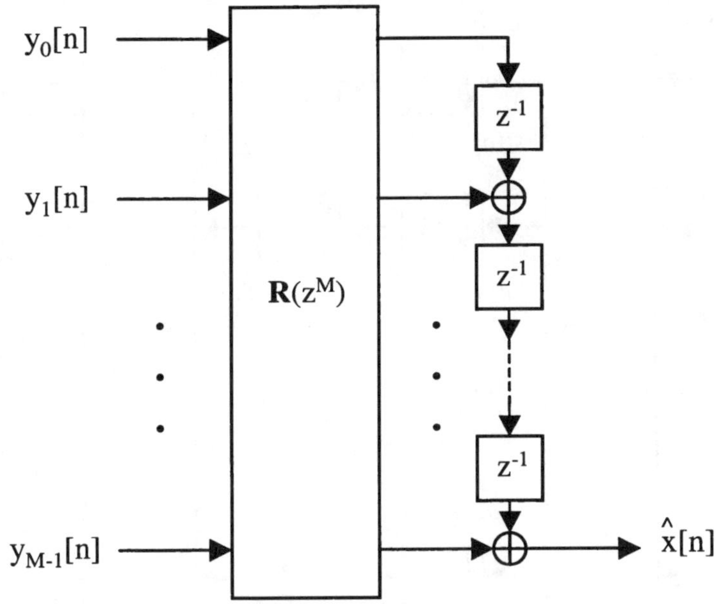

Figure 7.34 The polyphase representation of a synthesis filter bank.

and

$$\mathbf{R}(z) = \begin{bmatrix} R_{00}(z) & R_{01}(z) & \cdots & R_{0,M-1}(z) \\ R_{10}(z) & R_{11}(z) & \cdots & R_{1,M-1}(z) \\ \vdots & \vdots & \ddots & \vdots \\ R_{M-1,0}(z) & R_{M-1,1}(z) & \cdots & R_{M-1,M-1}(z) \end{bmatrix} \tag{7.63}$$

are referred to as the synthesis filter bank, the delay chain, and the polyphase matrix for the synthesis filter bank. Figure 7.34 shows the polyphase representation of a synthesis filter bank.

Now, the M-channel QMF bank in Figure 7.32 can be redrawn as shown in Figure 7.35 by using two polyphase representations of the analysis filter and synthesis filter banks in Figures 7.33 and 7.34, respectively. Furthermore, by using noble identities as discussed in Section 7.3.1, we can obtain the equivalent form of the M-channel

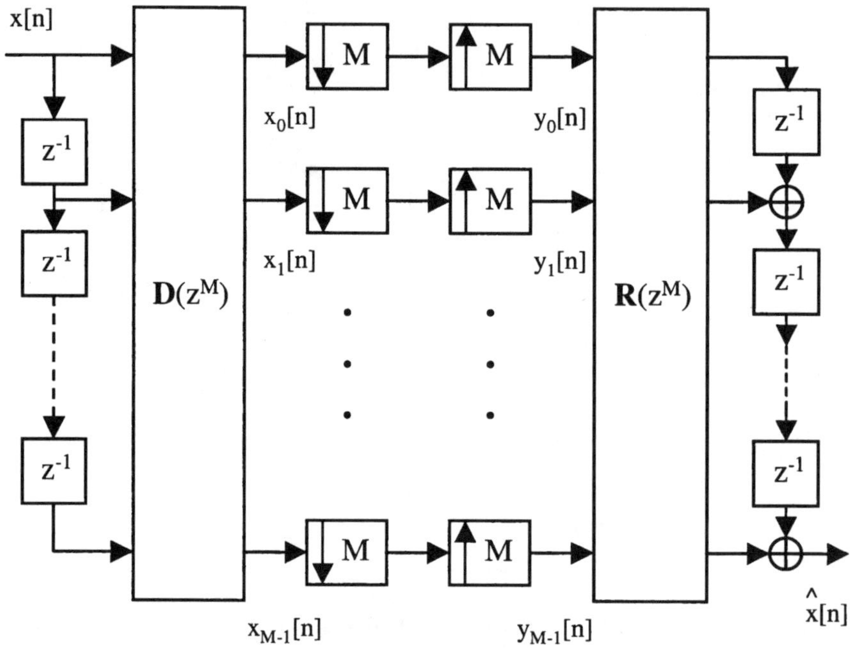

Figure 7.35 Representation of the M-channel QMF bank using two polyphase decompositions of the analysis filter and synthesis filter banks.

QMF bank as shown in Figure 7.36(a), which is referred to as the polyphase decomposition of the M-channel QMF bank. In addition, if two polyphase matrices, $\mathbf{D}(z)$ and $\mathbf{R}(z)$, are combined into the matrix $\mathbf{C}(z)$,

$$\mathbf{C}(z) = \mathbf{D}(z)\mathbf{R}(z), \tag{7.64}$$

then the polyphase decomposition of the M-channel QMF bank can be redrawn for simplification as shown in Figure 7.36(b).

Note, these equivalent representation forms of polyphase decomposition are extremely useful in the analysis and design of efficient M-channel QMF banks. As a result, the polyphase decomposition of the M-channel QMF bank enables the analysis and

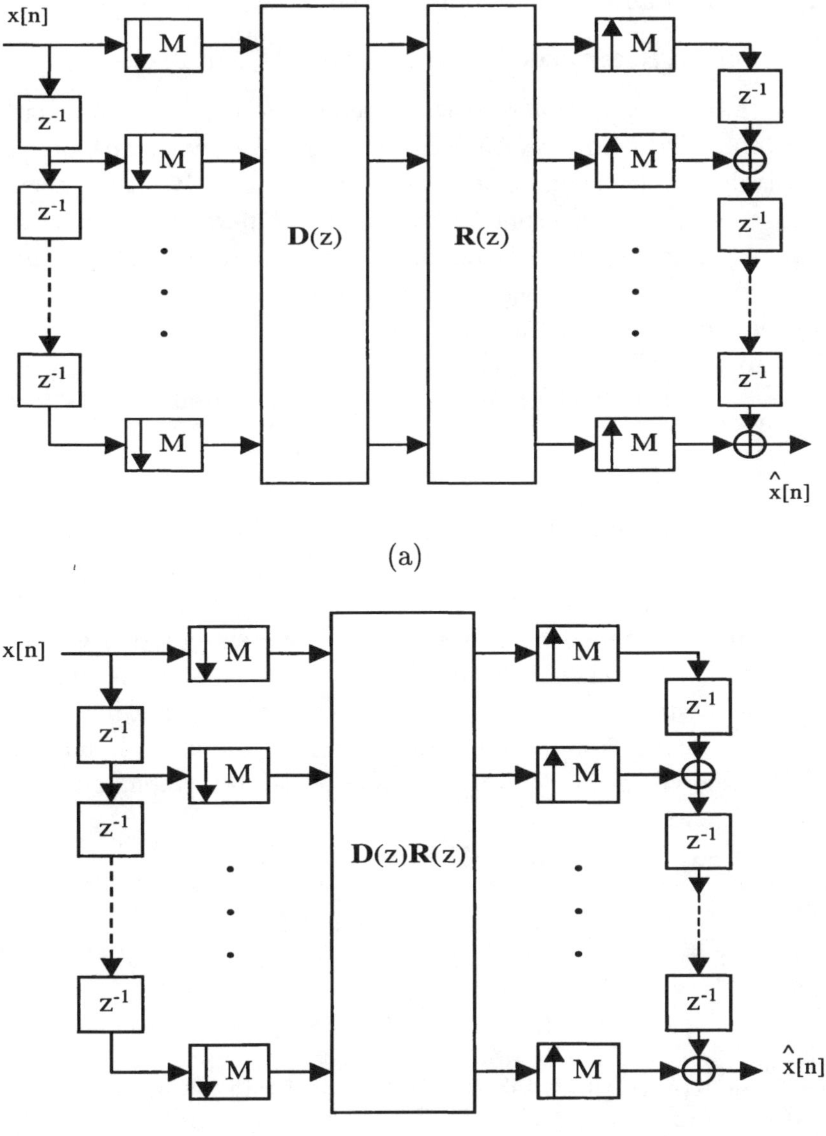

Figure 7.36 (a) Polyphase representation of an M-channel QMF bank using noble identities. (b) Simplification of polyphase representation of an M-channel QMF bank with $\mathbf{C}(z) = \mathbf{D}(z)\mathbf{R}(z)$.

synthesis filters to operate in such a way that the filters (polyphase matrices $\mathbf{D}(z)$ and $\mathbf{R}(z)$) operate at the lower sampling rate.

7.6.3 Perfect Reconstruction for M-channel QMF Banks

Recall that the condition of perfect reconstruction for an M-channel QMF bank is given by (7.53) in Section 7.6.1, with aliasing, and amplitude and phase distortions canceling completely. In this section, we discuss the more general conditions for perfect reconstruction of an M-channel QMF bank based on the results of polyphase decomposition as described in Section 7.6.2.

The simplification of the polyphase representation of an M-channel QMF bank with $\mathbf{C}(z) = \mathbf{D}(z)\mathbf{R}(z)$ is shown in Figure 7.36(b). It is clear that if two polyphase matrices, $\mathbf{D}(z)$ and $\mathbf{R}(z)$, satisfy

$$\mathbf{D}(z)\mathbf{R}(z) = \mathbf{I}, \tag{7.65}$$

then the output signal $\hat{x}[n]$ is equal to the input signal $x[n]$, and perfect reconstruction of an M-channel QMF bank is achieved. This condition indicates that perfect reconstruction of an M-channel QMF bank can be obtained by choosing the synthesis polyphase matrix, $\mathbf{R}(z)$, to be the inverse of the analysis polyphase matrix $\mathbf{D}^{-1}(z)$; this, of course, assumes that the polyphase matrix $\mathbf{D}(z)$ can be inverted. The condition in (7.65) is sufficient for perfect reconstruction for both the FIR and IIR system. Thus, the synthesis filter coefficients can be derived from (7.55).

It is also clear that if (7.65) is rewritten as

$$\mathbf{D}(z)\mathbf{R}(z) = Kz^{-n_0}\mathbf{I} \tag{7.66}$$

or

$$\mathbf{R}(z) = Kz^{-n_0}\mathbf{D}^{-1}(z), \tag{7.67}$$

where n_0 is a positive integer and K is a constant, then perfect reconstruction of an M-channel QMF bank can still be achieved. As can be seen, the result in (7.66) is consistent with the result obtained in (7.53).

The QMF bank has been successfully used in many DSP applications. This is because the theory of perfect reconstruction enables the QMF bank to be alias-free. The proof of the theory of perfect reconstruction was contributed by Smith and Barnwell [18, 19] and Mintzer [20]. For polyphase decomposition of the QMF bank, we cite the reference papers by Vaidyanathan [21-23], Vetterli [24], and Yim and Coakley [25].

7.7 The Uniform DFT Filter Bank

The discrete Fourier transform can be presented by using the matrix form of the DFT as discussed in Section 4.2.1. The $(M \times M)$ matrix $\mathbf{W}$ of the DFT has $(M \times M)$ elements $[\mathbf{W}]_{km} = W^{km}$ in (4.8), where $W = e^{\frac{-j2\pi}{M}}$. Note that $\mathbf{W}^T = \mathbf{W}$ since the column vectors $\mathbf{w}_i$ of matrix $\mathbf{W}$ of the DFT are orthonormal. Then, the quantity $\mathbf{W}^*$ is equal to $\mathbf{W}$, where the matrix $\mathbf{W}^*$ is a conjugate of the matrix $\mathbf{W}$.

From the definition of the matrix $\mathbf{W}$ and the relationship between input and output in (4.7), we may write

$$X_k[n] = \sum_{i=0}^{M-1} x[n-i]W^{-ki}, \quad k = 0, 1, 2, ..., M-1. \tag{7.68}$$

Equation (7.68) indicates that we need to compute a set of M signals $X_k[n]$ from a set of M signals $x[n-i]$ for every discrete-time index n as shown in Figure 7.37. In the z-transform domain representation, we may rewrite (7.68) as follows:

$$\begin{aligned} X_k(z) &= \sum_{i=0}^{M-1} z^{-i}W^{-ki}X(z) \\ &= \sum_{i=0}^{M-1} \left(zW^k\right)^{-i} X(z). \end{aligned} \tag{7.69}$$

Thus, we can represent (7.69) as

$$X_k(z) = H_k(z)X(z), \tag{7.70}$$

where $H_k(z)$ is expressed as

$$H_k(z) = H_0(zW^k), \tag{7.71}$$

and

$$H_0(z) = \sum_{i=0}^{M-1} z^{-i}. \tag{7.72}$$

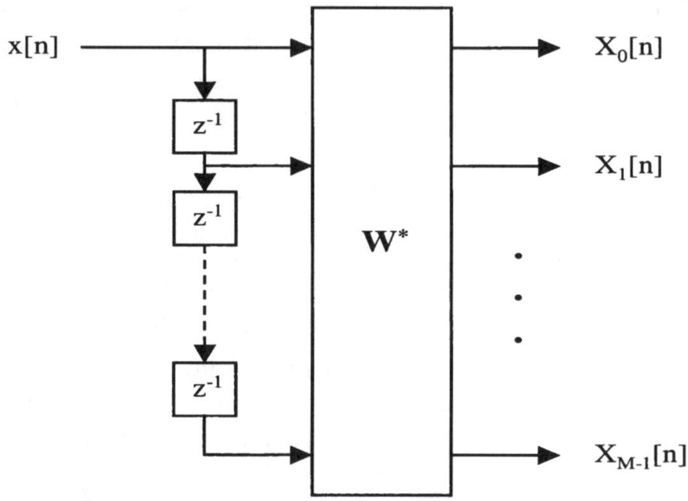

Figure 7.37 Representation of the uniform DFT filter bank.

Equation (7.72) is sometimes referred to as a *prototype filter*. As can be seen, (7.70) is equivalent to an analysis filter bank. Note that the filters have the filter length M, which is equal to the number of channels. Using the result of (2.124) in Example 2.8, the frequency response $H_0(e^{j\omega})$ in (7.72) is given by

$$H_0(e^{j\omega}) = e^{-j\omega(M-1)/2} \left[\frac{sin(M\omega/2)}{sin(\omega/2)} \right], \tag{7.73}$$

and the corresponding magnitude response $|H_0(e^{j\omega})|$ is obtained by

$$|H_0(e^{j\omega})| = \left| \frac{sin(M\omega/2)}{sin(\omega/2)} \right|. \tag{7.74}$$

Thus, the kth filter bank $H_k(z)$ has the frequency response

$$H_k(e^{j\omega}) = H_0(e^{j(\omega - \frac{2\pi k}{M})}). \tag{7.75}$$

Equation (7.75) is a uniformly shifted version of the prototype filter $H_0(e^{j\omega})$. Then, we have an analysis filter bank containing M filters.

Therefore, (7.71) or (7.75) is referred to as a *uniform DFT filter bank*.

Note that the analysis filter bank consisting of M filters is obtained from a single prototype filter $H_0(z)$ by uniformly shifting the frequency response based on the relation in (7.75). Those filters have wide transition bands and small stopband attenuation. In addition, the magnitude response of those filters has a large amount of overlap. This is because the prototype filter $H_0(z)$ in (7.72) itself is a very simple filter. However, those problems can be improved when we introduce the polyphase decomposition of the uniform DFT filter banks as discussed in the next section.

7.7.1 Polyphase Decomposition of the Uniform DFT Filter Bank

The polyphase decomposition can be used to implement a uniform DFT filter bank in a very efficient way. Given the prototype filter $H_0(z)$ in (7.72), the kth filter in (7.71) can be rewritten into the M-channel filter bank as

$$\begin{aligned} H_k(z) &= H_0(zW^k) \\ &= \sum_{i=0}^{M-1} (z^{-1}W^{-k})^i D_i(z^M), \end{aligned} \tag{7.76}$$

since $(zW^k)^M = z^M$. Then, substituting (7.76) into (7.70), we obtain

$$X_k(z) = \sum_{i=0}^{M-1} W^{-ki} \left[z^{-i} D_i(z^M) X(z) \right]. \tag{7.77}$$

Figure 7.38 shows an implementation structure of the uniform DFT filter bank by using polyphase decomposition in (7.77). Note that if $D_i(z^M) = 1$ for all i in (7.77), then (7.77) is equivalent to (7.69). This means that Figure 7.37 is a special case of Figure 7.38 when $D_i(z^M) = 1$ for all i. However, the presence of $D_i(z^M)$ in Figure 7.38 allows for increasing the length of the prototype filter $H_0(z)$. As a result, the analysis filter bank including M filters can have sharper cutoff and higher stopband attenuation in the uniform DFT filter bank.

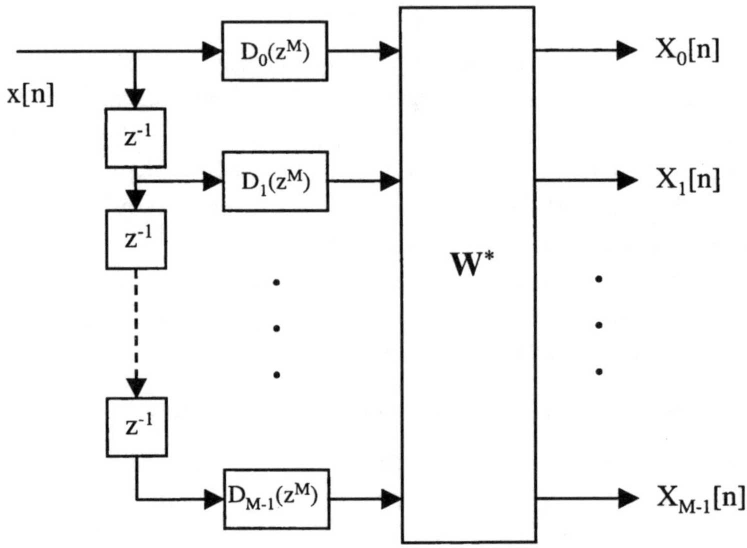

Figure 7.38 Polyphase decomposition of the uniform DFT filter bank.

7.7.2 Maximally Decimated Uniform DFT Filter Bank

In the M-channel QMF bank, the outputs of the analysis filter bank are decimated by a factor of M as shown in Figure 7.32. In a similar way, we are interested in decimating the outputs of the analysis filter bank in the uniform DFT filter bank by the factor of M. This can be done because each of those outputs has a bandwidth that is approximately M times narrower than that of the input signal $x[n]$. Thus, the polyphase decomposition of the uniform DFT filter bank with decimator functions is shown in Figure 7.39, by using noble identity. As a result, this polyphase structure of the uniform DFT filter bank is even more efficient for implementation.

7.7.3 Perfect Reconstruction of the Uniform DFT Filter Bank

In this section, we discuss the perfect reconstruction of the uniform DFT filter bank. Recall that the kth filters $H_k(z)$ in the analysis

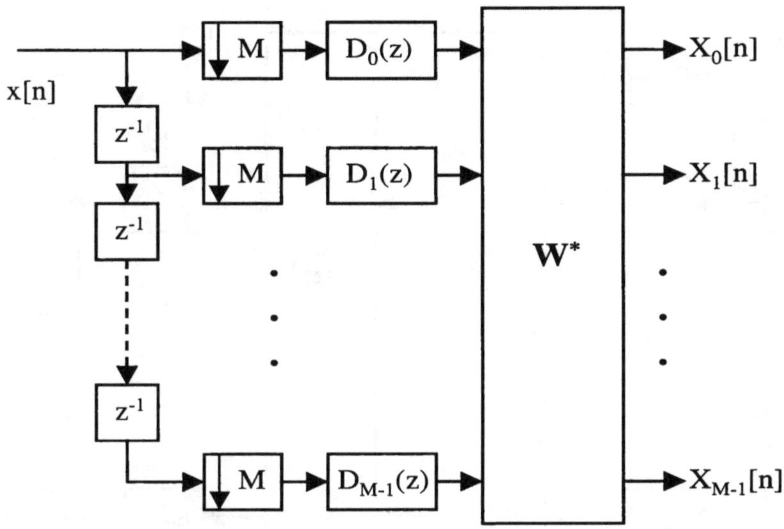

Figure 7.39 Polyphase decomposition of the uniform DFT filter bank with decimation by a factor of M.

filter bank are

$$H_k(z) = H_0(zW^k). \tag{7.78}$$

The kth filters $R_k(z)$ in the synthesis filter bank are given by

$$R_k(z) = W^{-k}R_0(zW^k), \tag{7.79}$$

where $R_0(z) = H_0(z)$. This was shown by Vaidyanathan [15]. Note that each filter in the synthesis filter bank has precisely the same magnitude response as the corresponding filter in the analysis filter bank.

Thus, it is clear that the polyphase matrix $\mathbf{D}(z)$ of the analysis filter bank is obtained by $\mathbf{D}(z) = \mathbf{W}^*$. For the perfect reconstruction of the uniform DFT filter bank, we obtain the polyphase matrix $\mathbf{R}(z)$ of the synthesis filter bank by taking $\mathbf{R}(z) = \mathbf{W}$. Under those conditions, the combined polyphase matrix $\mathbf{C}(z)$ is obtained by

$$\mathbf{C}(z) = \mathbf{D}(z)\mathbf{R}(z) = \mathbf{W}\mathbf{W}^* = M\mathbf{I}. \tag{7.80}$$

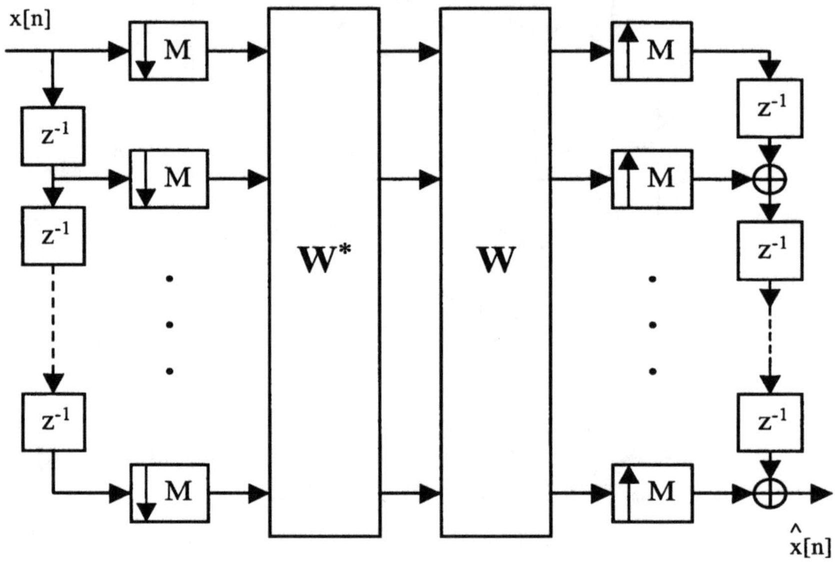

Figure 7.40 Perfect reconstruction of the analysis and synthesis banks in the uniform DFT filter bank.

As a result, the reconstructed signal $\hat{x}[n]$ satisfies the perfect reconstruction by

$$\hat{x}[n] = Mx[n - M + 1]. \tag{7.81}$$

Therefore, the aliasing problem due to overlap of the filters in the analysis filter bank can be completely canceled. Figure 7.40 shows the perfect reconstruction of the analysis and synthesis filter banks in the uniform DFT filter bank.

So far, we have presented the QMF bank and the uniform DFT bank with perfect reconstruction properties. The filter banks with perfect reconstruction concepts can also be extended to other transforms, such as discrete cosine transform (DCT), discrete sine transform (DST), and even discrete Hartley transform (DHT). In the technical literature, we cite the reference papers by Malvar [26], Ramstad [27], Koilpillai and Vaidyanathan [28], and Merched and Sayed [29]. For discrete-time wavelet-transform-based filter banks, we suggest the papers by Rioul and Vetterli [30], Vetterli and Herley [31], and Strang and Nguyen [32]. Recently, the filter banks based on principal components transform have been developed with applications in multiresolution analysis, image coding, and colored noise suppression by Tsatsanis and Giannakis [33], Xuan and Bamberger [34], and Akkarakaram and Vaidyanathan [35]. For the theory of principal components transform, we cite the reference paper by Miao [36].

7.8 Multirate Adaptive Filter Banks

The fundamental theory of conventionally adaptive filters was introduced in Chapter 6. Adaptive filtering is widely used for various applications such as adaptive system identification, adaptive prediction, adaptive equalization, adaptive echo cancellation, and adaptive noise cancellation. However, some applications that need adaptive filters with a large number of filter coefficients suffer from its heavy computational complexities and slow convergence (see Widrow and Stearns [37] and Morgan [38]). To overcome these problems, multirate adaptive filter banks (also known as subband adaptive filters) have been studied to reduce the computational complexity and to increase the convergence speed as presented in the reference papers by Gilloire and Vetterli [39], Kellermann [40], Somayazulu, Mitra, and Shynk [41], and Weiss et al. [42].

In this section, we introduce two approaches to multirate adaptive filter banks: (1) subband adaptive filters and (2) transform-based multirate adaptive filter banks using polyphase matrices.

7.8.1 Subband Approaches

The general structure of a multirate adaptive filter bank for system identification configuration is shown in Figure 7.41. The objective is to identify the unknown system, $U(z)$, by applying a multirate adaptive filter bank to the input signal $x[n]$ and the desired signal $d[n]$. In the multirate adaptive filter bank system, the input signal $x[n]$ and the desired signal $d[n]$ are first split into decimated frequency bands by analysis filter banks. Then the resulting subband signal $x_k[n]$ in the kth subband is supplied to an adaptive filter

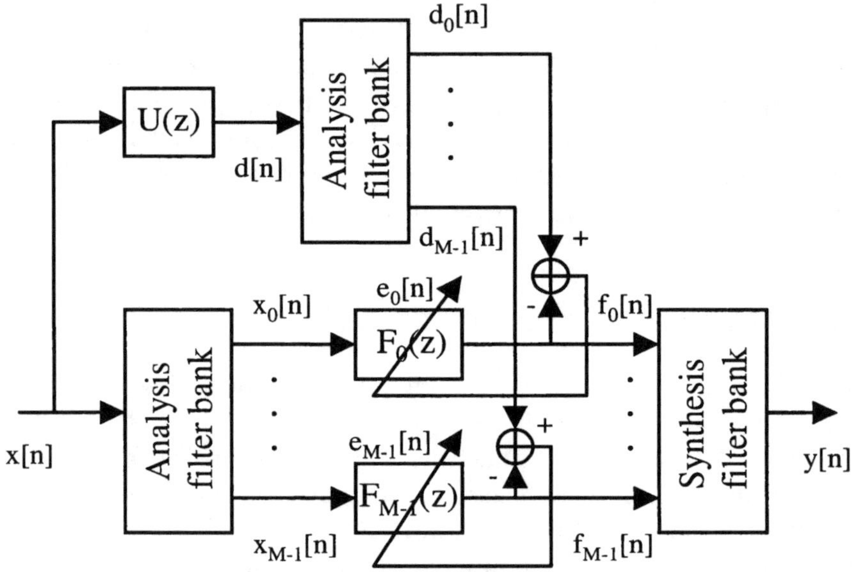

Figure 7.41 Generalized structure of multirate adaptive filter bank in system identification for an unknown system.

bank $F_k(z)$, which operates independently of those in adjacent bands. Thus, the subband error signals $e_k[n]$ are formed by using $d_k[n]$ while subtracting the output signal $f_k[n]$ of $F_k(z)$. Finally, the produced subband error signals $e_k[n]$ are used to update the filter coefficients by an adaptive filter algorithm in the multirate adaptive filter bank $F_k(z)$ for each iteration. Subsequently, the subband signals $f_k[n]$ can be recombined to form a full-band signal $y[n]$ by using a synthesis filter bank.

The benefits of the generalized structure of the multirate adaptive filter bank are as follows: (1) Since adaptive filter banks are adapted in each subband, the computational complexity per input sample is reduced by decimating a factor of M; (2) the multirate adaptive filter bank has a better convergence property because the condition number of correlation matrices of subband input signals is smaller than for the full-band case. However, the subband approaches have shown degradation in performance due to the introduction of undesirable aliasing components in the output of nonideal filter banks. This is because the filter banks are overlapping, and downsampling is used in the analysis filter banks. To solve these problems, one approach employing one or more "auxiliary" subbands, which are not decimated, in the multirate adaptive filter bank was reported by Somayazulu, Mitra, and Shynk [41]. Gilloire and Vetterli [39] presented another approach using cross filters to compensate the aliasing effects, but the convergence speed was slow. In addition, minimizing the error signal in the multirate adaptive filter bank by using statistically optimum approaches was proposed by Higa, Ochi, and Kinjo [43]. It has also been recognized that aliasing in the subband signals caused by the limited stopband attenuation in the analysis filter bank influences the minimum mean-square error (MMSE). Therefore, attempts to quantify an MMSE in the analysis filter bank due to aliasing have been studied by Petraglia and Mitra [44], Morgan and Thi [45], Ohno and Sakai [46], and Weiss et al. [47].

7.8.2 The Polyphase Matrix Approach

The multirate filter bank with adaptive filtering can result in a structure that unifies a direct form and a transform-based architecture. For a transform-based multirate adaptive filter bank, the polyphase matrices are particularly useful in the performance and implementation. Figure 7.42 illustrates a polyphase matrix approach to the transform-based multirate adaptive filter bank, while each of the subband filters $G_k(z)$ ($k = 0, 1, 2, ..., M - 1$) consists of a single filter coefficient or several filter coefficients, along with the LMS algorithm.

In this approach, the input signal $x[n]$ is first decimated by a factor of M to form subband signals, which are passed to the kth polyphase filters $D_k(z)$. The polyphase filters $D_k(z)$ are used to enhance the stopband attenuation of the prototype filter based on the transform used. Then the output signals of polyphase filter $D_k(z)$ are processed by a fixed orthogonal transform $\mathbf{W}^*$ of arbitrary length M. Next, the resulting transformed signals are filtered by subband adaptive filters $G_k(z)$. The subband adaptive filters are updated by

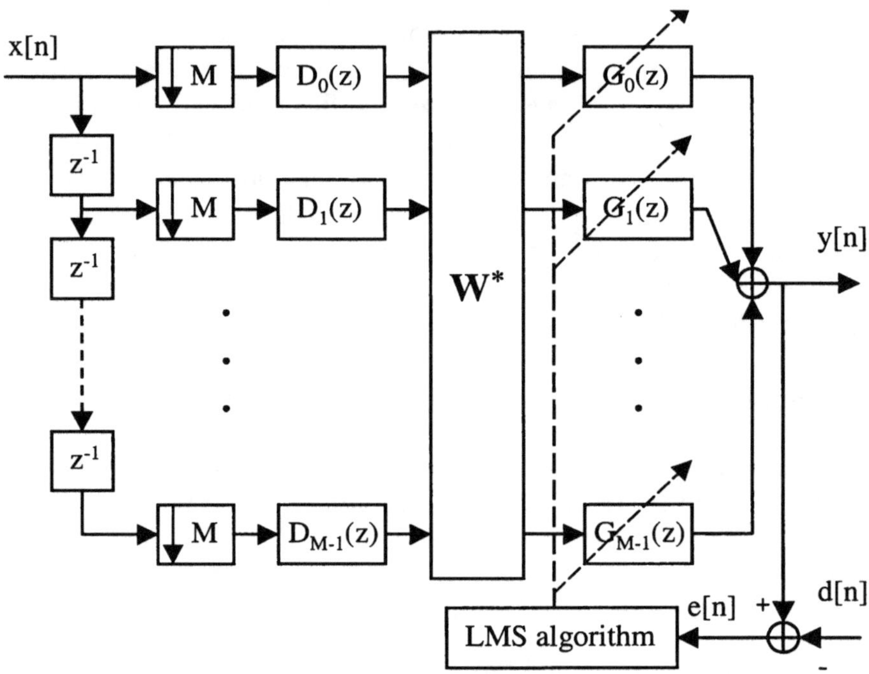

Figure 7.42 Transform-based multirate adaptive filter bank using the polyphase matrix approach.

using the LMS algorithm based on the error signal $e[n]$ for each iteration.

The choice of a transform $\mathbf{W}^*$ with good frequency selection decreases the correlation among the transformed signals. This can be used to obtain a significant improvement in the convergence speed of the LMS algorithm even for colored input signals. The discrete Fourier transform (DFT) or the discrete cosine transform (DCT) has been found to be useful in the multirate adaptive filter bank. Even the discrete Hartley transform (DHT) has been shown to be useful in the multirate adaptive filter bank in the reference paper by Merched and Sayed [29]. In addition, for the discrete Walsh transform-based multirate adaptive filter bank, we cite the reference paper by Wan and Ching [48].

Other adaptive methods, such as the RLS algorithm, can also be used to update the filter coefficients in the multirate adaptive filter bank. In the case of a subband decomposition with polyphase matrices using a good transform, the subband signals are approximately orthogonal. Therefore, each of the subband adaptive filters $G_k(z)$ can be independently adapted in parallel by using the RLS algorithm. This results in an improvement in performance and a reduction in the computational complexity and speeding up of the convergence.

7.9 Summary

In this chapter, we first reviewed the fundamental theory of discrete-time multirate signal processing, including decimation and interpolation as well as interconnection. For efficient implementation of these multirate systems, we next introduced the basic operations of multirate identity blocks, noble identities, and polyphase architecture for multirate systems. These basic methods for determining the implementation architecture of multirate systems are the key elements for ensuring correct and efficient operation in any practical application. To reduce the computational complexity further, the design of halfband filters, Mth-band filters, narrowband filters, and wideband filters was discussed. We then presented efficient filter design techniques as well as multistage implementations of multirate signal processing. These included decimation and interpolation systems. Finally, we introduced multirate filter banks, focusing on the two-channel and M-channel QMF banks. Perfect reconstruction of QMF banks with the alias-free condition was covered. We also discussed polyphase decomposition and representation for an M-channel QMF bank. This leads to a general form of perfect reconstruction for an M-channel QMF bank based on the polyphase representation. Furthermore, we extended concepts of the M-channel QMF bank to development of the uniform DFT filter bank, including polyphase decomposition, and maximally decimated and perfect-reconstruction representation. In addition, multirate adaptive filter banks were addressed, along with the subband adaptive filter and polyphase matrix approaches. The multirate adaptive filter bank has shown encouraging results in reducing computational complexity and increasing the speed of convergence.

As we know, in some cases the computational complexity in terms of multiplications and additions (MACs) is not always a fair measure of comparison for all the techniques in multirate signal processing systems. Therefore, we encourage the reader to always take into account the architecture of implementation on the target platform to achieve efficient and optimum solutions for any application.

References

[1] Schafer, R. W., and L. R. Rabiner, "A Digital Signal Processing Approach to Interpolation," *Proc. of IEEE*, Vol. 61, pp. 692-702, June 1973.

[2] Crochiere, R. E., and L. R. Rabiner, "Interpolation and Decimation of Digital Signals: a Tutorial Review," *Proc. of IEEE*, Vol. 69, pp. 300-331, March 1981.

[3] Crochiere, R. E., and L. R. Rabiner, *Multirate Digital Signal Processing*, Prentice Hall, Englewood Cliffs, New Jersey, 1983.

[4] Oppenheim, A. V., and R. W. Schafer, *Discrete-Time Signal Processing*, Prentice Hall, Englewood Cliffs, New Jersey, 1989.

[5] Bellanger, M., G. Bonnerot, and M. Coudreuse, "Digital Filter by Polyphase Network: Application to Sample Rate Alteration and Filter Banks," *IEEE Trans. on Acoustics, Speech and Signal Processing*, Vol. ASSP-24, pp. 109-114, April 1976.

[6] Vaidyanathan, P. P., "Multirate Digital Filters, Filter Banks, Polyphase Networks, and Applications: a Tutorial," *Proc. of IEEE*, Vol. 78, pp. 56-93, January 1990.

[7] Hsiao, C. C., "Polyphase Filter Matrix for Rational Sampling Rate Conversions," *Proceedings of IEEE International Conference on Acoustics, Speech and Signal Processing*, pp. 2173-2176, Dallas, Texas, April 1987.

[8] Bellanger, M., "On Computational Complexity in Digital Transmultiplexer Filters," *IEEE Transactions on communications*, Vol. COM-30, pp. 1461-1465, July 1982.

[9] Treichler, J. R., I. Fijalkow, and C. R. Johnson, "Fractionally Spaced Equalizers: How Long Should They Be?" *IEEE Signal Processing Magazine*, pp. 65-81, May 1996.

[10] Miao, G. J., and T. Acharya, "Dual Model Filter for Mobile Telecommunications," USA Patent Pending, December 1999.

[11] Neuvo, Y., C. Y. Dong, and S. K. Mitra, "Interpolated Finite Impluse Response Filters," *IEEE Trans. on Acoustics, Speech and Signal Processing*, Vol. ASSP-32, pp. 563-570, June 1984.

[12] Kovačevia, J., and M. Vetterli, "The Commutativity of Up/Downsampling in Two Dimensions," *IEEE Trans. on Information Theory*, Vol. IT-37, pp. 695-698, May 1991.

[13] Kalker, A. A. C. M., "Commutativity of Up/Down Sampling," *Electronics Letters*, Vol. 28, pp. 567-569, March 1992.

[14] Miao, G. J., J. Ho, and D. Brookshire, "Pre-Interpolation Filters for Upsampling," *MOTOROLA Journal of Technical Developments*, 1996.

[15] Vaidyanathan, P. P., *Multirate Systems and Filter Banks*, Prentice Hall, Englewood Cliffs, New Jersey, 1993.

[16] Mintzer, F., "On Half-Band, Third-Band and Nth Band FIR Filters and Their Design," *IEEE Trans. on Acoustics, Speech and Signal Processing*, Vol. ASSP-30, pp. 734-738, October 1982.

[17] Vaidyanathan, P. P., and T. Q. Nguyen, "A Trick for the Design of FIR Half-Band Filter," *IEEE Trans. on Circuits and Systems*, Vol. CAS-34, pp. 297-300, March 1987.

[18] Smith, M. J. T., and T. P. Barnwell, "A Procedure for Designing Exact Reconstruction Filter Banks for Tree Structed Subband Coders," *Proc. of IEEE Int. Conf. Acoustics, Speech and Signal Processing*, pp. 521-524, Tampa, Florida, March 1985.

[19] Smith, M. J. T., and T. P. Barnwell, "Exact Reconstruction Techniques for Tree-Structed Subband Coders," *IEEE Trans. on Acoustics, Speech and Signal Processing*, pp. 434-441, June 1986.

[20] Mintzer, F., "Filters for Distortion-Free Two-Band Multirate Filter Banks," *IEEE Trans. on Acoustics, Speech and Signal Processing*, Vol. ASSP-33, pp. 626-630, June 1985.

[21] Vaidyanathan, P. P., "Theory and Design of M-Channel Maximally Decimated Quadrature Mirror Filters with Arbitrary M, Having Perfect Reconstruction Property," *IEEE Trans. on Acoustics, Speech and Signal Processing*, Vol. ASSP-35, pp. 476-492, April 1987.

[22] Vaidyanathan, P. P., "Quadrature Mirror Filter Banks, M-band Extensions and Perfect-Reconstruction Techniques," *IEEE Signal Processing Magazine*, Vol. 4, pp. 4-20, July 1987.

[23] Vaidyanathan, P. P., and V. C. Liu, "Classical Sampling Theorems in the Context of Multirate and Polyphase Digital Filter Bank Structures," *IEEE Trans. on Acoustics, Speech and Signal Processing*, Vol. 37, No. 9, pp. 1480-1495, September 1988.

[24] Vetterli, M., "A Theory of Multirate Filter Banks," *IEEE Transactions on Acoustics, Speech and Signal Processing*, Vol. ASSP-35, pp. 356-372, March 1987.

[25] Yim, W., and Coakley, F, "Polyphase Matrix and Lattice Decomposition for Multirate Filters and Filter Banks," *Proc. IEEE Int. Conf. on Acoustics, Speech and Signal Processing*, Vol. 4, pp. 625-627, San Francisco, California, March 1992.

[26] Malvar, H. S., "Modulated QMF Filter Bank with Perfect Reconstruction," *Electronic Letters*, Vol. 26, pp. 906-907, June 1990.

[27] Ramstad, T. A., "Cosine Modulated Analysis-Synthesis Filter Bank with Critical Sampling and Perfect Reconstruction," *Proc. IEEE of Int. conf. on*

Acoustics, Speech and Signal Processing, pp. 1789-1792, Toronto, Canada, May 1991.

[28] Koilpillai, R. D., and P. P. Vaidyanathan, "Cosine-Modulated FIR Filter Banks Satisfying Perfect Reconstruction," *IEEE Trans. on Signal Processing*, Vol. SP-40, pp. 770-783, April 1992.

[29] Merched, R., and Sayed, A. H., "An Embedding Approach to Frequency-Domain and Subband Adaptive Filtering," *IEEE Trans. on Signal Processing*, Vol. 48, No. 9, pp. 2607-2619, September 2000.

[30] Rioul, O., and M. Vetterli, "Wavelets and Signal Processing," *IEEE Signal Processing Magazine*, pp. 14-38, October 1991.

[31] Vetterli, M., and C. Herley, "Wavelets and Filter Banks: Theory and Design," *IEEE Trans. on Signal Processing*, Vol. SP-40, No. 9, pp. 2207-2232, 1992.

[32] Strang, G., and T. Nguyen, *Wavelets and Filter Banks*, Wellesley-Cambridge Press, Wellesley, Massachusetts, 1996.

[33] Tsatsanis, M. K., and G. B. Giannakis, "Principal Component Filter Banks for Optimal Multiresolution Analysis," *IEEE Trans. on Signal Processing*, Vol. 43, pp. 1766-1777, August 1995.

[34] Xuan, B., and R. H. Bamberger, "2D Factorizable FIR Principal Component Filter Banks," *Proc. IEEE of Int. conf. Acoustics, Speech and Signal Processing*, Vol. 3, pp. 1538-1541, Atlanta, Georgia, May 1996.

[35] Akkarakaran, S., and P. P. Vaidyanathan, "Results on Principal Component Filter Banks: Colored Noise Suppression and Existence Issues," *IEEE Trans. on Information Theory*, Vol. 47, No. 3, pp. 1003-1020, March 2001.

[36] Miao, G. J., "Maximum Principal Components for Blind Frequency Estimation of Multiple Complex Exponential Signals in White Noise," *Proc. of Int. Conf. on Signal Processing Applications and Technology*, Vol. 2, pp. 1179-1183, Toronto, Canada, September 1998.

[37] Widrow, B., and S. D. Stearns, *Adaptive Signal Processing*, Prentice-Hall, Englewood Cliffs, New Jersey, 1985.

[38] Morgan, D. R., "Slow Asymptotic Convergence LMS Acoustic Echo Cancelers," *IEEE Trans. on Speech and Audio Processing*, Vol. 3, No. 2, pp. 126-136, March 1995.

[39] Gilloire, A., and M. Vetterli, "Adaptive Filtering in Subband with Critical Sampling: Analysis, Experiments and Application to Acoustic Echo Cancellation," *IEEE Trans. on Signal Processing*, Vol. 40, No. 8, pp. 1862-1875, August 1992.

[40] Kellermann, W., "Analysis and Design of Multirate System for Cancellation of Acoustical Echoes," *Proc. of IEEE Int. Conf. on Acoustics, Speech and Signal Processing*, Vol. 5, pp. 2570-2573, April 1988.

[41] Somayazulu, V. S., S. K. Mitra, and J. J. Shynk, "Adaptive Line Enhancement Using Multirate Techniques," *Proc. of IEEE Int. Conf. on Acoustics, Speech and Signal Processing*, pp. 928-931, May 1989.

[42] Weiss, S., S. R. Dooley, R. W. Stewart, and A. K. Nandi, "Adaptive

Equalization in Oversampled Subbands," *Electronic Letters*, Vol. 34, No. 15, pp. 1452-1453, July 1998.

[43] Higa, Y., H. Ochi, and S. Kinjo, "A Subband Adaptive Filter with the Statistically Optimum Analysis Filter Bank," *IEEE Trans. on Circuits and Systems II: Analog and Digital Signal Processing*, Vol. 45, No. 8, pp. 1150-1154, August 1998.

[44] Petraglia, M. R., and S. K. Mitra, "Performance Analysis of Adaptive Filter Structures Based on Subband Decomposition," *Proc. IEEE Int. Symp. on Circuits and Systems*, Vol. 1, pp. 60-63, May 1993.

[45] Morgan, D. R., and J. C. Thi, "A Delayless Subband Adaptive Filter Architecture," *IEEE Trans. on Signal Processing*, Vol. 43, pp. 1819-1830, August 1995.

[46] Ohno, S., and H. Sakai, "Spectral Analysis of Subband Adaptive Digital Filters," *IEEE Trans. on Signal Processing*, Vol. 48, No. 1, pp. 254-257, January 2000.

[47] Weiss, S., A. Stenger, R. W. Stewart, and R. Rabenstein, "Steady-State Performance Limitations of Subband Adaptive Filters," *IEEE Trans. on Signal Processing*, Vol. 49, No. 9, pp. 1982-1991, September 2001.

[48] Wan, K. F., and P. C. Ching, "Multilevel Split-Path Adaptive Filtering and Its Unification with Discrete Walsh Transform Adaptation," *IEEE Trans. on Circuits and System II: Analog and Digital Signal Processing*, Vol. 44, No. 2, pp. 147-151, February 1997.

8

Multirate Data Converters

8.1 Introduction

Data converters provide the link between the analog world of transducers and the digital world of signal processing, computing, and data processing systems. For a computer or a digital signal processor to process analog signals, these signals must first be converted from the analog domain to the digital domain. This type of data conversion is performed by an *analog-to-digital* (A/D) converter. The reverse operation, conversion of a digital signal into an analog form, is performed by a *digital-to-analog* (D/A) converter.

Systems including both analog and digital functions have been developed with higher levels of functionality and integration by incorporating mixed-signal integrated circuits. In most cases, such mixed-signal-based systems-on-chip (SOCs) perform inherently multirate signal processing because of the different sampling rates employed at various stages in the entire system. Applications of multirate signal processing in SOCs are also emerging, both for converting between analog signals and digital samples, and for implementing processing functions in an easier and more economical way than would be possible using purely digital or analog techniques. A bridge function between multirate analog signal processing and multirate digital signal processing involves the *multirate data*

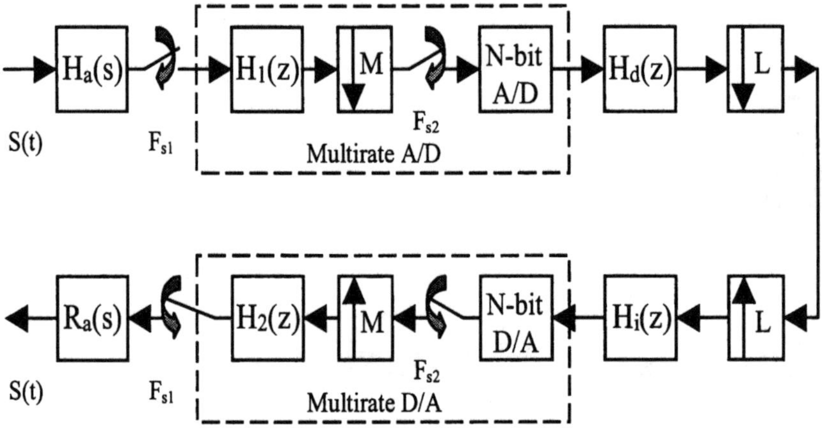

Figure 8.1 Generalized multirate architecture for mixed-signal processing of analog signals.

converter.

The multirate data converter can be embedded into generalized SOC multirate architectures for digital processing of analog signals as shown in Figure 8.1. In this system, $H_a(s)$ is a continuous-time analog filter; $H_1(z)$, with downsampling by M, is an analog discrete-time decimator; $H_d(z)$, with downsampling by L, represents a discrete-time decimator; $H_i(z)$, with upsampling by L, represents a discrete-time interpolator; and $H_2(z)$, with upsampling by M, is an analog discrete-time interpolator. We denote the subsystem of an analog discrete-time decimator and a Nyquist N-bit A/D converter as a *multirate A/D converter*, and the subsystem of an analog discrete time interpolator and a Nyquist N-bit D/A converter as a *multirate D/A converter*. Many other multirate architectures can be developed by using the building blocks as shown in Figure 8.1. Careful design trade-offs must be implemented in order to achieve high performance specifications for a given complexity.

In this chapter, we primarily focus on the system concepts for conventional data convertion, and multirate data convertion and their applications rather than design of the data converters for circuits. We begin in Section 8.2 by presenting the basic A/D converter, with an emphasis on sampling, quantization, and coding. The D/A converter operation is introduced in Section 8.3, with a focus on the amplitude distortion due to the zero-order-hold function and methods to reduce that distortion. In Section 8.4, we introduce the basic concept of the multirate A/D converter, including the multirate front-end subsystem used to provide the required filtering conditions for converting between continuous-time and discrete-time signals as well as the converters, both single-rate and multirate, that transform analog samples into digital signals. The multirate D/A converter that transforms the digital signals back into analog samples is given in Section 8.5. The oversampling A/D converter is presented in Section 8.6. In Section 8.7, another class of A/D converters and D/A converters, which are usually designated as oversampling sigma-delta converters, are described. Unlike Nyquist data converters, these converters relate input and output sequences with different frequencies and are, therefore, inherently multirate. Subsequently, the system architecture based on the quadrature mirror filter (QMF) banks based A/D converters and D/A converters is introduced in Section 8.8. Finally, a brief summary and conclusions are presented in Section 8.9.

8.2 Analog-to-Digital Converter

An A/D converter is a physical device that converts an input voltage or current into a binary code. The binary code represents a quantized amplitude value close to the amplitude of the input signal. Conceptually, the A/D converter can be viewed as a three-step process including sampling, quantization, and coding as shown in Figure 8.2.

8.2.1 Sampling

We briefly introduced the sampling of a continuous-time signal in Section 2.2. Sampling in Figure 8.2 can be considered as the conversion of a continuous-time signal into a discrete-time signal obtained by taking samples of the continuous-time signal at discrete-time instants. There are many ways to sample a continuous-time signal into the discrete-time signal. However, in this section we limit

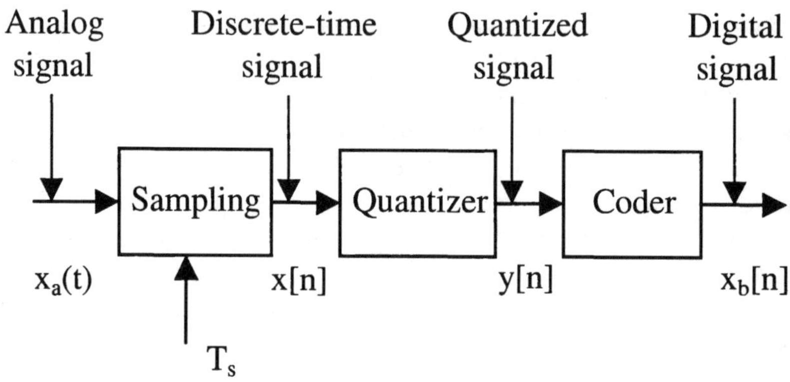

Figure 8.2 Basic representation of an analog-to-digital (A/D) converter.

our discussion to *uniform sampling*, which is the type of sampling most used in practical applications.

The relationship of converting a continuous-time signal into the discrete-time signal is described by

$$x[n] = x_a(nT_s), \quad -\infty < n < \infty, \tag{8.1}$$

where $x[n]$ is the discrete-time signal obtained by taking samples of the continuous-time signal $x_a(t)$, $t = nT_s$, every T_s seconds. T_s is called the *sampling period* or *sample interval*. Its reciprocal, $F_s = \frac{1}{T_s}$, is called the *sampling rate* or the *sampling frequency*.

Given any continuous-time signal, it is important to determine the sampling rate F_s so that the continuous-time signal can be exactly recovered from its discrete-time values $x[n]$. The *sampling theorem* provides the solution to this problem.

The sampling theorem states that if the highest frequency contained in the continuous-time signal $x_a(t)$ is F_{max} and a sampling rate $F_s \geq 2F_{max}$ is used, then the continuous-time $x_a(t)$ can be reconstructed exactly from the discrete-time signal $x[n]$ using the interpolation function

$$r(t) = \frac{sin(\pi F_s t)}{\pi F_s t}. \tag{8.2}$$

Thus, the continuous-time signal $x_a(t)$ can be expressed as

$$x_a(t) = \sum_{n=-\infty}^{\infty} x[n] r\left(t - \frac{n}{F_s}\right). \tag{8.3}$$

Note that if the minimum sampling rate $F_s = 2F_{max}$ is used, then (8.3) yields

$$x_a(t) = \sum_{n=-\infty}^{\infty} x[n] \left\{ \frac{sin\left[2\pi F_{max}\left(t - \frac{n}{2F_{max}}\right)\right]}{2\pi F_{max}\left(t - \frac{n}{2F_{max}}\right)} \right\}. \tag{8.4}$$

The sampling rate $F_s = 2F_{max}$ is known as the *Nyquist sampling rate.*

In practice, sampling of a continuous-time signal $x_a(t)$ is performed by using a sample-and-hold (S/H) circuit, since the sampling conversion is not instantaneous. The sampled signal $x[n]$ is then quantized and converted into a digital signal. The ideal S/H system has a system output described by

$$x_{S/H}(t) = \sum_{-\infty}^{\infty} x[n] h_0(t - nT_s), \tag{8.5}$$

where $x[n]$ is the sample of the continuous-time signal $x_a(t)$ at $t = nT_s$, and $h_0(t)$ is the impulse response of the zero-order-hold system defined by

$$h_0(t) = \begin{cases} 1, & 0 < t < T_s \\ 0, & \text{otherwise.} \end{cases} \tag{8.6}$$

As can be seen from (8.5) and (8.6), the output of the zero-order-hold $h_0(t)$ is a staircase waveform, where the sample values are held constant during each interval of T_s seconds as shown in Figure 8.3.

The goal of the S/H circuit, which is a digitally controlled analog circuit, is to continuously sample the input continuous-time signal

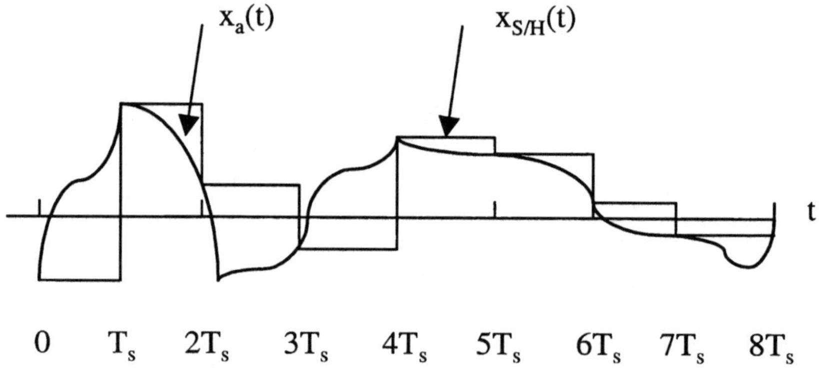

Figure 8.3 Time-domain response of an ideal sample-and-hold circiut.

$x_a(t)$ and then to hold that sample value constant for as long as it takes for the quantizer and coder to obtain its digital signal. This allows an A/D converter to operate more slowly as compared to the time it takes to actually acquire the sample. Such an ideal S/H circuit does not introduce distortion into the conversion process during the sample and hold modes.

In practice, however, an S/H circuit introduces time-related degradations such as errors or *jitter* due to sampling-time uncertainty and sampling-clock-time uncertainty. In addition, an S/H circuit produces nonlinear variations in the duration of the sampling aperture. Such problems of distortion in the S/H circuit have been considered in the books by Plassche [1] and Proakis and Manolakis [2].

In the following section, we assume that the S/H circuit is an ideal model and does not introduce errors. We then focus on the quantization noise along with coding.

8.2.2 Quantization and Coding

Quantization and sampling in time are at the heart of data conversion. Periodic sampling at frequency rates more than twice the signal frequency bandwidth does not introduce distortion. However, since the signal must be represented digitally for a computer or a digital signal processor to process it, roundoff, or quantization distortion, is introduced. The goal in designing quantization methods for data converters is to limit such distortion. In this section, we will discuss some properties of uniform quantization that are useful for determining the noise power for data converters.

Figure 8.4(a) shows a uniform quantizer, which rounds off an input signal $x[n]$ by a staircase function, a nonlinear relation between the quantized output $y[n]$ and the input $x[n]$. Such quantization can be represented as

$$y[n] = Q\{x[n]\}. \tag{8.7}$$

Also, if we treat the quantizer error $e[n]$ as an additive noise signal, then we can represent the quantizated signal $y[n]$ by a linear function $Gx[n]$ with an error $e[n]$ as follows:

$$y[n] = Gx[n] + e[n], \tag{8.8}$$

where gain G is the slope of the straight line that passes through the center of the uniform quantization levels as shown in Figure 8.4(a), (b), and (c). Note that, for $x[n]$ in some range, the error signal $e[n]$ is bounded by $\pm\Delta/2$ if gain G is set to unity.

For statistical analysis of quantization error, we assume that the error signal $e[n]$ has the following properties: (1) The error signal $e[n]$ is uncorrelated with all values of the input signal $x[n]$; (2) the error signal $e[n]$ has equal probability of lying anywhere in the uniform range of $\pm\Delta/2$; (3) the random process, $e[n]$, is uncorrelated with itself and can be considered white noise; and (4) the error signal $e[n]$ can be treated as a sample sequence of a stationary random process. Therefore, the mean value of error signal $e[n]$ is zero and the variance of the error signal $e[n]$ is

$$\begin{aligned}\sigma_e^2 &= \frac{1}{\Delta}\int_{-\Delta/2}^{+\Delta/2} e^2[n]de \\ &= \frac{\Delta^2}{12}.\end{aligned} \tag{8.9}$$

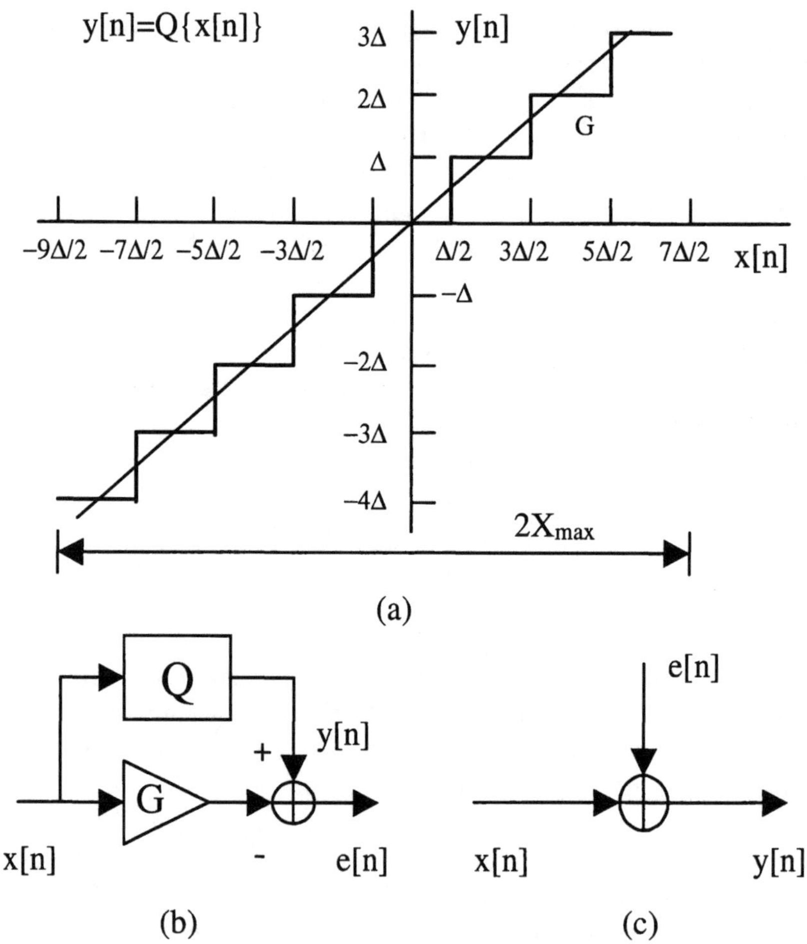

Figure 8.4 A generalized representation of uniform quantization. (a) Input and output characteristic of uniform quantization. (b) The relation of the quantization error $e[n]$ and input signal $x[n]$ is explicitly shown. (c) The relation of the quantization error $e[n]$ and input signal $x[n]$ is not presented.

Coding in an A/D converter assigns a unique binary number to each quantization level. Given Q levels of quantization, we need at least Q different binary numbers. If we have a word length of B bits, we can represent 2^B distinct binary numbers. Thus, we usually require $2^B \geq Q$.

There are many binary coding schemes. The two's-complement representation is most often used in digital signal processing. In general, in a two's-complement coding system, a B-bit binary fraction of the form $b_0 b_1 ... b_{B-1}$ has the value

$$-b_0 2^0 + b_1 2^1 + \cdots + b_{B-1} 2^{B-1}. \tag{8.10}$$

The leftmost bit in (8.10) is the sign bit or the most significant bit (MSB). The remaining bits represent either binary integers or fractions. The term b_{B-1} in (8.10) is also called the least significant bit (LSB). Thus, in the two's-complement coding system, the binary symbols have the following meaning, as shown in Table 8.1, when B is 4.

Therefore, with full-scale signal value X_{max} as shown in Figure 8.4(a), the step size $\triangle$ or the *resolution* of the quantization can be expressed as

$$\triangle = \frac{2X_{max}}{2^B}. \tag{8.11}$$

Substituting (8.11) into (8.9), the variance of the error signal $e[n]$ is therefore

$$\sigma_e^2 = \frac{2^{-2B} X_{max}^2}{3}. \tag{8.12}$$

The *signal-to-noise ratio* (SNR), which is a common measure of the fidelity of data converters, is defined as the ratio of signal power (variance) σ_x^2 to noise power (variance) σ_e^2 as follows:

$$\begin{aligned} SNR(\text{dB}) &= 10log_{10}\left(\frac{\sigma_x^2}{\sigma_e^2}\right) \\ &= 10log_{10}\left(\frac{3 \cdot 2^{2B}\sigma_x^2}{X_{max}^2}\right) \\ &= 6.0206B + 4.7712 + 20log_{10}\left(\frac{\sigma_x}{X_{max}}\right). \end{aligned} \tag{8.13}$$

Table 8.1 The Relation of Numeric Value and Binary Symbol

Number	Numeric Value	Two's-Complement
+7	+7/8	0111
+6	+6/8	0110
+5	+5/8	0101
+4	+4/8	0100
+3	+3/8	0011
+2	+2/8	0010
+1	+1/8	0001
0	0	0000
-1	-1/8	1111
-2	-2/8	1110
-3	-3/8	1101
-4	-4/8	1100
-5	-5/8	1011
-6	-6/8	1010
-7	-7/8	1001
-8	-8/8	1000

Equation (8.13) indicates that SNR (dB) increases 6 dB for every bit added to the A/D converter. For example, given an SNR requirement of 70 dB, we need approximately 12-bit uniform quantization in the A/D converter.

Note that the third term in (8.13) suggests that the root mean square (RMS) value σ_x of the signal amplitude should be less than the peak amplitude of the signal. In practice, if the RMS value σ_x is too large, the peak signal amplitude will exceed the full-scale amplitude of the A/D converter. In this case, (8.13) has a severe distortion known as *clipping* or *hard limiting*. If the value σ_x is too small, (8.13) will be negative. In this case, the SNR decreases since most of the available range is unused. Therefore, the signal amplitude should be carefully matched to the full-scale amplitude X_{max} of the A/D converter.

Detailed analyses and mathematical treatments of quantization noise can be found in the references contributed by Oppenheim and Schafer [3], Gray, Chou, and Wong [4], Gray [5], Blumberg [6], Widrow, Kollal, and Liu [7], as well as Gray and Neuhoff [8].

8.3 Digital-to-Analog Converter

In order to convert a digital signal into the continuous-time signal, we need a digital-to-analog (D/A) converter in most practical applications. Recall that a bandlimited lowpass analog signal $x_a(t)$, which has been sampled at $F_s \geq 2F_{max}$, can be perfectly recovered from its discrete-time signal $x[n]$ based on the sampling theorem as presented in Section 8.2.1. The ideal reconstruction formula derived from (8.4) is as follows:

$$x_a(t) = \sum_{n=-\infty}^{\infty} x[n] \left\{ \frac{sin[\pi(t - nT_s)/T_s]}{\pi(t - nT_s)/T_s} \right\}, \tag{8.14}$$

where the sampling interval $T_s = 1/F_s$, and F_s is the sampling rate. Equation (8.14) indicates that the system takes the discrete-time signal $x[n]$ as input and produces the analog signal $x_a(t)$ as output. Such a system is referred to as an *ideal D/A converter.* However, the direct realization of an ideal D/A converter based on (8.14) is impractical for most applications since (8.14) is both noncausal and physically nonrealizable.

8.3.1 Operation of a D/A Converter

A block diagram of a D/A converter followed by a sample-and-hold (S/H) amplifier and an approximate analog-reconstructed lowpass filter is shown in Figure 8.5. First, digital signals are applied to the D/A converter as parallel signals. Then the D/A converter converts the digital signal into an analog signal. The relationship between the digital signal and the analog signal can be expressed by the following equation:

$$V_a = \sum_{n=0}^{B-1} b_n 2^n R_{ref}, \tag{8.15}$$

where R_{ref} is a reference value such as a reference voltage or current, 2^n indicates the binary weighting of the bit values as a function of the variable n, and V_a represents the output of the analog signal.

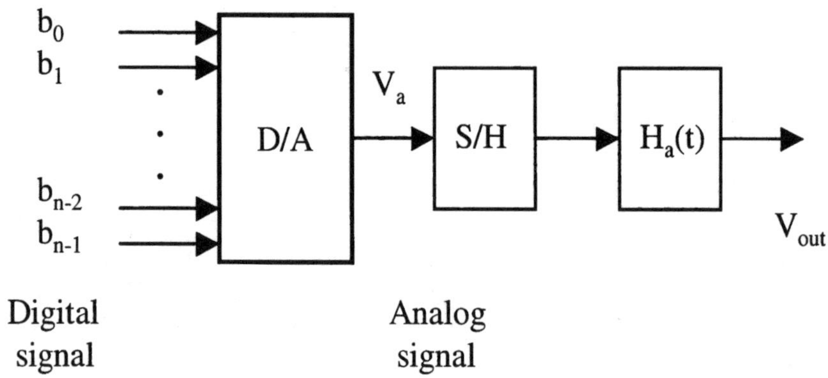

Figure 8.5 Block diagram of a digital-to-analog converter followed by an S/H amplifier and an approximate analog-reconstructed lowpass filter.

During the operation in Figure 8.5, the D/A converter acts as a sample-and-hold function where the digital signal is converted into an analog signal. This sample-and-hold in the D/A converter is referred to as the *zero-order-hold*; it introduces the amplitude distortion for a D/A converter. The frequency response of the zero-order-hold function for a D/A converter is

$$\begin{aligned} H_0(j\Omega) &= \frac{1-e^{-j\Omega t_h}}{j\Omega} \\ &= t_h \left[\frac{sin(\Omega t_h/2)}{\Omega t_h/2}\right] e^{-j\Omega t_h/2}, \end{aligned} \tag{8.16}$$

where t_h is the time duration of the zero-order-hold circuit in the hold mode. Thus, the amplitude characteristic of the zero-order-hold system function is obtained by

$$|H_0(j\Omega)| = t_h \left[\frac{sin(\Omega t_h/2)}{\Omega t_h/2}\right]. \tag{8.17}$$

Equation (8.17) is a well-known $\frac{sin(x)}{x}$ distortion function for amplitude reduction, which is introduced by the zero-order-hold during the D/A operation. The $\frac{sin(x)}{x}$ function is sometimes referred to as the *sinc* function.

Note that if the hold time t_h is equal to the sampling interval T_s, such as $t_h = T_s = 1/F_s$, then in every clock cycle, input data changes are held constant over that clock cycle. Substituting $\Omega = 2\pi F_{max}$ into (8.17), (8.17) can be rewritten as

$$|H_0(j2\pi F_{max})| = \frac{1}{F_s}\left[\frac{sin(\pi F_{max}/F_s)}{\pi F_{max}/F_s}\right]. \tag{8.18}$$

Furthermore, if the Nyquist sampling rate is used, $F_s = 2F_{max}$, in (8.18), then we obtain

$$\begin{aligned} |H_0(j2\pi F_{max})| &= \frac{1}{F_s}\left[\frac{sin(\pi/2)}{\pi/2}\right] \\ &= \frac{2}{\pi}\left(\frac{1}{F_s}\right). \end{aligned} \tag{8.19}$$

Note that (8.19) yields an amplitude reduction of $\frac{2}{\pi}$. This is equivalent to -3.9224 dB.

Figure 8.6 shows the amplitude response of the distortion function $\frac{sin(x)}{x}$ operation when using the Nyquist sampling rate. On the other hand, such a function has a great benefit in that it acts as an approximate lowpass filter to eliminate the spurious images. Note that the narrowband signals are repeated with regard to the sampling rate F_s. However, the operation of the $\frac{sin(x)}{x}$ function can produce sufficient suppression of the images at $f = nF_s$, for $n \neq 0$. In addition, the amplitude response $|H_0(j2\pi f)|$ of the images decreases as the frequency f increases, achieving a simple lowpass filter.

The sample-and-hold (S/H) amplifier as shown in Figure 8.5 is used to hold the output of the D/A converter constant at the previous output value until the new sample at the output of the D/A reaches steady state. The S/H amplifier samples and holds the new value in the next sampling interval. Therefore, the S/H amplifier can approximately represent the analog signal as a series of rectangular pulses whose heights are equal to the corresponding values of the signal. The S/H amplifier transmits very short output pulses to the reconstruction analog lowpass filter $H_a(t)$. As a result, the S/H amplifier does not introduce significant amplitude distortion at output frequencies close to half the sampling rate.

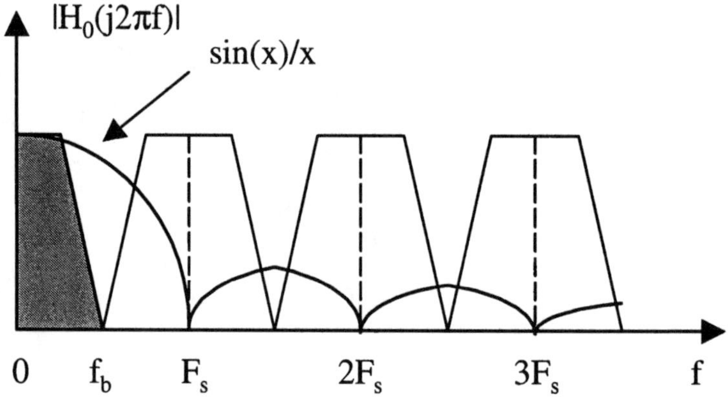

Figure 8.6 The amplitude response of a zero-order-hold function when using the Nyquist sampling rate.

Recall that the frequency response of the ideal reconstruction filter from (8.14) is

$$H_r(j\Omega) = \begin{cases} T, & |\Omega| < \pi/T_s \\ 0, & |\Omega| > \pi/T_s. \end{cases} \tag{8.20}$$

The compensated reconstruction analog lowpass filter in the frequency domain is obtained by

$$H_a(j\Omega) = \frac{H_r(j\Omega)}{H_0(j\Omega)H_{S/H}(j\Omega)}, \tag{8.21}$$

where $H_{S/H}(j\Omega)$ represents the transfer function of the S/H amplifier in the frequency domain. For a mathematical simplification, we assume that the S/H amplifier in Figure 8.5 is an ideal system. Thus, the compensated reconstruction analog lowpass filter in (8.21) becomes

$$H_a(j\Omega) = \frac{H_r(j\Omega)}{H_0(j\Omega)}. \tag{8.22}$$

Substituting (8.16) and (8.20), with $t_h = T_s$, into (8.22), then the compensated reconstruction analog lowpass filter in (8.22) can be rewritten by

$$H_a(j\Omega) = \begin{cases} \left[\frac{\Omega T_s/2}{sin(\Omega T_s/2)}\right] e^{j\Omega T_s/2}, & |\Omega| < \pi/T_s \\ 0, & |\Omega| > \pi/T_s. \end{cases} \tag{8.23}$$

Equation (8.23) indicates that we need to design a compensated reconstruction analog lowpass filter in such a way that the amplitude reduction of the distortion function $\frac{sin(x)}{x}$ is compensated by an increase in the amplitude characteristic of this filter at the high-frequency band edge. In addition, the phase response in (8.23) ideally corresponds to a time advance of $T_s/2$ to compensate for the delay introduced by the zero-order-hold in the D/A converter. However, it is difficult to realize such a time advance in practical applications. Moreover, the compensation of the magnitude response in (8.23) is impossible because the gain of the zero-order-hold, $H_0(j\Omega)$, drops to -3.9224 dB at $f = F_s$. Therefore, in the following section, we discuss several other methods to reduce the amplitude distortion of the zero-order-hold in the D/A converter.

8.3.2 Methods of Reducing Amplitude Distortion

The first method of reducing amplitude distortion for a D/A converter is to consider reducing the hold time t_h in (8.17). If the hold time is $t_h = \beta T_s = \beta \frac{1}{F_s}$, with $\Omega = 2\pi F_{max}$, then (8.17) yields as follows,

$$|H_0(j2\pi F_{max})| = \frac{\beta}{F_s}\left[\frac{sin(\beta\pi F_{max}/F_s)}{\beta\pi F_{max}/F_s}\right], \tag{8.24}$$

where $0 \leq \beta \leq 1$. As can be seen, the scale factor β is used to control the amplitude distortion of the $\frac{sin(x)}{x}$ function. For example, if $\beta = 0.25$ and $F_s = 2F_{max}$, (8.24) becomes

$$|H_0(j2\pi F_{max})| = \frac{\beta}{F_s}\left(\frac{3.0615}{\pi}\right). \tag{8.25}$$

This indicates that the amplitude distortion is reduced to $\frac{3.0615}{\pi}$ or -0.2244 dB when the hold time t_h is equal to $\frac{T_s}{4}$.

With a uniform sampling rate F_s, the reduction of hold time t_h can be achieved by using a switch at the output of the D/A converter.

The output signal of the D/A converter is used to pass through the compensated reconstruction analog lowpass filter during the hold time t_h. On the other hand, a zero output value is applied to the compensated construction analog lowpass filter during the time of $T_s - t_h$ [1].

The second method of reducing amplitude distortion for a D/A converter is to increase the sampling rate F_s. Note that the ratio of $\frac{F_{max}}{F_s}$ decreases when the sampling rate F_s increases. As a result, the amplitude distortion of a D/A converter can be reduced by using (8.18). For example, if we use $4F_s$ or $8F_s$ for the sampling rate in (8.18), then the amplitude distortion of a D/A converter is reduced to -0.2244 dB or -0.0559 dB, respectively, which can be compensated by using a digital interpolation filter preceding the D/A converter or the analog reconstruction filter. A D/A converter with oversampling for reducing the amplitude distortion is sometimes referred to as an *oversampling D/A converter.*

8.4 Multirate A/D Converter

The multirate A/D converter is developed based on the emergence of signal processing technologies, such as charge transport devices and switched capacitors as reported by Gersho [9], Gregorian, Martin, and Temes [10], and Dooley [11]. These can also be operated in the discrete-time domain. Although the signal amplitudes are still represented in analog form, the technologies leading to the development of analog multirate signal processing have found very important application areas in the realization of high-speed interfaces between analog and digital forms. The rationale for developing a multirate A/D converter is to achieve high-speed conversion using low-speed and low-cost devices. It also seems to be to reduce the requirement of high-order analog antialiasing filters upfront.

Figure 8.7 shows the generalized architecture of a multirate A/D converter, including a continuous-time antialiasing lowpass filter, discrete-time sampled-data decimation filter, and a Nyquist A/D converter. Sampled continuous-time signals represent the simplest case of analog multirate signal processing for transforming a continuous-time signal into a sequence of analog samples at a frequency F_s. The continuous-time antialiasing filter $H_a(s)$ band-

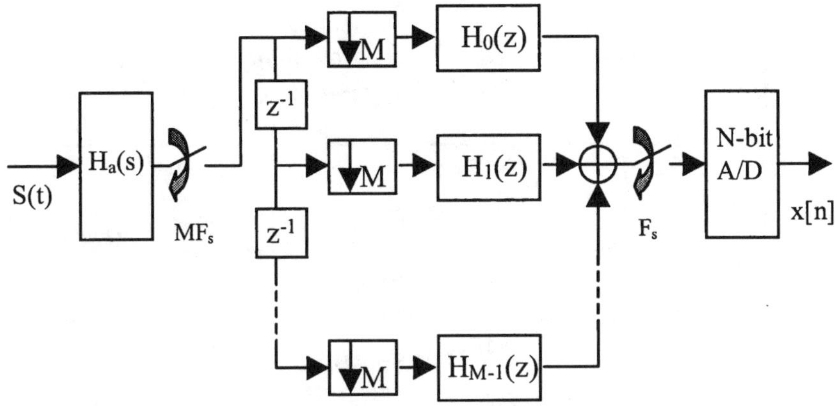

Figure 8.7 The generalized architecture of the multirate A/D converter contains three stages: (1) $H_a(s)$, an anti-aliasing analog filter; (2) $H_i(z)$ $(i = 0, 1, ..., M - 1)$, an analog discrete-time sampled-data decimator with polyphase structure; and (3) an N-bit Nyquist A/D converter.

limits the input analog signal in such a way that small aliasing effects can initially be tolerated. Usually, the output of a continuous-time antialiasing filter is initially oversampled at a rate factor M greater than the desired Nyquist sampling rate F_s so that a low-order continuous-time antialiasing filter $H_a(s)$ can be used. Then, the resulting sequence of samples is processed by a discrete-time sampled-data decimation filter $H(z)$ with a downsampling factor of M to reduce the frequency from MF_s to F_s and remove the aliased components. Finally, the Nyquist A/D converter transfers the discrete-time sequence into digital sample data with the Nyquist sampling frequency. This process obtains high-speed A/D conversion using low-speed and low-cost A/D converters. Figure 8.8 illustrates the entire process of a multirate A/D converter in the frequency domain.

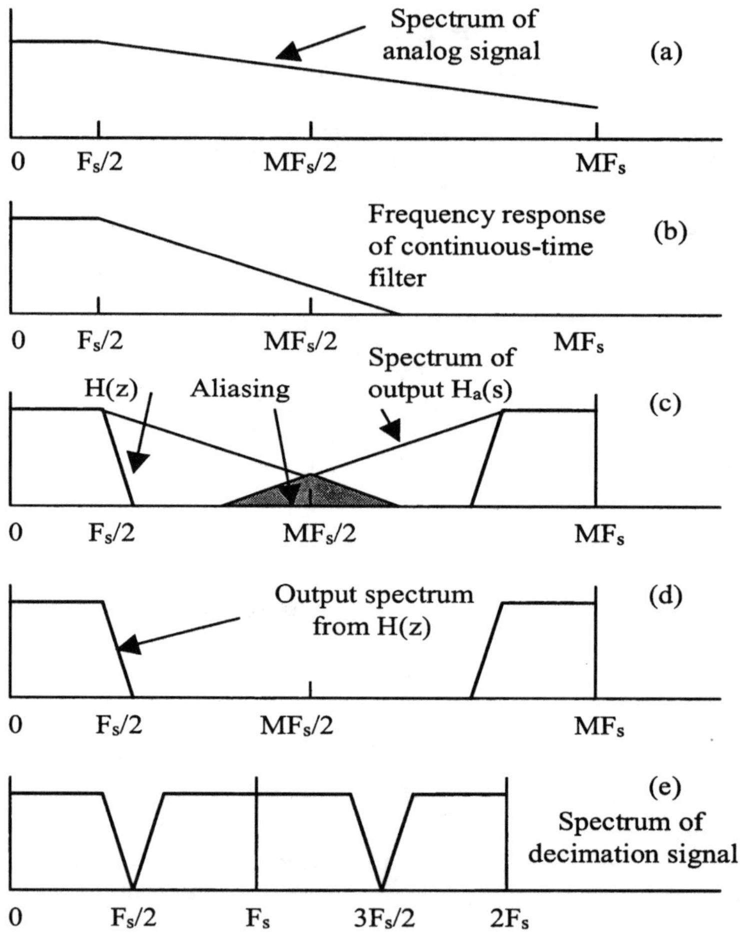

Figure 8.8 Frequency operation of the continuous-time filter $H_a(s)$ and discrete-time sampled-data decimator in the multirate A/D converter. (a) Spectrum of analog input; (b) frequency response of continuous-time filter $H_a(z)$; (c) overlay frequency responses of an input signal for $H(z)$, discrete-time sampled-data decimation filter $H(z)$, and an aliasing component; (d) spectrum of output $H(z)$; and (e) spectrum of decimation signal with a sampling frequency at F_s.

Eventually, the combined amplitude characteristics of the continuous-time filter $H_a(s)$ and discrete-time sampled-data decimation filter $H(z)$ can be considered the overall antialiasing filter function. This enables transfer of a continuous-time signal to a digital sequence with the Nyquist sampling frequency F_s. In addition, this allows the use of a low-order analog filter as an antialiasing filter followed by an efficient polyphase architecture of the discrete-time decimation filter $H(z)$ or even a discrete-time Mth-band decimation filter for implementation.

The decimation filter $H(z)$ is an analog decimation filter with downsampling by M. In fact, this filter can be built using charge transport devices and switched capacitors, which are also discrete-time. The reason this is still analog is that the signal samples have not been quantized. Many successes have been reported using this strategy since 1980. There are two kinds of decimation filters: *IIR switched-capacitor* and *FIR switched-capacitor*. The circuit design of a discrete-time sampled-data decimation filter can be found in the references by Franca and Santos [12], Martins and Franca [13], Franca and Martins [14], Uehara and Gray [15], Martins and Franca [16], and Neves et al. [17].

Employing a conventional polyphase architecture for the FIR decimation filter with switched-capacitor technology is not practical in application because of the long impulse responses required. In other words, this would require a rather large number of switched-capacitor branches and switching waveforms. To reduce the implementation complexity of the filter, it is desirable to develop equivalent terms that only utilize one delay element of an active switched capacitor in the various polyphase filters. For this purpose, we present a design example to illustrate an alternative polyphase architecture, which leads to an FIR switched-capacitor decimation filter using significantly fewer branches and switching waveforms than in conventional polyphase structures. It is thus more attractive for mixed-signal integrated circuit implementation.

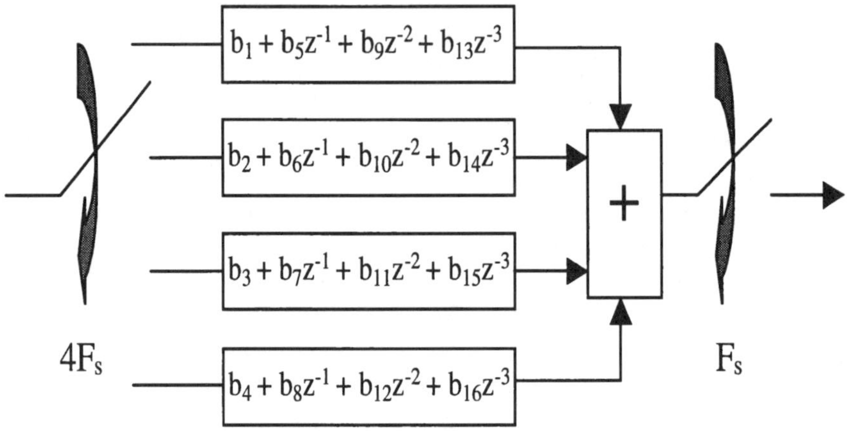

Figure 8.9 The common polyphase architecture of discrete-time switched-capacitor decimation filter in a multirate A/D converter. The polyphase filters contain the following: $H_0(z) = b_1 + b_5z^{-1} + b_9z^{-2} + b_{13}z^{-3}$, $H_1(z) = b_2 + b_6z^{-1} + b_{10}z^{-2} + b_{14}z^{-3}$, $H_2(z) = b_3 + b_7z^{-1} + b_{11}z^{-2} + b_{15}z^{-3}$, and $H_3(z) = b_4 + b_8z^{-1} + b_{12}z^{-2} + b_{16}z^{-3}$.

Example 8.1 Minimum-Hardware Form of a Polyphase Architecture Implementation for a Discrete-Time Switched-Capacitor Decimator

Consider a discrete-time FIR switched-capacitor decimator with a filter length $N = 16$ and a downsampling factor of $M = 4$ in a multirate A/D converter. The common polyphase architecture for implementing this FIR switched-capacitor decimator is shown in Figure 8.9. The polyphase filters $H_i(z)$ $(i = 0, 1, 2, 3)$ are implemented using canonical switched-capacitor elements, which can be controlled by nonoverlapping switching waveforms and operated by two switched waveforms with switching period $1/F_s$. The output accumulator is a block common to all polyphase filters $H_i(z)$ and is implemented by using one operational amplifier. However, such a common polyphase architecture still leads to the switched-capacitor decimator circuit having a rather large number of switched-capacitor branches and switching waveforms in the filter system. As a result, circuits are more sensitive to errors in terms of the capacitance ratios and switch timing. In addition, the architecture's computational complexity may be beyond practically acceptable limits.

An alternative polyphase architecture with a minimum number of active-delayed blocks for implementation of a discrete-time FIR switched-capacitor decimator is illustrated in Figure 8.10. This filter has a parallel form that is developed based on the delay element z^{-i} ($i = 0, 1, 2, 3$) as shown in Figure 8.9. First, the filter coefficients with delay z^0 are directly implemented in the parallel form of the polyphase structure. Second, for the filter coefficients with delay z^{-1}, we use an additional delay element of an active switched-capacitor. Third, the filter coefficients with delay z^{-2} are obtained by cascading two of the delay elements of the active switched-capacitor. Finally, the filter coefficients with delay z^{-3} are formed by cascading three of the delay elements of the active switched-capacitor. Each delay element can be obtained by an input operational amplifier operating as a charge-to-voltage converter, which is followed by an output switched-capacitor branch operating as a voltage-to-charge converter. Using this minimum hardware form of polyphase architecture, the computational complexity can be reduced for the multirate A/D converter.

8.5 Multirate D/A Converter

A generalized architecture of a multirate D/A converter is shown in Figure 8.11, which contains the Nyquist D/A converter, the discrete-time sampled-data interpolation filter with upsampling factor of L, and the continuous-time reconstruction lowpass filter. Efficient implementation of the interpolation system is realized by means of a polyphase structure. The basic operation of the multirate D/A converter is to achieve the complementary step of recovering a continuous-time signal from a sequence of digital samples. Such a multirate D/A converter can obtain the result of high-speed D/A conversion by using a low-speed, low-power, and low-cost D/A

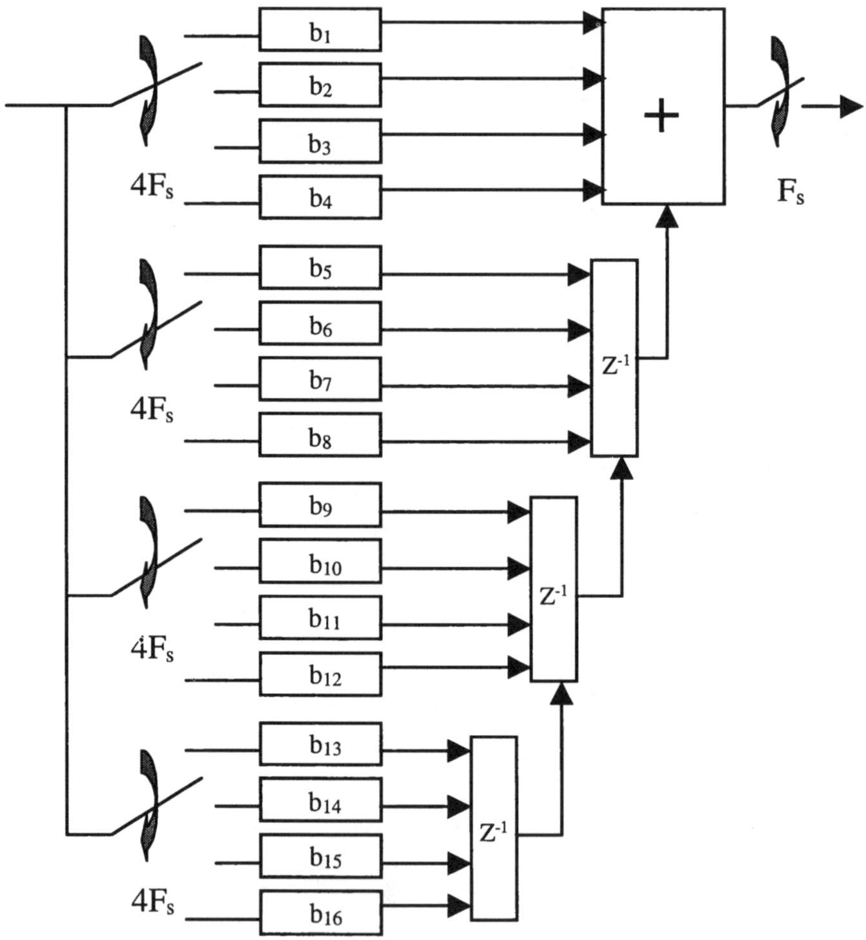

Figure 8.10 The minimum-hardware form of polyphase architecture for a discrete-time switched-capacitor decimation filter in the multirate A/D converter.

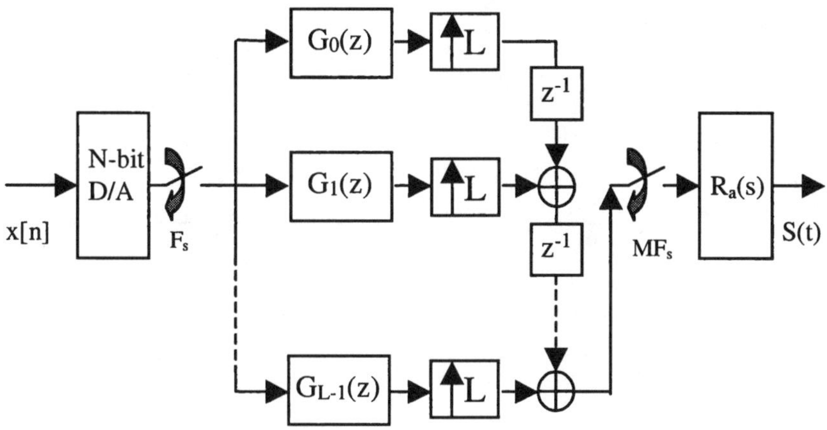

Figure 8.11 A generalized architecture of a multirate D/A converter including an N-bit D/A converter, a discrete-time sampled-data interpolation filter with upsampling by L, and a continuous-time reconstruction lowpass filter.

converter.

To illustrate, Figure 8.12 shows the frequency domain representation of the multirate D/A converter operation. The discrete-time sampled-data interpolation subsystem increases the frequency of the input samples by L-fold and, at the same time, removes the associated $L-1$ frequency band images as shown in Figures 8.12(a) and (b). Then, the images with sampling frequency LF_s at the center frequencies of kLF_s, $k = 1, 2, ...$, are filtered by the subsequent continuous-time reconstruction lowpass filter, as shown in Figure 8.12(c). As we can see, the combined amplitude characteristics of the interpolator and continuous-time reconstruction lowpass filters produce the overall anti-imaging filter function, which is needed to allow a continuous-time signal to be recovered from a sequence of analog samples with frequency F_s.

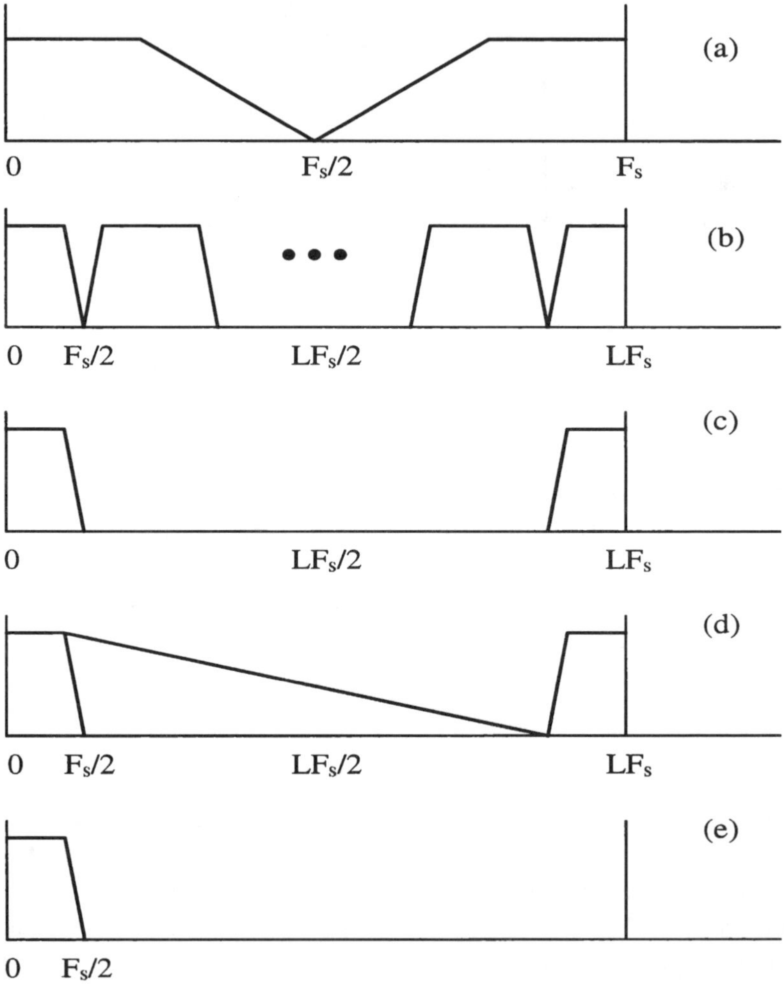

Figure 8.12 The frequency response of a multirate D/A converter operation: (a) Input spectrum of interpolator sampled at F_s. (b) Output spectrum of upsampling by L with sampling at LF_s. (c) Spectrum output of filter $G(z)$ with sampling at LF_s. (d) Frequency response of filter $R_a(s)$ and frequency response of filter $G(z)$ together. (e) Spectrum of output of filter $R_a(s)$.

8.6 Oversampling A/D Converter

The oversampling A/D converter is a technique that improves the resolution obtained from straightforward Nyquist sampling rate A/D conversion. The improvement is achieved by oversampling the signal in such a way that samples are acquired from the analog waveform at a rate significantly faster than the Nyquist sampling rate. Such A/D conversion has been shown to be an alternative to the implementation of high-resolution low-speed A/D convertion. An oversampling A/D converter can also be treated as a multirate A/D converter because it relates an input of analog samples with a high sampling rate to an output of digital data operating at a much lower sampling rate.

The quantization noise power $P_e(f)$ obtained from the oversampling A/D converter is exactly the same quantization noise power produced by a Nyquist sampling rate A/D converter, but the frequency distribution is different because of the higher sampling rate. The quantization noise power $P_e(f)$ is uniformly distributed between $-F_s/2$ to $+F_s/2$, where F_s is the sampling frequency as shown in Figure 8.13. For the Nyquist sampling rate, all the quantization noise power $P_e(f)$ is represented by the area of the tall rectangle occurring across the signal bandwidth $f_b = f_{sN}/2$. For the oversampled rate, the same quantization noise power $P_e(f)$ that was represented by the area of the wide rectangle has been spread over a bandwidth of sampling frequency F_s, which is much greater than the signal bandwidth f_b. As can be seen, only a relatively small fraction of the total quantization noise power falls within the frequency bandwidth from $-f_b$ to f_b. Therefore, the quantization noise power $P_e(f)$ outside the frequency band of the signal can be greatly attenuated with a digital lowpass filter following the A/D converter.

Figure 8.14 shows a generalized block diagram of an oversampled A/D converter including oversampling, the A/D converter model, and the discrete-time decimation filter with downsampling by factor M. The discrete-time decimator is realized by using a common polyphase structure. The digital signal can be downsampled to the Nyquist sampling rate without affecting the SNR (dB) after using the discrete-time decimation system.

In Figure 8.14, for a mathematical simplification we assume that the signal $x[n]$ and the quantization error $e[n]$ are statistically

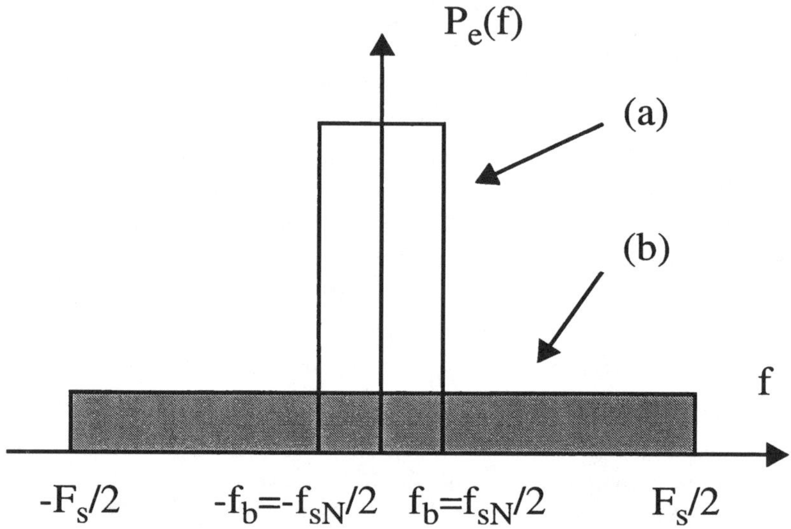

Figure 8.13 The power spectral density of quantization noise. (a) A Nyquist sampling rate A/D converter. (b) An oversampling rate A/D converter.

independent. Then the output $Y(z)$ of the oversampling A/D converter can be expressed in z-domain as follows:

$$Y(z) = X(z)H_X(z) + E(z)H_E(z), \tag{8.26}$$

where $H_X(z)$ is the signal transfer function and $H_E(z)$ is the noise transfer function. Assume that the random process is stationary and $H_X(f) = H_E(f) = 1$. Then the power spectral density $P_{xy}(f)$ at output of the oversampling A/D converter is

$$\begin{aligned} P_{xy}(f) &= P_x|H_X(f)|^2 \\ &= \sigma_x^2. \end{aligned} \tag{8.27}$$

The signal variance σ_{xy}^2 is the same as the input signal variance σ_x^2, since the signal power is assumed to occur within the signal frequency

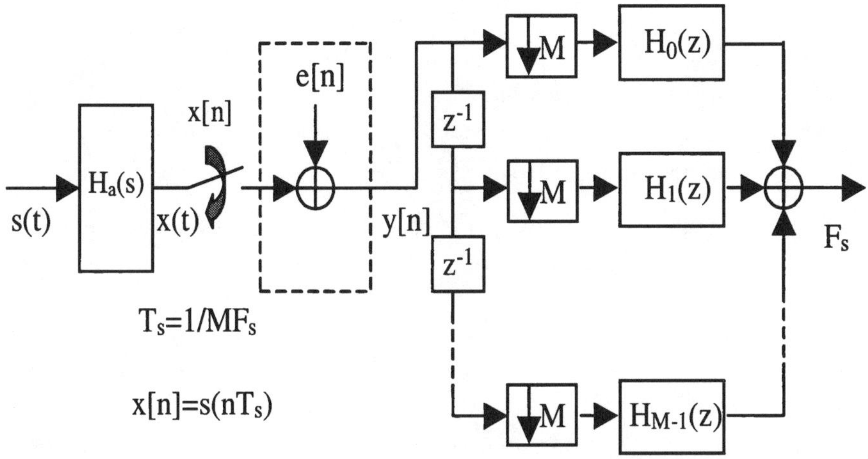

Figure 8.14 A generalized oversampling A/D converter system including the antialiasing filter, the A/D converter model, and the decimation subsystem.

band only. With white noise assumption for the quantization error $e[n]$, the noise power $P_e(f)$ is equal to σ_e^2/F_s. Then the power spectral density $P_{ey}(f)$ is

$$
\begin{aligned}
P_{ey}(f) &= P_e|H_E(f)|^2 \\
&= \frac{\sigma_e^2}{F_s}.
\end{aligned}
\tag{8.28}
$$

This indicates that the power spectral density $P_{ey}(f)$ is also equal to σ_e^2/F_s.

With an ideal discrete-time decimation filter with a cutoff of f_b following the oversampling A/D quantizer, the quantization noise variance σ_{ey}^2 within the signal frequency band from $-f_b$ to f_b is

$$
\begin{aligned}
\sigma_{ey}^2 &= \int_{-f_b}^{f_b} P_{ey}(f)df \\
&= \int_{-f_b}^{f_b} \frac{\sigma_e^2}{F_s} df
\end{aligned}
$$

$$= \sigma_e^2 \left(\frac{2f_b}{F_s} \right). \tag{8.29}$$

Therefore, the SNR (dB) of the output from the entire oversampling A/D converter is as follows:

$$\begin{aligned} SNR(\text{dB}) &= 10log_{10} \left(\frac{\sigma_{xy}^2}{\sigma_{ey}^2} \right) \\ &= 10log_{10} \left[\frac{\sigma_x^2}{\sigma_e^2 (2f_b/F_s)} \right] \\ &= 10log_{10} \left(\frac{\sigma_x^2}{\sigma_e^2} \right) + 10log_{10} \left(\frac{F_s}{2f_b} \right). \end{aligned} \tag{8.30}$$

Note that SNR (dB) in (8.30) is the same as SNR (dB) in (8.13) for the Nyquist sampling rate A/D converter if $F_s = 2f_b$. For the oversampling rate, let $2^B = F_s/(2f_b)$, where B is the number of bits. Then SNR (dB) in (8.30) can be expressed as

$$SNR(\text{dB}) = 10log_{10} \left(\frac{\sigma_x^2}{\sigma_e^2} \right) + 3.0103B. \tag{8.31}$$

As a result, SNR (dB) increases about 3 dB for every doubling of the oversampling ratio. In other words, this is equivalent to improving the resolution of an A/D converter by one-half bit. In addition, the analog antialiasing filter does not need sharp cutoff frequencies because of the oversampling. Therefore, the benefits of oversampling in the A/D converter are to reduce quantization noise in the frequency band of interest and to decrease the requirement of sharp cutoffs for the antialiasing and reconstruction analog filters.

A vast technical literature exists on the topic of oversampling A/D converters. For a tutorial discussion, we suggest the papers by Uchimura et al. [18], Temes and Candy [19], Candy and Temes [20, 21], and Aziz, Sorensen, and Spiegel [22].

8.7 Sigma-Delta A/D Converter

Sigma-delta A/D converters have become quite popular to achieve high-resolution, because the recent advances in very large scale

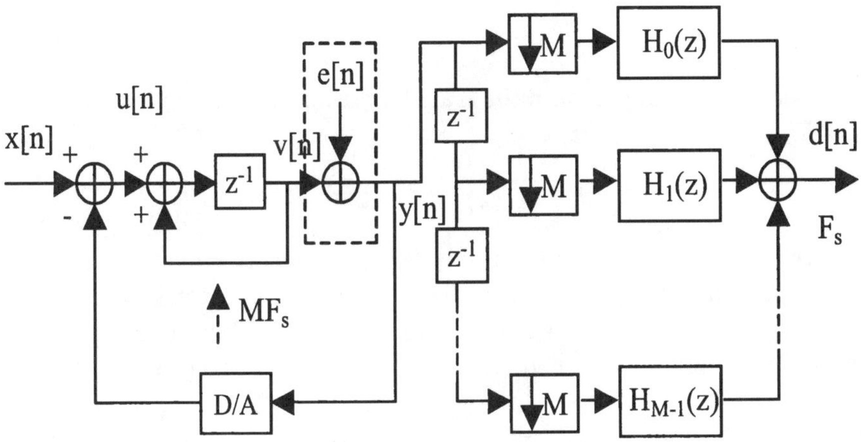

Figure 8.15 Block diagram of a first-order sigma-delta A/D converter system.

integration (VLSI) technology have made feasible complex digital processing of the bit stream. Such A/D converters use the technique of noise shaping in addition to oversampling to allow high-resolution conversion of relatively low bandwidth signals. One significant advantage of the method is that analog signals are converted using only a few bits or even a one-bit A/D converter so that the precision of analog signal processing circuits can be much less than the resolution of the overall A/D converter.

8.7.1 First-Order Sigma-Delta Modulator

Figure 8.15 shows a block diagram of a first-order sigma-delta A/D converter system. The system consists of an analog sigma-delta modulator followed by a digital decimation subsystem with a polyphase structure for implementation. The sigma-delta modulator contains an integrator, an internal A/D converter or a quantizer, and a D/A converter used in the feedback path.

The discrete-time integrator, which is often called a *feed-forward loop filter*, has a time-domain version of the differentiator as follows:

$$v[n] = u[n-1] + v[n-1]. \tag{8.32}$$

The corresponding z-domain transfer function is given by

$$\frac{V(z)}{U(z)} = \frac{z^{-1}}{1-z^{-1}}. \tag{8.33}$$

It is common to use a one-bit D/A converter since it is perfectly linear. In this case, we assume that the D/A converter is ideal so that it is replaced by a unity gain transfer function. Then, the sigma-delta modulator output $Y(z)$ in the z-domain is expressed as

$$\begin{aligned} Y(z) &= [X(z) - Y(z)]\frac{z^{-1}}{1-z^{-1}} + E(z) \\ &= z^{-1}X(z) + (1-z^{-1})E(z), \end{aligned} \tag{8.34}$$

where $E(z)$ is the z-domain transform of the noise sequence $e[n]$ generated by the quantizer in the A/D converter. Therefore, $H_X(z) = z^{-1}$ is the signal transfer function (STF), and $H_E(z) = 1 - z^{-1}$ is the noise transfer function (NTF). In a sigma-delta modulator, $H_X(f)$ will have a flat frequency response in the interval $|f| \leq f_b$. In contrast, $H_E(f)$ will have a high attenuation in the frequency band $|f| \leq f_b$ and a no-interest region in the frequency interval $f_b < |f| \leq MF_s/2$. Note that the input signal $x[n]$ is not distorted in any way by the network, and simply experiences a delay from input to output. The performance of the system is determined by the noise transfer function, $H_E(z)$. Thus, the noise transfer function in the frequency domain turns out to be

$$\begin{aligned} |H_E(e^{j\omega})|^2 &= |(1-e^{-j\omega})|^2 \\ &= |(e^{j\omega/2} - e^{-j\omega/2})e^{-j\omega/2}|^2 \\ &= 4sin^2(\omega/2) \\ &= 2(1-cos\omega). \end{aligned} \tag{8.35}$$

Equation (8.35) is also referred to as a *noise shaping* function in the sigma-delta modulator. Then, the lowpass decimation filter $H(z)$ with downsampling factor M eliminates the noise from the signal band for $\omega > \pi/M$. Therefore, with an ideal lowpass decimation filter $H(z)$, the power of the noise sequence at the output of the sigma-delta A/D converter system as shown in Figure 8.15 becomes

$$
\begin{aligned}
\sigma_{ed}^2 &= \int_0^{\pi/M} P_{ed}(e^{j\omega})d\omega \\
&= \int_0^{\pi/M} P_e(e^{j\omega})|H_E(e^{j\omega})|^2|H(e^{j\omega})|^2 d\omega \\
&= \frac{\sigma_e^2}{\pi}\int_0^{\pi/M} |H_E(e^{j\omega})|^2 d\omega \\
&= \frac{\sigma_e^2}{\pi}\int_0^{\pi/M} |2(1-cos\omega)|^2 d\omega \\
&= \frac{2\sigma_e^2}{M}\left[1-\frac{sin(\pi/M)}{\pi/M}\right].
\end{aligned} \tag{8.36}
$$

Substituting (8.12) into (8.36), (8.36) can be rewritten,

$$
\sigma_{ed}^2 = 2\left(\frac{2^{-2B}X_{max}^2}{3M}\right)\left[1-\frac{sin(\pi/M)}{\pi/M}\right]. \tag{8.37}
$$

The noise shaping as described in (8.35) yields a further reduction of the noise by a factor of $2\left[1-\frac{sin(\pi/M)}{\pi/M}\right]$. As can be seen, the expression in (8.36) goes to zero when M goes to ∞.

When the sampling frequency is high enough or $F_s \gg f_b$, the sigma-delta A/D converter is able to achieve high resolution overall. Using $H_X(z) = z^{-1}$ and $H_E(z) = 1-z^{-1}$, the noise power within the signal frequency ranges from $-f_b$ to f_b at the output of a first-order sigma-delta modulator is defined by

$$
\begin{aligned}
\sigma_{ey}^2 &= \int_{-f_b}^{f_b} P_{ey}(f)df \\
&= \int_{-f_b}^{f_b} P_e(f)|H_E(z)|^2 df \\
&= \int_{-f_b}^{f_b} \frac{\sigma_e^2}{F_s}\left[4sin^2\left(\frac{\pi f}{F_s}\right)\right] df \\
&= \int_{-f_b}^{f_b} 2\frac{\sigma_e^2}{F_s}\left[1-cos\left(\frac{2\pi f}{F_s}\right)\right] df \\
&= \frac{2\sigma_e^2}{\pi}\left[\frac{2\pi f_b}{F_s} - sin\left(\frac{2\pi f_b}{F_s}\right)\right].
\end{aligned} \tag{8.38}
$$

Note that, for small frequency f, we may use the approximation $sin(x) \approx x$ to obtain

$$sin^2\left(\frac{\pi f}{F_s}\right) \approx \left(\frac{\pi f}{F_s}\right)^2, \tag{8.39}$$

where $F_s >> f_b$. Thus, the noise power within the signal frequency range from $-f_b$ to f_b is given by

$$\begin{aligned} \sigma_{ey}^2 &= \int_{-f_b}^{f_b} \frac{\sigma_e^2}{F_s}\left[4sin^2\left(\frac{\pi f}{F_s}\right)\right] df \\ &\approx 4\int_{-f_b}^{f_b} \frac{\sigma_e^2}{F_s}\left(\frac{\pi f}{F_s}\right)^2 df \\ &\approx \frac{\pi^2\sigma_e^2}{3}\left(\frac{2f_b}{F_s}\right)^3. \end{aligned} \tag{8.40}$$

Therefore, the SNR (dB) from the output of a first-order sigma-delta modulator is obtained by:

$$\begin{aligned} SNR(\text{dB}) &= 10log_{10}\left(\frac{\sigma_{xy}^2}{\sigma_{ey}^2}\right) \\ &= 10log_{10}\left(\frac{\sigma_x^2}{\sigma_{ey}^2}\right) \\ &= 10log_{10}\left(\frac{\sigma_x^2}{\sigma_e^2}\right) - 10log_{10}\left(\frac{\pi^2}{3}\right) \\ &\quad +30log_{10}\left(\frac{F_s}{2f_b}\right). \end{aligned} \tag{8.41}$$

If we assume that the oversampling rate is $\left(\frac{F_s}{2f_b}\right) = 2^B$, then (8.41) can be rewritten as

$$SNR(\text{dB}) = 10log_{10}\left(\frac{\sigma_x^2}{\sigma_e^2}\right) - 10log_{10}\left(\frac{\pi^2}{3}\right) + 9.031B. \tag{8.42}$$

Equation (8.42) indicates that SNR increases 9 dB for every doubling of the oversampling frequency in the first-order sigma-delta A/D modulator. The 3 dB of the SNR in (8.42) is due to the reduction in quantization noise power $P_e(f)$, and a further 6 dB of SNR is due to the noise-shaping filter characteristic $H_E(f)$. This is equivalent to improving the A/D converter resolution by 1.5 bits. As a result, the noise power is reduced by increasing the sampling rate to spread the quantization noise over a large frequency bandwidth, and then by shaping the noise power spectrum using a discrete-time lowpass decimation filter.

8.7.2 Second-Order Sigma-Delta Modulator

The structure of a second-order sigma-delta A/D converter, which is widely used, as shown in Figure 8.16, now contains two integrators. The transfer functions of the first one and the second one are $1/(1-z^{-1})$ and $z^{-1}/(1-z^{-1})$, respectively. The output $Y(z)$ in the z-domain of the second-order sigma-delta A/D modulator is given by

$$Y(z) = z^{-2}X(z) + (1 - z^{-1})^2 E(z). \tag{8.43}$$

Thus, the sigma-delta modulator has the signal transfer function $H_X(z) = z^{-2}$ and the noise transfer function $H_E(z) = (1 - z^{-1})^2$.

The frequency response of the noise transfer function, $H_E(e^{j\omega}) = (1 - e^{-j\omega})^2$, in the second-order sigma-delta modulator is shown in Figure 8.17, along with the noise transfer functions of the first-order, the third-order, and the fourth-order sigma-delta modulator. The vertical line in Figure 8.17 indicates the signal bandwidth, f_b, where, in this case, $f_b = \frac{F_s}{20}$. In general, the frequency response of Nth order of the noise transfer function is expressed as

$$H_E(e^{j\omega}) = (1 - e^{-j\omega})^N. \tag{8.44}$$

As we can see, compared with the noise transfer function in the first-order sigma-delta modulator, the noise transfer function in the second-order sigma-delta modulator provides more quantization noise suppression within the low-frequency signal band, and more amplification of the noise outside the signal band. Moreover, more noise power is pushed outside the frequency signal band when the order of sigma-delta modulator increases.

At the output of a second-order sigma-delta modulator, the in-band noise power within the frequency range $[-f_b, f_b]$ is given by

$$\sigma_{ey}^2 = \int_{-f_b}^{f_b} P_{ey}(f)df$$

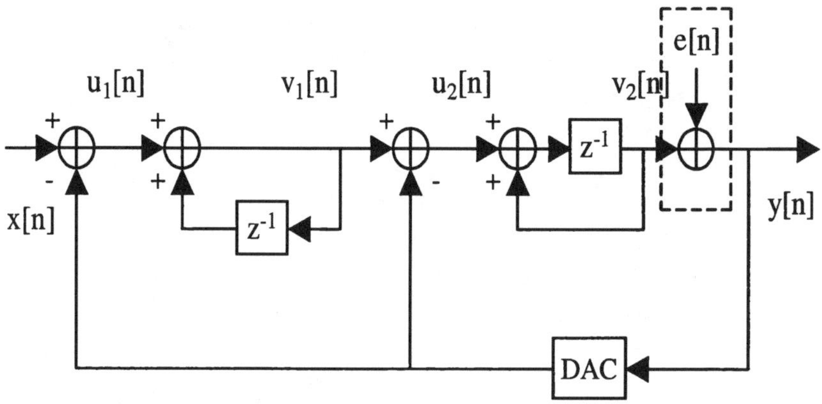

Figure 8.16 Block diagram of a second-order sigma-delta A/D modulator without the discrete-time decimation subsystem.

$$\begin{aligned}
&= \int_{-f_b}^{f_b} P_e(f)|H_E(e^{j\omega})|^2 df \\
&= \int_{-f_b}^{f_b} \frac{\sigma_e^2}{F_s}|4sin^2(\pi f/F_s)|^2 df \\
&= \frac{16\sigma_e^2}{F_s}\int_{-f_b}^{f_b}[sin(\pi f/F_s)]^4 df. \qquad (8.45)
\end{aligned}$$

For small frequency f, we may use the approximation $sin(x) \approx x$ to obtain

$$[sin(\pi f/F_s)]^4 \approx \left(\frac{\pi f}{F_s}\right)^4. \qquad (8.46)$$

Thus, (8.45) can be rewritten as

$$\begin{aligned}
\sigma_{ey}^2 &= \frac{16\sigma_e^2}{F_s}\int_{-f_b}^{f_b}[sin(\pi f/F_s)]^4 df \\
&\approx \frac{16\sigma_e^2}{F_s}\int_{-f_b}^{f_b}\left(\frac{\pi f}{F_s}\right)^4 df
\end{aligned}$$

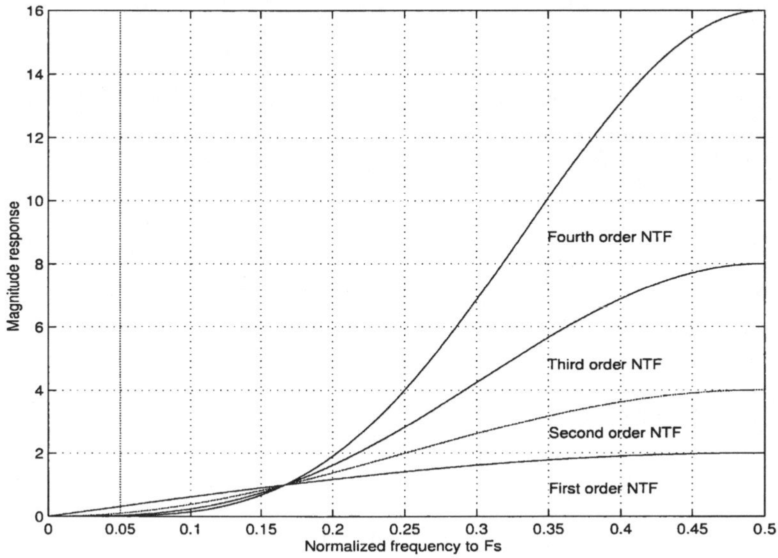

(a)

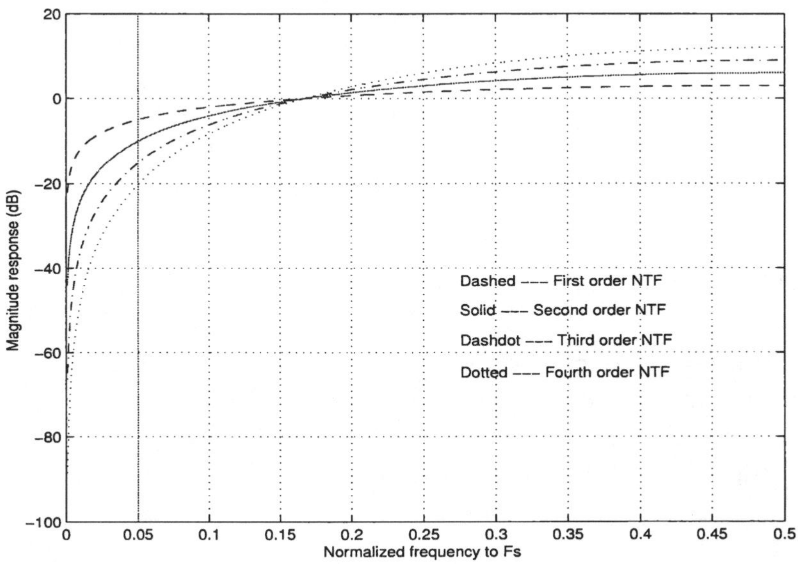

(b)

Figure 8.17 The noise transfer function for the first, second, third and fourth-order sigma-delta modulators. (a) Magnitude response with linear scale. (b) Magnitude response in dB scale.

$$\approx \left(\frac{16\sigma_e^2}{F_s}\right)\left[\frac{2F_s}{5\pi}\left(\frac{\pi f_b}{F_s}\right)^5\right]. \tag{8.47}$$

Therefore, in-band noise power σ_{ey}^2 in the second-order sigma-delta A/D modulator is

$$\sigma_{ey}^2 \approx \frac{\pi^4\sigma_e^2}{5}\left(\frac{2f_b}{F_s}\right)^5, \tag{8.48}$$

where $F_s \gg f_b$.

Using (8.48), the SNR (dB) for the second-order sigma-delta modulator yields the following:

$$\begin{aligned} SNR(\text{dB}) &= 10log_{10}\left(\frac{\sigma_{xy}^2}{\sigma_{ey}^2}\right) \\ &= 10log_{10}\left(\frac{\sigma_x^2}{\sigma_{ey}^2}\right) \\ &= 10log_{10}\left(\frac{\sigma_x^2}{\sigma_e^2}\right) \\ &\quad - 10log_{10}\left(\frac{\pi^4}{5}\right) + 50log_{10}\left(\frac{F_s}{2f_b}\right). \end{aligned} \tag{8.49}$$

Now, let $2^B = \frac{F_s}{2f_b}$, substituting it into (8.49), we obtain

$$SNR(\text{dB}) = 10log_{10}\left(\frac{\sigma_x^2}{\sigma_e^2}\right) - 10log_{10}\left(\frac{\pi^4}{5}\right) + 15.051B. \tag{8.50}$$

Thus, for every doubling of the oversampling ratio, the SNR improves about 15 dB in the second-order sigma-delta A/D modulator. In other words, the resolution of second-order sigma-delta A/D modulator is equivalently increased by 2.5 bits, which is 1 bit better than the improvement achieved by the first-order sigma-delta A/D modulator.

8.7.3 *N*th-Order Sigma-Delta Modulator

For an Nth-order sigma-delta A/D modulator, the structure contains N integrators based on a straightforward extension of the second-order sigma-delta A/D modulator. In this case, the signal transfer function in Nth-order case is given by $H_X(z) = z^{-N}$ and the noise transfer function is obtained by $H_E(z) = (1 - z^{-1})^N$, which contains N zeros at $z = 1$ or at DC frequency on the unit circle. Then, the output $Y(z)$ can be expressed by

$$Y(z) = z^{-N}X(z) + (1 - z^{-1})^N E(z). \tag{8.51}$$

Therefore, within the frequency range $[-f_b, f_b]$, the inband quantization noise power at the output of a Nth-order sigma-delta A/D modulator is calculated by

$$\begin{aligned}
\sigma_{ey}^2 &= \int_{-f_b}^{f_b} P_{ey}(f)df \\
&= \int_{-f_b}^{f_b} P_e(f)|H_E(e^{j\omega})|^2 df \\
&= \int_{-f_b}^{f_b} \frac{\sigma_e^2}{F_s}[4sin^2(\pi f/F_s)]^N df.
\end{aligned} \tag{8.52}$$

With the approximation, $sin(x) \approx x$, for small value x, it can be shown that (8.52) becomes

$$\sigma_{ey}^2 \approx \frac{\pi^{2N}}{2N+1}\sigma_e^2 \left(\frac{2f_b}{F_s}\right)^{2N+1}, \tag{8.53}$$

where $F_s \gg f_b$.

Using (8.53), the SNR (dB) for the Nth-order sigma-delta A/D modulator is then:

$$\begin{aligned}
SNR(\text{dB}) &= 10log_{10}\left(\frac{\sigma_{xy}^2}{\sigma_{ey}^2}\right) \\
&= 10log_{10}\left(\frac{\sigma_x^2}{\sigma_{ey}^2}\right) \\
&= 10log_{10}\left(\frac{\sigma_x^2}{\sigma_e^2}\right) - 10log_{10}\left(\frac{\pi^{2N}}{2N+1}\right) \\
&\quad + (20N+10)log_{10}\left(\frac{F_s}{2f_b}\right).
\end{aligned} \tag{8.54}$$

Again, assume that the oversampling rate is $\left(\frac{F_s}{2f_b}\right) = 2^B$. Then, (8.54) can be rewritten as follows:

$$SNR(\text{dB}) = 10log_{10}\left(\frac{\sigma_x^2}{\sigma_e^2}\right) - 10log_{10}\left(\frac{\pi^{2N}}{2N+1}\right) + (6.02B + 3.01)B. \quad (8.55)$$

Therefore, for every doubling of the oversampling frequency ratio, the Nth-order sigma-delta A/D modulator can provide ($6.02B$+3.01) dB in SNR. In other words, the resolution of sigma-delta A/D modulator can be increased by approximately ($B + 0.5$) bits.

The sampling theorem was initially introduced by Nyquist [23] and later contributed to in the classic paper by Shannon [24]. Sigma-delta A/D converters, which are developed on the sampling theorem along with oversampling concepts, have been treated in many technical sources. Specifically, we cite the references by Gray, Chou, and Wong [4], Candy and Temes [21], and Aziz, Sorensen, and Spiegel [22]. In the context of the multiband sigma-delta, and multistage sigma-delta modulations, we recommend the papers by Chou, Wong, and Gray [25], Wong and Gray [26], Aziz, Sorensen, and Spiegel [27], Joseph, Tarassenko, and Collins [28], and Dick and Harris [29]. The high-order sigma-delta A/D converters were reported on in the papers by Ribner [30] and Rio et al. [31].

Example 8.2 First- and Second-Order Sigma-Delta A/D Converters with a 1-Bit Quantizer

Suppose that the input signal is a sinusoid with a full-scale amplitude voltage $V = 1$ and corresponding signal power $V^2/2 = 0.5$. The input signal has a $f_b = 100$ kHz signal band. In this example, we want to achieve a final resolution of 16 bits from the output of the sigma-delta modulator. This corresponds to 96 dB in SNR.

For a first-order sigma-delta modulator with a 1-bit internal A/D converter, (8.41) indicates that the oversampling frequency F_s requires at least 297.5 MHz to obtain 16-bit resolution. With such high operation speed, it is not easy for the sampled-data analog-switched capacitor integrator to implement this sigma-delta A/D modulator.

For a second-order sigma-delta modulator with a 1-bit internal

A/D converter based on (8.49), we only need the required oversampling frequency $F_s = 22.9$ MHz to achieve 16-bit resolution. As can be seen, in contrast to the oversampling frequency of 297.5 MHz required by the first-order sigma-delta modulator, it is very reasonable for the sampled-data analog-switched capacitor integrator to achieve the necessary circuit speed of the second-order case in current complementary metal-oxide semiconductor (CMOS) technology.

Figure 8.18 shows a power spectral density estimate from the output of the second-order sigma-delta A/D modulator using 1-bit quantizer with oversampling frequency of $F_s = 22.9$ MHz. The tall peak is a sinusoidal input frequency of 100 kHz. It is clear that the noise shaping is greatly attenuated when the noise power is at the low frequency range, $f < f_b$. In addition, such power spectra can also be used to compute the in-band SNR (dB) over the entire frequency band of the 100-kHz input signal.

8.8 Hybrid QMF Bank A/D and D/A Converter

In this section, we discuss a specific architecture based on hybrid QMF bank A/D and D/A converters. Hybrid QMF-bank-based converters realize high-speed conversion using a number of low-speed and low-cost converters. Such hybrid converters have many applications in video, wireless, and wireline communications, as well as network communications.

8.8.1 QMF-Based A/D Converter

A QMF-based A/D converter as shown in Figure 8.19 is developed based on the immediate idea of the PRP QMF bank as described in Section 7.6, but using analog analysis filters, digital synthesis filters, and a set of low-speed A/D converters. Such a QMF-based A/D converter can achieve a very high speed using M low-speed A/D subconverters operating at M-fold lower sampling rate.

The basic operation idea of a QMF-based A/D converter is to first decompose the analog signal $x_a(t)$ into subband signals by using the analog filter bank $H_{ak}(z)$, $k = 0, 1, ..., M - 1$. These analog signals are passed through downsampling and sampled by a set of A/D subconverters at a sampling rate of F_s/M, and then converted

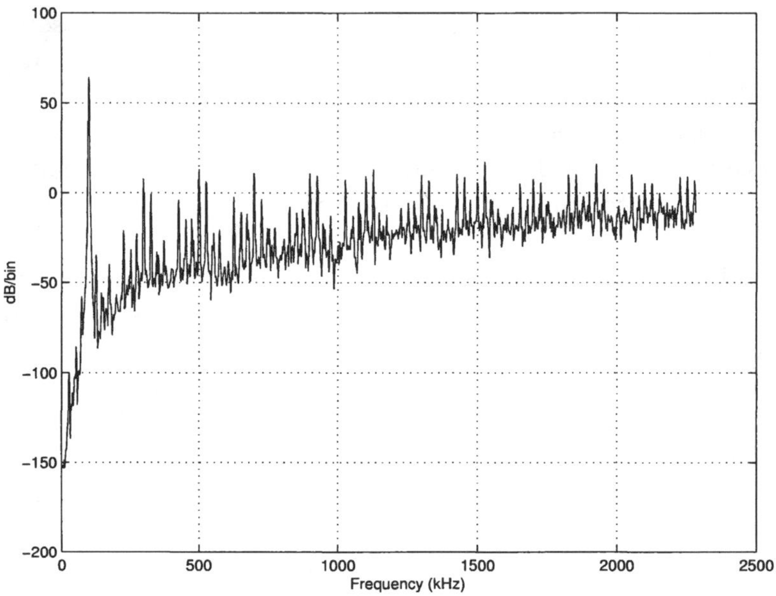

Figure 8.18 The portion of frequency plot from 0 to 2,260 kHz of output power spectrum of the second-order sigma-delta A/D modulator using a 1-bit quantizer without the implementation of decimation subsystem.

into digital subband signals, with nonoverlapping frequency bands of bandwidth $2\pi/M$. The digitized subband signals are sequentially passed through an upsampler to recover the desired sampling rate, and are recombined through the synthesis filter bank $R_k(z)$ $(k = 0, 1, ..., M-1)$ to obtain the digitally reconstructed signal $\hat{x}[n]$.

Obviously, the goal here is to design the PRP QMF bank such that the digitized reconstructed signal $\hat{x}[n]$ closely represents the digital signal $x[n]$ that would have been captured by direct A/D conversion of analog signal $x_a(t)$ at the very high sampling rate F_s. However, it is clear that there is aliasing at the output of the analog analysis bank since the analog analysis filter bank $H_{ak}(z)$ is not ideal. Therefore, a digital synthesis filter bank should be designed to minimize the problem of aliasing as well as amplitude and phase

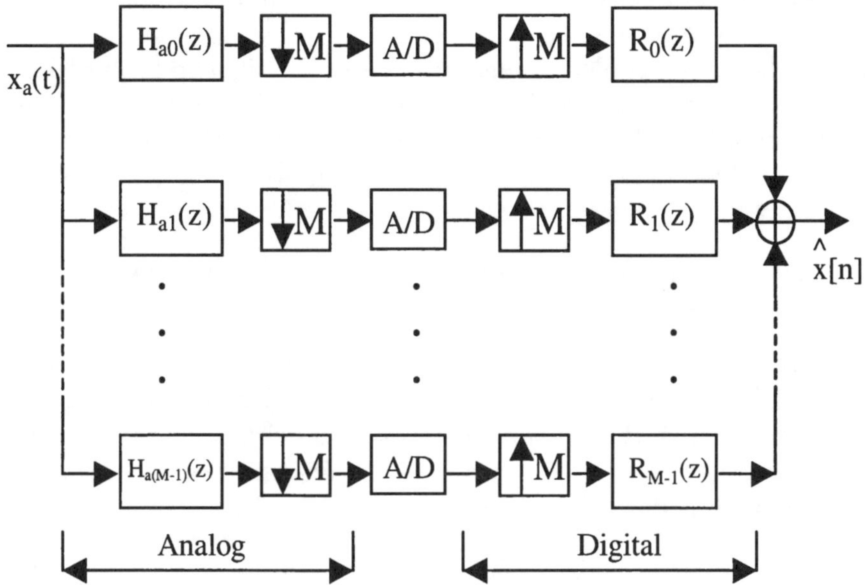

Figure 8.19 The implementation of single very high-speed A/D converter by using QMF-based low-speed A/D subconverters.

distortions.

Another issue in the QMF-based A/D converter is the harmonic distortion due to mismatches among a set of M low-speed A/D subconverters, and the jitter problem due to uneven sample timing of high-frequency inputs preceding each downsampling and A/D converter when switching from sampling to the hold mode. Fortunately, the harmonic distortion can be substantially eliminated by a digital synthesis filter bank at little additional cost, and the jitter problem of uneven sample timing of high-frequency inputs can be reduced by the decimation stage in the analog analysis filter bank. This is because each downsampling function acts as a second sample-and-hold (S/H) architecture after each S/H output of each $H_{ak}(z)$, which is built by the switch-capacitor filter. In other words, the input signal does not change when the downsampling function occurs as a transition from sample to hold.

The more detailed discussions of and solutions to problems posed by the QMF-based A/D converter can be found in the references by Petraglia, Maloberti, and Mitra [32], Petraglia and Mitra [33], Franca, Petraglia, and Mitra [34], and Lowenborg, Johansson, and Wanhammar [35].

Example 8.3 Polyphase Representation of a Two-Channel QMF-Based A/D Converter

In this example, we show an efficient architecture for implementation of a two-channel QMF-based A/D converter using polyphase representation. The generalized two-channel QMF-based A/D converter is shown in Figure 8.20(a). The discrete-time analog analysis filter bank is designed to have the following form:

$$H_{a1}(z) = H_{a0}(-z). \tag{8.56}$$

The above equation indicates that $H_{a1}(z)$ is a good discrete-time analog highpass filter if $H_{a0}(z)$ is a good discrete-time analog lowpass filter. Using polyphase decomposition theory, we let

$$H_{a0}(z) = E_0(z^2) + z^{-1}E_1(z^2). \tag{8.57}$$

Based on (8.56), we can write $H_{a1}(z)$ as follows:

$$H_{a1}(z) = E_0(z^2) - z^{-1}E_1(z^2). \tag{8.58}$$

To cancel the aliasing, the digital synthesis filters are chosen to be

$$R_0(z) = H_{a1}(-z), \qquad R_1(z) = -H_{a0}(-z). \tag{8.59}$$

Then, the digital synthesis filters $R_0(z)$ and $R_1(z)$, which satisfy (8.59), can be represented in terms of $E_0(z^2)$ and $E_1(z^2)$ as follows:

$$R_0(z) = E_0(z^2) + z^{-1}E_1(z^2), \tag{8.60}$$

and

$$R_1(z) = z^{-1}E_1(z^2) - E_0(z^2). \tag{8.61}$$

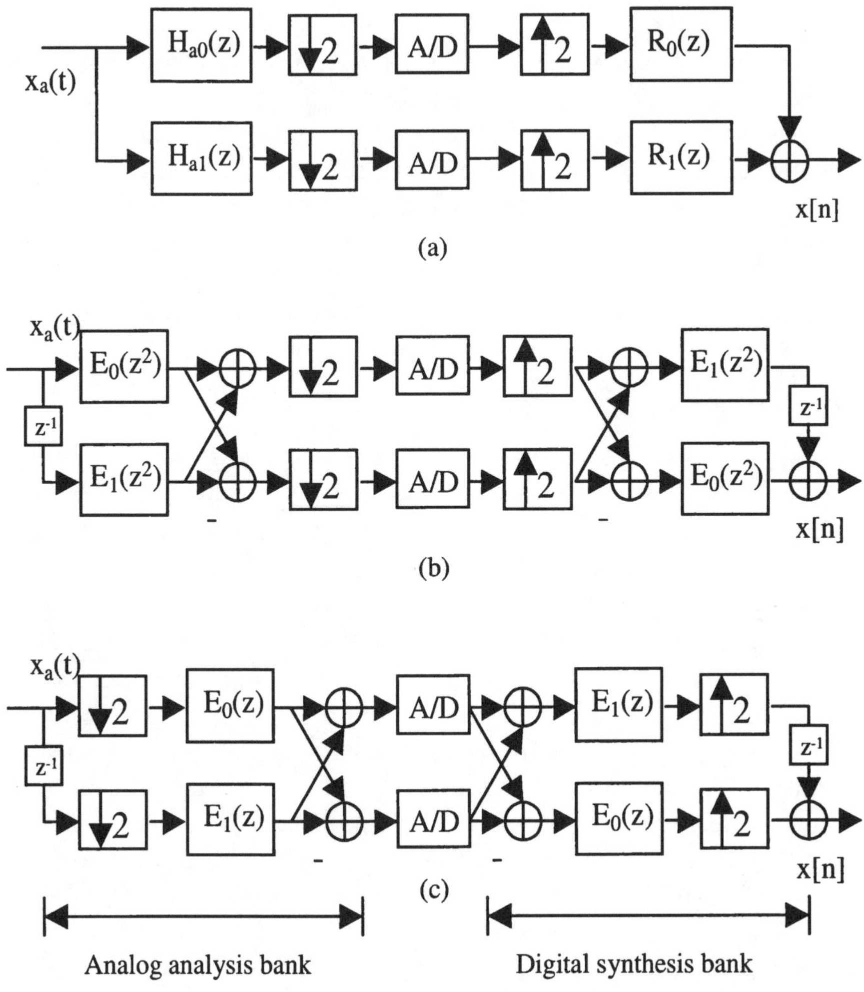

Figure 8.20 Efficient architecture for implementation of the two-channel QMF-based A/D converter: (a) the generalized structure of two-channel QMF-based A/D converter; (b) the representation of two-channel QMF-based A/D converter in polyphase structure; and (c) the efficient implementation structure of a two-channel QMF-based A/D converter using the noble identities method.

Using (8.57), (8.58), (8.60), and (8.61), we draw the QMF-based A/D converter in Figure 8.20(b). Furthermore, using noble identities theory, we move the downsampling to the left of the polyphase subfilter in the analog analysis filter bank and shift the upsampling to the right of the polyphase subfilter in the digital synthesis filter bank as indicated in Figure 8.20(c). Therefore, the polyphase representation of the analog analysis filter bank and the digital synthesis filter bank are now operating at the lowest sampling rate so that the minimum number of MACs per unit for implementation is achieved. As a result, a very-high-speed A/D converter can be substituted by using two low-speed A/D subconverters based on the QMF approach.

8.8.2 QMF-Based D/A Converter

The architecture of a QMF-based D/A converter in Figure 8.21 is introduced by substituting a set of M low-speed D/A subconverters in the synthesis branches for the single very-high-speed D/A converter in the common analog signal. The digital signal $y[n]$ is first split into subband signals with no-overlap in frequency by using a digital analysis bank $H_k(z)$ $(k = 0, 1, ..., M - 1)$. Then, the D/A subconverter is implemented at the low sampling rate of F_s/M, eventually based on the decimated subband signals $d_k[n]$ $(k = 0, 1, ..., M - 1)$. After that, the analog subband signals are upsampled by an integer factor M as well as passed through the discrete-time analog synthesis filter bank $R_{ak}(z)$ $(k = 0, 1, ..., M - 1)$ and recombined to form the desired analog signals, $y_a(t)$.

The design goal here is to choose the digital analysis filters and analog synthesis filters in such a way that the aliasing can be canceled completely and the amplitude distortion and the phase distortion can be reduced to the desired level at the output. Such a system has a great advantage with respect to implementation of D/A subconverters at an M-fold lower sampling rate. This leads to great savings in power consumption, die area size, and the cost of the SOC product.

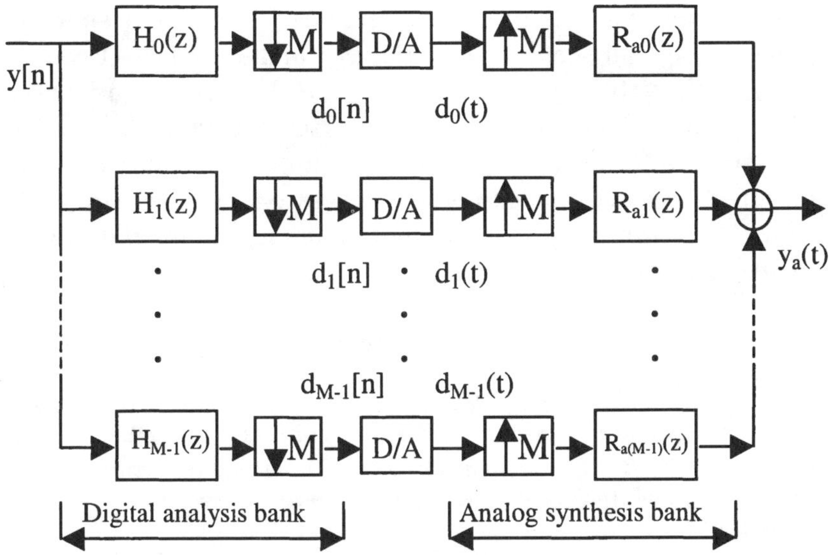

Figure 8.21 The implementation of a single very-high-speed D/A converter by using QMF-based low-speed D/A subconverters.

8.9 Summary

In this chapter, we introduced recent developments in the area of multirate data converters, which act as a bridge between continuous-time analog multirate signal processing and discrete-time multirate signal processing, based on the fundamental theory of multirate signal processing. We first presented the conventional A/D and D/A converters. Then, the concepts of multirate A/D conversion and multirate D/A conversion were discussed in detail, leading to the presentation of efficient system architectures for implementation. The quantization noise problem associated with A/D conversion, and state-of-the-art high-resolution technologies including oversampling A/D converters and sigma-delta A/D converters were presented.

These methods can produce high-quality digital signals for mixed-signal circuit integration. Furthermore, the hybrid QMF bank-based A/D converter and D/A converter were described. It was shown how to achieve the equivalent of a single very-high-speed A/D converter and D/A converter by using a set of M low-speed and low-cost A/D subconverters and D/A subconverters. This approach shows great potential for future expansions, where the benefits of multirate data converters and analog-digital integrated systems can be fully exploited for wireless and wireline communication, and for SOC implementation using multirate signal processing.

References

[1] Plassche, R. V. D., *Integrated Analog-To-Digital and Digital-To-Analog Converters*, Kluwer Academic Publishers, Dordrecht, The Netherlands, 1994.

[2] Proakis, J. G., and D. G. Manolakis, *Digital Signal Processing: Principles, Algorithms, and Applications*, 3d ed., Prentice Hall, Upper Saddle River, New Jersey, 1996.

[3] Oppenheim, A. V., and R. W. Schafer, *Discrete-Time Signal Processing*, Prentice Hall, Englewood Cliffs, New Jersey, 1989.

[4] Gray, R. M., W. Chou, and P. W. Wong, "Quantization Noise in Single-Loop Sigma-Delta Modulation with Sinusoidal Inputs," *IEEE Trans. on Communications*, Vol. 37, No. 9, pp. 956-968, September 1989.

[5] Gray, R. M., "Quantization Noise Spectra," *IEEE Trans. on Information Theory*, Vol. 36, No. 6, pp. 1220-1244, November 1990.

[6] Blumberg, L. M., "Time-Domain Properties of Instantaneous Uniform Quantization," *IEEE Trans. on Circuits and Systems II: Analog and Digital Signal Processing*, Vol. 40, No. 12, pp. 767-776, December 1993.

[7] Widrow, B., I. Kollar, and M. C. Liu, "Statistical Theory of Quantization," *IEEE Trans. on Instrumentation and Measurement*, Vol. 45, No. 2, pp. 353-361, April 1996.

[8] Gray, R. M., and D. L. Neuhoff, "Quantization," *IEEE Trans. on Information Theory*, Vol. 44, No. 6, pp. 2325-2383, October 1998.

[9] Gersho, A., "Charge-Transfer Filtering," *Proc. of IEEE*, Vol. 67, pp. 196-218, February 1979.

[10] Gregorian, A., K. Martin, and G. C. Temes, "Switched-Capacitor Circuit Design," *Proc. of IEEE*, Vol. 71, pp. 941-966, August 1979.

[11] Dooley, D. J. (eds.), *Data Conversion Integrated Circuits*, IEEE Press, New York, 1980.

[12] Franca, J., and Santos, S., "FIR Switched-Capacitor Decimators with Active-Delayed Block Polyphase Structures," *IEEE Trans. on Circuits and Systems*, Vol. 35, No. 8, pp. 1033-1037, August 1988.

[13] Martins, R. P., and J. E. Franca, "A Novel N-th Order IIR Switched-Capacitor Decimator Building Block with Optimum Implementation," *Proc. IEEE Int. Symp. on Circuits and Systems*, pp. 1471-1474, May 1989.

[14] Franca, J., and R. P. Martins, "IIR Switched-Capacitor Decimator Building Blocks with Optimum Implementation," *IEEE Trans. on Circuits and Systems*, Vol. 37, No. 1, pp. 81-90, January 1990.

[15] Uehara, G., and P. Gray, "Practical Aspects of High-Speed Decimation Filter Implementation," *Proc. IEEE Int. Symp. Circuits and Systems*, pp. 2308-2311, San Diego, California, May 1992.

[16] Martins, R. P., and J. E. Franca, "Cascade Switched-Capacitor IIR Decimating Filters," *IEEE Trans. on Circuits and Systems I: Fundamental Theory and Applications*, Vol. 42, No. 7, pp. 367-376, July 1995.

[17] Neves, R. F., A. Petraglia, S. K. Mitra, and J. E. Franca, "A New Mixed Analog-Digital Architecture for High-Frequency Narrow Bandwidth Channel Digitization," *Proc. IEEE Int. Symp. Circuits and Systems*, Vol. 1, pp. 437-440, Atlanta, Georgia, May 1996.

[18] Uchimura, K., T. Hayashi, T. Kimura, and A. Iwata, "Oversampling A-to-D and D-to-A Converters with Multistage Noise Shaping Modulators," *IEEE Trans. on Acoustics, Speech and Signal Processing*, Vol. 36, No. 12, pp. 1899-1905, December 1988.

[19] Temes, G. C., and J. C. Candy, "A Tutorial Discussion of the Oversampling Method for A/D and D/A Conversion," *Proc. IEEE Int. Symp. Circuits and Systems*, Vol. 2, pp. 910-913, 1990.

[20] Candy, J. C., and G. C. Temes, "Oversampling Methods for Data Conversion," *IEEE Pacific Rim Conference on Communications, Computers and Signal Processing*, pp. 498-502, May 1991.

[21] Candy, J. C., and G. C. Temes (eds.), *Oversampling Delta-Sigma Data Converters: Theory, Design and Simulation*, IEEE Press, New York, 1992.

[22] Aziz, P. M., H. V. Sorensen, and J. V. D. Spiegel, "An Overview of Sigma-Delta Converters," *IEEE Signal Processing Magazine*, pp. 61-84, January 1996.

[23] Nyquist, H., "Certain Topics in Telegraph Transmission Theory," *Transactions on AIEE*, Vol. 47, pp. 617-644, April 1928.

[24] Shannon, C. E., "Communication in the Presence of Noise," *Proc. of IRE*, pp. 10-21, January 1949.

[25] Chou, W., P. W. Wong, and R. M. Gray, "Multistage Sigma-Delta Modulation," *IEEE Trans. on Information Theory*, Vol. 35, No. 4, pp. 784-796, July 1989.

[26] Wong, P. W., and R. M. Gray, "Two-Stage Sigma-Delta Modulation," *IEEE Trans. on Acoustics, Speech and Signal Processing*, Vol. 38, No. 11, pp. 1937-1952, November 1990.

[27] Aziz, P. M., H. V. Sorensen, and J. V. D. Spiegel, "Multiband Sigma-Delta Modulation," *Electronics Letters*, Vol. 29, No. 9, pp. 760-762, April 1993.

[28] Joseph, D., L. Tarassenko, and S. Collins, "Analysis and Simulation of a Cascaded Delta-Sigma Modulator," *Conf. Publ. of IEE Advanced A/D and D/A Conversion Techniques and Their Applications,* pp. 54-57, July 1999.

[29] Dick, C., and F. Harris, "FPGA Signal Processing Using Sigma-Delta Modulation," *IEEE Signal Processing Magazine*, pp. 20-35, January 2000.

[30] Ribner, D. B., "A Comparison of Modulator Networks for High-Order Oversampled $\sum\Delta$ Analog-to-Digital Converters," *IEEE Trans. on Circuits and Systems,*, Vol. 38, No. 2, pp. 145-159, February 1991.

[31] Rio, R. D., F. Medeiro, B. Perez-Verdu, and A. Rodriguez-Vazquez, "High-Order Cascade Multibit $\sum\Delta$ Modulators for xDSL Applications," *Proc. IEEE Int. Symp. on Circuits and Systems*, pp. 37-40, Geneva, Switzerland, May 2000.

[32] Petraglia, A., F. Maloberti, and S. K. Mitra, "QMF-Based A/D Converters: Overview and New Results," *Proc. Int. Conf. Analog to Digital and Digital to Analog Conversion*, pp. 112-117, Wales, U.K., September 1991.

[33] Petraglia, A., and S. K. Mitra, "High Speed A/D Conversion Incorporating a QMF Bank," *IEEE Trans. on Instrumentation and Measurement*, Vol. 41, pp. 427-431, June 1992.

[34] Franca, J., A. Petraglia, and S. K. Mitra, "Multirate Analog-Digital Systems for Signal Processing and Conversion," *Proc. of IEEE*, Vol. 85, No. 2, pp. 242-262, February 1997.

[35] Lowenborg, P., H. Johansson, and L. Wanhammar, "A Class of Two-Channel Approximately Perfect Reconstruction Hybrid Analog/Digital Filter Banks," *Proc. IEEE Int. Symp. on Circuits and Systems,* pp. 579-581, Geneva, Switzerland, May 2000.

9

Statistical Signal Classification

9.1 Introduction

In this chapter, we introduce the statistical signal classification, which is built on the the fields of knowledge in probability and random processes, matrix algebra, and discrete-time signal processing. Statistical signal classification has developed into a widely practiced field with multidisciplinary applications. The role of statistical signal classification as an important component of automated intelligent digital systems continues to emerge as digital machine intelligence becomes a dominant technology in the twenty-first century. Statistical signal classification is used for digital signal preprocessing, description of measurements, and classification or decision making. An important aspect of the statistical signal classification system is that it can be built into the application specific integrated circuit (ASIC) chip (see Sastry, Ranganathan, and Remedios [1], Cheng and Xia [2]) or implemented by using a DSP chip.

Applications of statistical signal classification include image analysis and processing (see Gong and Kulikowski [3]), speech recognition (see Rabiner and Juang [4]), optical character recognition (see Blue et al. [5]), handwriting analysis (see Lu and Shridhar [6]), object recognition (see Olson [7]), face recognition (see Jia and Nixon

[8]), fingerprint identification (see Blue et al. [5]), signal analysis and classification (see Chen [9], Miao and Yu [10], Miao and Clements [11]), seismic signal analysis, medical diagnosis (see Miao, Sun, and Yu [12], Miao, Benkeser, and Nichols [13]), radar target classification (see Novak, Owirk, and Netishen [14]), human language translation, electrocardiograph signal understanding (see Schalkoff [15]), neural signals of the brain (see Echard et al. [16]), and computer vision. Recently, as digital signal processors rapidly develop, a statistical signal classification system using DSP-based system-on-chip (SOC) makes it possible for biomatrics identification and classification in real-time applications. Such SOC intelligence systems are useful for security purposes and can be built into a notebook personal computer or a wireless cellular phone to protect private data and information.

Statistical signal classification can be defined as the categorization of input digital data into identifiable digital pattern classes using the extraction of significant features or attributes of the digital data from a background of impertinent details. Generally, this can be referred to as a digital pattern recognition problem, briefly introduced in Section 1.3. Depending on whether the correct classification of some available digital signal pattern vectors is known, the process of statistical signal classification can be divided into supervised classification and unsupervised classification. Unsupervised classification, a very challenging problem, is in many cases the only alternative.

In supervised statistical signal classification, a set of training samples representing the type of features and underlying classes is labeled as to correct class. This yields a training or learning problem. If the form of the densities is known, the parameter estimation approaches can be used. However, in some cases, a potential problem of the parameter estimation approaches is that we require the underlying class-conditional pattern distribution in both form and parameters. Unfortunately, in practice we are not able to determine a specific form for the distributions. In addition, the form of the distributions chosen does not fit one of the estimate formulations. In this case, we need to use the nonparametric learning techniques.

The problem of unsupervised statistical signal classification can be viewed as the problem of estimating parameters of the density functions from the unlabeled digital signal samples drawn from an unknown distribution such as a mixture density.[1] The class-dependent density functions can be recovered from the mixture density, under certain conditions, by using a declustering procedure. The declustering process can be achieved by using *cluster analysis.* In the early research, use of histograms for declustering of mixture densities was suggested by Patrick and Hancock [17], Leboucher and Lowitz [18], and Yarman-Vural and Ataman [19].

Cluster analysis is unsupervised classification in which a priori information about the digital-signal sampled data is unavailable. The structure of the digital-signal sampled data as illustrated by clustering analysis frequently provides practical physical insights of the measured quantity. Cluster techniques including the *partitioning* and *hierarchical* methods successfully apply to various studies in different areas.

In order to improve the performance of the statistical signal classification, designing a reliable set of characteristic measurements and their associations for the representation of digital signal pattern vectors is essential. So, the study of statistical signal classification contains both the analysis of digital signal pattern characteristics and design methods for implementing the system.

In this chapter, we present a generalized structure of supervised and unsupervised statistical signal classification systems and the methods to implement these systems. The design of supervised or unsupervised statistical signal classification methods consists of several major stages: digital signal preprocessing, selection of measurement features and feature extraction, cluster analysis, specification of the optimum decision function, and estimation of the probability of misclassification. A block diagram of a supervised or unsupervised statistical signal classification system is shown in Figure 9.1. The upper block of the diagram represents the implementation system and the lower block of the diagram indicates the training process of the supervised or unsupervised statistical signal classification system. Note that herein, the cluster analysis as shown in Figure 9.1 is used only in the unsupervised statistical signal classification systems.

---

1. A mixture density is a weighted sum of individual probability density functions.

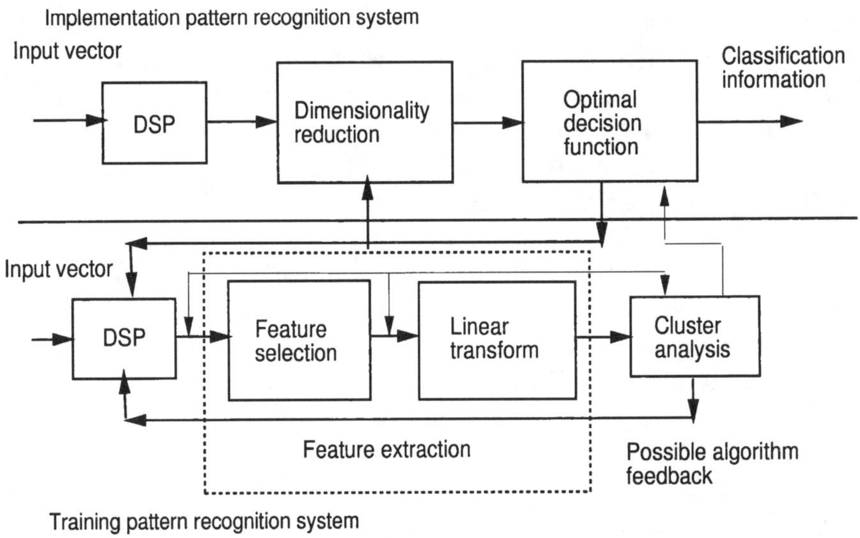

Figure 9.1 General block diagram of a supervised or unsupervised statistical signal classification system (see Miao [20]).

This chapter is organized as follows. Section 9.2 introduces statistical pattern representation, the definitions of terminologies, and the general digital signal preprocessor. Section 9.3 presents the feature extraction theory including selection of measurement features, feature selection, and dimensionality reduction. We also present two technical methods, including the *$\mathcal{F}$-test* and the *stepwise discriminat analysis*. These two methods are use to choose a subset of discriminatory measurement features from initial measurement features. For dimensional reduction, the *Karhunen-Loève* transform is presented. Substantial unsupervised learning and cluster analysis including four cluster techniques, *partitioning*, *hierarchical*, *density seeking*, and *clumping*, are given in Section 9.4. Furthermore, the partitioning and hierarchical approaches as well as characteristic methods for clustering evaluation are described in detail. Section 9.5 introduces signal classification theory including nonparametric signal classification and parametric signal classification. For parametric signal classification, we concentrate on the statistical signal classifiers based on *Gaussian* models and the methods to implement the classifier. Estimation of probability of misclassification error based on the parametric approach and the nonparametric approach is given in Section 9.6. Finally, this chapter is summarized in Section 9.7.

9.2 Statistical Pattern Representation

An abstract view of the statistical pattern representation is shown in Figure 9.2. We assume that there is a mapping between class-membership space H and pattern space P. The mapping is achieved via a relation function T_i for each class, ω_i, and may be probabilistic. Each class ω_i produces a subset of patterns in pattern space P, where the ith pattern is denoted by p_i. In most cases, however, these subspaces overlap to allow patterns p_i from different classes ω_i to share attributes. Then another mapping function M transforms patterns p_i from subspaces of P into observation or measured features m_i. This mapping M reflects the choice of measurement features for the statistical signal classification. Note that, in Figure 9.2, identical measurements or observations may come from different patterns p_i of pattern space P that correspond to different underlying classes ω_i. This suggests a potential problem of the statistical signal classification with ambiguity. However, the abstract view in Figure 9.2 is well suited for modeling in statistical signal classification.

9.2.1 Definition of Terminologies

For mathematical simplification, we define some terminologies in the statistical pattern representation for the statistical signal classification as follows:

Patterns

A *pattern* or a *pattern vector* is a set of measurements or observations. These measurements or observations can be such things as amplitude, weight, line, voltage, image, digital waveform, and so on. In addition, a pattern or a pattern vector may be converted from one representation to another. Naturally, a statistical signal classification is established based on patterns.

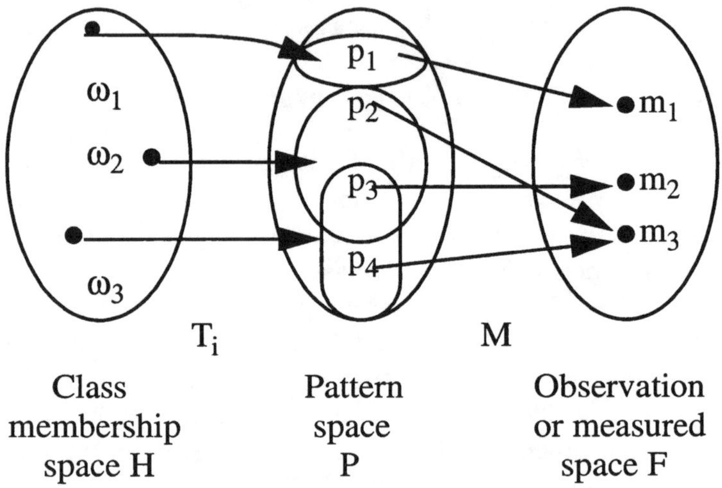

Figure 9.2 The mapping relationship in the abstract view of the statistical pattern representation.

Features

A *feature* is any extractable measurement and may be symbolic, numerical, or both. Features may be represented by using continuous, discrete-time, or even discrete binary variables. In addition, features may result from applying a feature extraction or a mapping to the input data. The key to extracting features is to ensure that all the features are computationally feasible and lead to good results for the statistical signal classification.

Feature Vector and Feature Space

Features are arranged in a d-dimensional vector that is referred to as the *feature vector* and denoted by $\mathbf{x}$. The feature vector $\mathbf{x}$ yields a multidimensional measurement *feature space*. A statistical signal classifier partitions feature space into regions for each class with labeled information.

Measure of Similarity

The *measure of similarity* is a quantity determining the likeness between one pattern and another pattern. For example, if one pattern, $\mathbf{x}$, is very similar to another pattern that belongs to pattern class ω_i, then we intuitively classify $\mathbf{x}$ into the pattern class ω_i.

The Euclidean distance, which is one measure of vector similarity between two patterns $\mathbf{x}$ and $\mathbf{y}$, is given by

$$\begin{aligned} D &= \|\mathbf{x}-\mathbf{y}\| \\ &= (\mathbf{x}-\mathbf{y})^T(\mathbf{x}-\mathbf{y}) \\ &= \sum_{i=1}^{d}(x_i - y_i), \end{aligned} \tag{9.1}$$

where d is a d-dimensional vector. Equation (9.1) is a measure of their similarity. If the distance is smaller, then the similarity is greater. However, there are other meaningful distance measures that are sometimes useful. For example, weighted distance measures are defined by

$$D = (\mathbf{x}-\mathbf{y})^T\mathbf{C}^{-1}(\mathbf{x}-\mathbf{y}), \tag{9.2}$$

where $\mathbf{C}$ is the covariance matrix with positive definite and symmetric. Another example, the *Mahalanobis distance* from a pattern $\mathbf{x}$ to the mean vector $\mathbf{m}$

$$D = (\mathbf{x}-\mathbf{m})^T\mathbf{C}^{-1}(\mathbf{x}-\mathbf{m}) \tag{9.3}$$

is also a useful measure of similarity, where $\mathbf{C}$ is the covariance matrix of a pattern class. The distance between two mean vectors $\mathbf{m}_i$ and $\mathbf{m}_j$

$$D = (\mathbf{m}_i-\mathbf{m}_j)^T\mathbf{C}^{-1}(\mathbf{m}_i-\mathbf{m}_j) \tag{9.4}$$

is often referred to as the Mahalanobis distance as well.

Measures of similarity can use a nondistance function such as a cosine of the angle between two pattern vectors $\mathbf{x}$ and $\mathbf{y}$,

$$cos\phi = \frac{\mathbf{x}^T\mathbf{y}}{\|\mathbf{x}\|\|\mathbf{y}\|}, \tag{9.5}$$

when $\mathbf{x}$ and $\mathbf{y}$ are oriented in the same direction with respect to the origin. For example, this $cos\phi$ function is useful when the regions

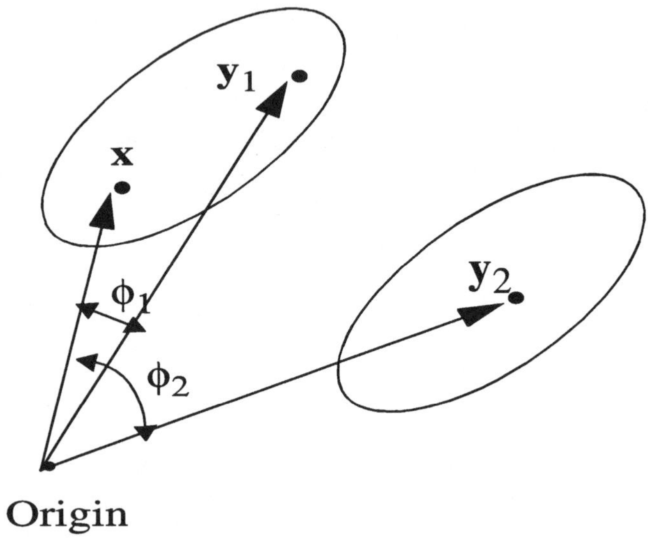

Figure 9.3 Illustration of a similarity measure.

of the pattern spaces tend to develop along with principal axes as shown in Figure 9.3. As we can see, the pattern vector $\mathbf{y}_1$ is more similar to the pattern vector $\mathbf{x}$ then the pattern vector $\mathbf{y}_2$ because $cos\phi_1$ is greater than $cos\phi_2$.

Pattern Classes

A *pattern class* is a set of patterns that share some common attributes and are known to originate from the same source in $\mathcal{H}$ as shown in Figure 9.2. The goal in the statistical signal classification is to identify suitable attributes and to form a measure of similarity along with an associated matching process.

Linear Separability

If a decision boundary that correctly classifies all the training pattern vectors for a class problem exists, the pattern vectors are said to be *linearly separable.* However, a linear separability does not exist in many applications. This leads to misclassification error.

Training

Training in the statistical signal classification uses labeled samples of types of patterns to be used in the actual application. Supervised training assumes a labeled data set with respect to pattern class, while unsupervised training must determine the partitions of the sample data, since the training data set does not have labeled pattern classes.

9.2.2 Digital Signal Preprocessor

A major function of the *digital signal preprocessor* is to filter out unwanted and multiple digital signals, and eliminate them from the aggregate statistics. In addition, the digital signal preprocessor transforms the raw input digital-signal sampled data to aid computational visual feasibility, and feature extraction and to minimize noise.[2]

The digital signal preprocessor extracts the measurement features from the digital sampled signal. Each of the resulting measurement features indicates a specific measurable characteristic of the signal. These measurement features do not share the same units, because they describe the same physical objects from various aspects. To enable the application of subsystems in the statistical signal classification as shown in Figure 9.1, the values of the output measurement features from the digital signal preprocessor are normalized. The normalization is accomplished by dividing each parameter by its sample standard deviation (or maximum, or median, or mean value), making the entire digital parameter space roughly "circular" in spread.

The digital measurement features contain two kinds of signal pattern features: the *intraset measurement features* and the *interset measurement features*. The intra-set measurement features are the characterizing attributes common to all the signal pattern

---

2. Noise is generalized to represent a number of imperfect circumstances. Noise includes nonlinear distortions from the analog-to-digital (A/D) converter or errors in input signal measurement, errors in digital signal preprocessing, errors in selection of measurement features and feature extraction, and errors in training digital-signal sampled data.

classes under consideration. Those signal measurement features carry no discriminatory information and should be ignored. The interset measurement features are used to represent the differences between signal pattern classes. The extraction of the interset signal measurement features has been recognized as an important problem in the design of the statistical signal classification system. However, in practice, the determination of a complete set of discriminatory measurement features is extremely difficult in most problems of statistical signal classification. Fortunately, in some cases, we are able to find some of the discriminatory measurement features based on the observed sampled signals.

The digital signal preprocessor is treated in the general area of digital signal processing (DSP). More detailed mathematical treatment of DSP can be found in previous chapters of this book or in one of the many excellent textbooks including Dudgeon and Mersereau [21], Quackenbush, Barnwell, and Clements [22], Oppenheim and Schafer [23], and Proakis and Manolakis [24].

Example 9.1 Extraction of Digital Measurement Features in Character Classification

In this example, two characters "A" and "H" in the grid-measuring scheme can be identified using digital measurement features as shown in Figure 9.4. As we can see, the entire grid block has the 8×8 element. Then, digital signal measurements can be arranged to form a signal measurement vector, $\mathbf{x} = (x_1, x_2, ..., x_{64})^T$, where each element x_i is defined as having the value 1 if the ith cell contains part of the character, and otherwise is assigned the value 0. Therefore, the characters in the grid-measuring scheme can easily be classified by using the above signal measurement vector $\mathbf{x}$, since the difference between signal measurement vectors $\mathbf{x}$ is based on the characters "A" and "H."

Example 9.2 Extraction of the Measurement Features in Digital Waveform Classification

Figure 9.5(a) shows a plot of digital waveforms sampled at 100 MHz with 10-bit resolution from an A/D converter (see Miao and Clements [11]). The vertical axis is voltage and the horizontal axis

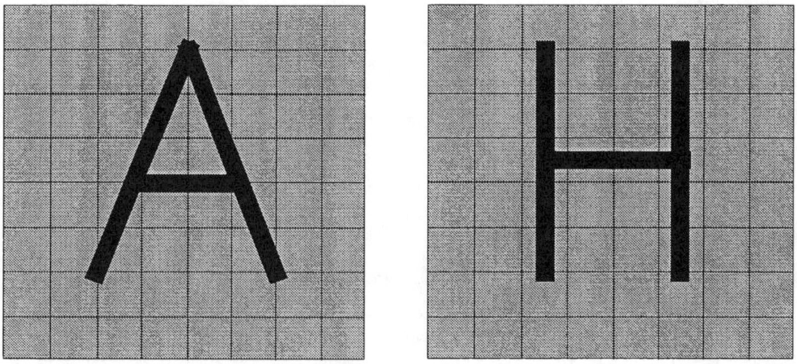

Figure 9.4 A simple scheme for the generation of digital signal measurement vectors in the character classification.

is discrete sampling time. The digital waveform is obtained from an alpha-constant of a digital air monitor with radon and plutonium sources.

To establish the statistical signal classifier, one of the most important tasks is to extract a set of measurement features. Considering Figure 9.5(b), the initial extraction of measurement features includes VT, height, area, rise time, decay time, mean time, V1, and V2. The extended signal measurement features contain:

1. Average acceleration rate (AAR) = (Height)/(Rise time)
2. Plurality index (PI) = (VT - V2)/(Mean time)
3. Resistance index (RI) = (VT - V2)/(VT)
4. Peak and valley ratio (PVR) = (VT)/(V2)
5. Rise and decay time ratio (RDTR) = (Rise time)/(Decay time)
6. Rise and mean time ratio (RMTR) = (Rise time)/(Mean time)
7. Energy rate (ER) = (Energy)/(Mean time)

Note that some of the above measurement features may not be useful for revealing differences among the pattern classes. Therefore, selection of a subset of discriminant measurement features is important. In the next section, we will introduce feature extraction theory, with a discussion of selecting a subset

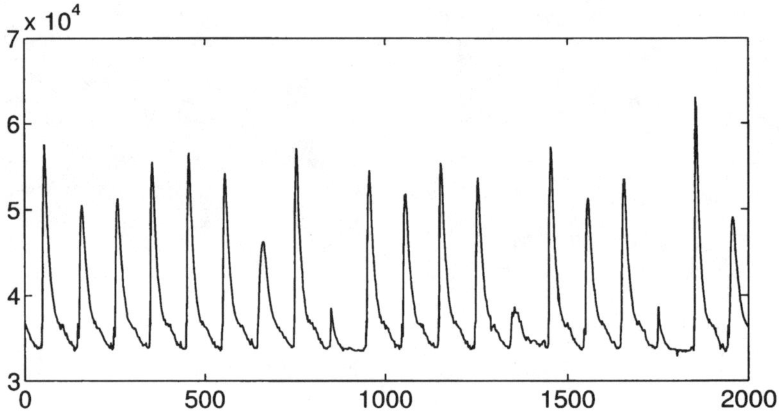

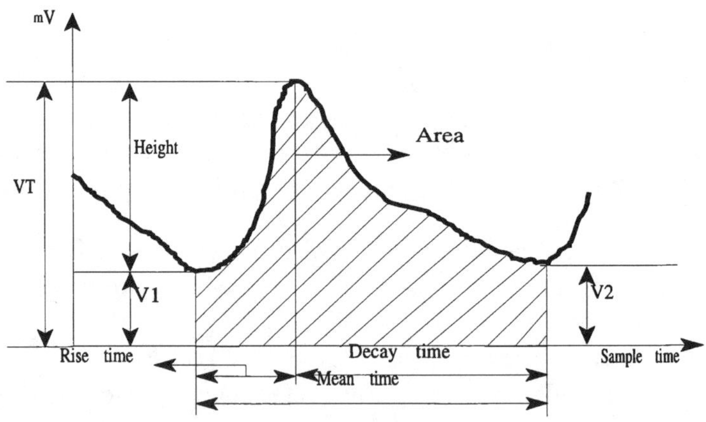

Figure 9.5 (a) The plot of originally digital waveforms with radon and plutonium sources sampled at 100 MHz with a 10-bit A/D converter from an alpha-constant of a digital air monitor system. (b) A sketch of extracting all the possibly digital measurement features in the statistical signal classification.

of discriminant measurement features, feature selection, and dimensionality reduction.

9.3 Feature Extraction Theory

The performance of a supervised or unsupervised statistical signal classification often depends crucially on the performance of the feature extractor. Usually, the methods by which digital measurement features are selected are intuitive and empirical, drawing upon the designer's knowledge and experience with the particular problem. One of the most important guides to feature selection is that the measurement features should be insensitive to irrelevant variations, such as limited amounts of translations, rotations, scale changes, and so on, while emphasizing differences that are important for distinguishing between different pattern classes.

The analysis of feature extraction is problem-dependent. In other words, the extraction of appropriate measurement features for statistical signal classification depends on the pattern vectors and the number of pattern classes under consideration. There are many approaches with respect to the analysis of feature extraction. In our presentation, analytical approaches to feature extractor design are restricted to *selection of measurement features*, *feature selection*, and *dimensionality reduction*. All of the approaches assume that a relatively large number of digital measurement features are available for analysis.

9.3.1 Selection of Measurement Features

Given the digital signal pattern sample of N observations on training data, an important question to ask is whether or not all of the signal measurement features are necessary for providing good discrimination. Often, the answer is no. The number of features measured on the pattern samples is called the dimensionality, Q, in the statistical signal classification problem. One would expect that an increase in the dimensionality of a particular problem would lead to an increase in the performance of a statistical signal classifier, since additional information becomes available. However, in practice, the addition of further signal measurement features can lead to a decrease in the classification performance. Because the number of features in the discrimination model increases with the dimensionality Q, the number of training objects N must be increased correspondingly. Otherwise, the increase in number of measurement features for a fixed data set will lead to poorly posed settings. In such cases, estimates of parameters have very large variances (see Aeberhard, Coomans, and Vel [25], Fukunaga and Hayes [26], Miao [20]). In extreme cases, no satisfactory training is possible by direct means.

It has been shown that there is a "peaking" phenomenon in the finite-sample-size case. With the addition of new signal measurement features, the probability of a misclassification error initially drops, attains a minimum, and then begins to increase. The number of measurement features at which the probability of misclassification error is minimum is called the optimal number. This depends on the pattern sample size, the type of classification function, and the distributions of the pattern vectors. Most importantly, the number relies on the effectiveness of features and their ordering. Even though there is no limitation on the number of measurement features, a small number will clearly lead to a more manageable and less time-consuming learning process. Furthermore, reducing the number of initial measurement features can lead to economies in the design of the digital hardware and software. Therefore, selecting a subset of discriminatory measurement features from the initial set that best reveals the difference among the pattern classes is necessary.

The selection of measurement features is very important in designing a supervised or unsupervised classification machine, and is a major factor in its success or failure. There is no general theory on selection of measurement features, because the selection of measurement features usually depends on the particular signal classification problem under consideration. However, the selection of measurement features can depend on prior knowledge of or expert experience with respect to the particular statistical signal classification problem. Then, the digital pattern space is defined by the selection of measurement features.

In this section, we introduce two statistical techniques, the *$\mathcal{F}$-test* and *stepwise discriminant analysis*, for the selection of discriminatory measurement features from the initial measurement features. In practice, the $\mathcal{F}$-test and the stepwise discriminant analysis are frequently used.

The $\mathcal{F}$-test chooses features based on the ratio of dispersion within each pattern class and squared distance of pattern class to the overall average of total pattern samples. When the $\mathcal{F}$-test ratio is large, the pattern classes have good separation. The stepwise discriminant analysis procedure can select a subset of measurement features that best reveals differences among the pattern classes. This method produces a good discrimination model using forward selection, backward elimination, and/or stepwise selection. The stepwise discriminant analysis is done under the assumption that the pattern classes are approximately multivariate Gaussian distributed with a common covariance matrix. If the $\mathcal{F}$-test or the stepwise discriminant analysis is used carefully with cross-validation, it can be a valuable tool in selecting a subset of key digital signal measurement features.

The $\mathcal{F}$-test

Consider a set of N digital signal pattern samples S_{ij} divided into M pattern classes and described by Q measurement features, where $i = 1, ..., N$, and $j = 1, ..., Q$. Let $G_T(j)$ be the average of the pattern samples. $G_k(j)$ is the average of the pattern samples belonging to the pattern class C_k with the number of pattern samples N_k (where $k = 1, ..., M; j = 1, ..., Q$):

$$G_T(j) = \frac{1}{N}\sum_{i=1}^{N} S_{ij} \qquad j = 1, ..., Q, \tag{9.6}$$

$$G_k(j) = \frac{1}{N_k}\sum_{i \in C_k}^{N_k} S_{ij} \quad k = 1, ..., M \text{ and } j = 1, ..., Q. \tag{9.7}$$

We also define $W(j)$ to be the dispersion within a jth pattern class,

$$W(j) = \sum_{k=1}^{M}\sum_{i \in C_k}^{N_k} \frac{1}{N}[S_{ik} - G_k(j)]^2 \qquad j = 1, ..., Q. \tag{9.8}$$

$D(j)$ is the squared distance of the jth pattern class to the overall average $G_T(j)$,

$$D(j) = \frac{1}{N} \sum_{k=1}^{M} N_k[G_k(j) - G_T(j)]^2 \qquad j = 1, ..., Q. \tag{9.9}$$

When $D(j)$ is large and $W(j)$ is small, pattern classes will be well separated. Therefore, selection of the jth signal measurement feature with discriminating power is determined by using the $\mathcal{F}$-test

$$\mathcal{F}(j) = \left(\frac{N-1-M}{M-1}\right) \left[\frac{D(j)}{W(j)}\right]. \tag{9.10}$$

Note that (9.10) is also referred to as the $\mathcal{F}$ distribution, as introduced in Section 3.3.8.

The detailed $\mathcal{F}$-test procedures for choosing the discriminate measurement feature may be described in the following steps:

1. State the null hypothesis H_0 and specify the significance level α under $\mathcal{F}$ distribution.
2. Initialize by selecting a signal measurement feature with the highest $\mathcal{F}$-test ratio based on (9.10).
3. Select the next signal measurement feature with the highest $\mathcal{F}$-test ratio at the significance level α and examine its correlation with signal measurement features previously selected.
4. If the correlation is high, then ignore this signal measurement feature. Otherwise, keep this signal measurement feature.
5. If this is the last signal measurement feature, then stop. Otherwise, go to step 3.

Stepwise Discriminant Analysis

The stepwise discriminant analysis is a modified version of the $\mathcal{F}$-test algorithm (see Costanza and Afifi [27], and Jennrich [28]). This analysis selects a subset of signal measurement features to produce a good discrimination model using forward selection, backward elimination, or stepwise selection.

As a basis for stepwise selection, the Wilks' Λ-statistic has been used to describe a within class (see Wilks [29]),

$$\Lambda(j) = \frac{| W(j) |}{| W(j) + D(j) |} \qquad j = 1, ..., Q. \tag{9.11}$$

In attempting to study the influence of one additional measurement feature, it is useful to define the partial Λ-statistic

$$\Lambda(j+1) = \frac{\Lambda(1, 2, ..., Q, Q+1)}{\Lambda(j)}. \tag{9.12}$$

The test procedure of stepwise discriminant analysis essentially compares two models: (1) the full discriminant model contains $X_1, X_2, ..., X_Q$ and X_{Q+1} as independent signal measurement features; (2) the reduced discriminant model includes $X_1, X_2, ..., X_Q$, but not X_{Q+1}. The goal of the test procedure is to decide which model is more appropriate based on how much additional information X_{Q+1} provides about the discriminant model over that already provided by $X_1, X_2, ..., X_Q$. The partial $\mathcal{F}$-test is usually used in the test procedure to determine the discrimination power of the above two models. The corresponding partial $\mathcal{F}$-test statistic based on (9.12) is given by (see Rao [30])

$$\mathcal{F} = \left(\frac{N - M - Q}{M - 1} \right) \left[\frac{1 - \Lambda(j+1)}{\Lambda(j+1)} \right]. \tag{9.13}$$

The partial $\mathcal{F}$-test is used to examine the significance level α of the change from $\Lambda(j)$ to $\Lambda(j+1)$ under the null hypothesis H_0, which is a test of the improvement in the discriminating power by introducing another measurement feature.

Stepwise selection in stepwise discriminant analysis begins with no measurement features in the model. At each step, if the feature fails to meet the null hypothesis criterion based on (9.13), then the feature is removed. Otherwise, the feature that contributes most to the discriminatory power of the model is added. The process of stepwise selection stops after examining all the signal measurement features.

The final set of measurement features selected by either the $\mathcal{F}$-test or the stepwise discriminant analysis is not guaranteed necessarily to be the best possible set of measurement features. However, the $\mathcal{F}$-test or the stepwise discriminant analysis is able to determine a subset of key measurement features based on the knowledge of the sampled signal pattern data and cross-validation.

9.3.2 Feature Selection

Feature selection may be considered as a special case of linear feature extraction in which some measurement features receive weights of zero. Feature selection differs from selection of measurement features, the latter being a selection procedure that defines the Q measurement features for pattern classes.

Optimal feature selection can only be done by examining all possible sets of P measurement features chosen from an initial Q measurement features. Exhaustive searching requires the order of $C_Q^P = \frac{Q!}{P!(Q-P)!}$ possible subsets to be examined (see Young and Calvert [31]). For even moderate values of P and Q, this approach often results in excessive computation. For example, select $P = 10$ out of $Q = 25$, then, $C_Q^P = 3.269 \times 10^6$.

The main goal of feature selection is to choose P features from a given set of Q ($P \ll Q$) without significantly degrading the performance of the statistical signal classification system. In other words, the probability of a misclassification error after applying feature selection should be kept as low as possible. Unfortunately, a direct determination of the probability of misclassification error for feature selection is usually not possible in practice. Therefore, many analytical approaches have been proposed to find indirect criteria to serve as a guide (see Tou and Gonzalez [32]). However, we will focus on the more tractable problem of dimensionality reduction (or linear transformation), since feature selection can be considered as a special case of linear transformation, in that some measurement features have zero weights.

9.3.3 Dimensionality Reduction

The method of *dimensionality reduction* attempts to obtain a smaller number of features by forming combinations of the original ones. Both linear and nonlinear transformations have been suggested as tools. Overall, the performance of nonlinear transformations is better than the performance of linear transformations for the enhancement of class separation (see Fu [33], Siedlecki, Siedlecka, and Sklansky [34]). However, the implementation of nonlinear transformations generally needs much more complex computation than that of linear transformations. We will only discuss linear transformation in this section.

Let $\mathbf{X} = \{\mathbf{x}_1, \mathbf{x}_2, ..., \mathbf{x}_N\}$ be an N-dimensional pattern space Ω_x and $\mathbf{Y} = \{\mathbf{y}_1, \mathbf{y}_2, ..., \mathbf{y}_M\}$ be an M-dimensional feature space extracted from $\mathbf{X}$ by applying a linear transformation, typically with $M \ll N$. A linear *feature extractor* is a linear transformation that maps an N-dimensional signal pattern vector $\mathbf{X}$ in Ω_x into an M-dimensional signal feature space $\mathbf{Y}$ in Ω_y,

$$\mathbf{Y} = \mathcal{T}(\mathbf{X}), \quad \mathbf{X} \text{ in } \Omega_x \text{ and } \mathbf{Y} \text{ in } \Omega_y. \tag{9.14}$$

It should be noted that since typically $M \ll N$, the mapping is not one to one, and the inverse $\mathcal{T}^{-1}(\mathbf{Y})$ is not unique anymore.

So far, we have noted that there are many families of transformation techniques with different philosophies. Among linear transformations, the best known and most useful feature extraction scheme is the Karhunen-Loève transform. It is a modified version of *principal components analysis* that will be introduced in Chapter 10.

Suppose that $\mathbf{X} = \{\mathbf{x}_i\}$, where $i = 1, 2, ..., N$ is an N-dimensional pattern vector to be transformed onto a $\mathbf{Y}$, where $\mathbf{Y}$ is M-dimensional feature space. Each feature $\mathbf{x}_i$ can be described by an expansion of the form

$$\begin{aligned} \mathbf{x}_i &= \sum_{j=1}^{M} \beta_{ij} u_j \\ &= \beta_{i1}u_1 + \beta_{i2}u_2 + \beta_{i3}u_3 + ... + \beta_{iM}u_M, \end{aligned} \tag{9.15}$$

where u_j is an orthonormal system of vectors, that is $u_i^T u_j = \delta_{ij}$ (where $\delta_{ij} = 1$ if $i = j$, otherwise $\delta_{ij} = 0$ if $i \neq j$) and then $\beta_{ij} = \mathbf{x}_i^T u_j$, for $j = 1, 2, ..., M$. Now, let a subsystem of vectors $\{u_1, u_2, u_3\}$ be a truncation of $\mathbf{x}_i$ to projections $\mathbf{z}_i$ such that the sum of squared error, $\mathcal{E}_{SSE}$, is minimum,

$$\mathbf{z}_i = \beta_{i1}u_1 + \beta_{i2}u_2 + \beta_{i3}u_3 + C, \tag{9.16}$$

where C is an N-dimensional vector. Then

$$\begin{aligned}\varepsilon_i &= \mathbf{x}_i - \mathbf{z}_i \\ &= \sum_{j=4}^{M} \beta_{ij} u_j - C \\ &= \sum_{j=4}^{M} (\mathbf{x}_i^T u_j) u_j - C. \end{aligned} \tag{9.17}$$

Therefore, the sum of squared error, $\mathcal{E}_{SSE}$, is

$$\begin{aligned}\mathcal{E}_{MSE} &= \sum_{i=1}^{N} \varepsilon_i^2 \\ &= \sum_{i=1}^{N} (\mathbf{x}_i - \mathbf{z}_i)^2. \end{aligned} \tag{9.18}$$

As we can see, the projection $\mathbf{z}_i$ can be displayed by using the orthonormal coordinates β_{i1}, β_{i2}, and β_{i3} when (9.18) is obtained. The sum of squared error $\mathcal{E}_{SSE}$ can be minimized under the constraint $u_i^T u_j = \delta_{ij}$. Then, the optimality criterion is

$$\mathcal{E} = \mathcal{E}_{SSE} + \sum_{j=4}^{M} \lambda_j (1 - \|u_j\|^2), \tag{9.19}$$

where λ_j are *Lagrange* multipliers imposing unit-length constraints on $\{u_j\}$. Differentiating (9.19) with respect to C and setting the result to zero yields

$$\begin{aligned}\frac{\partial \mathcal{E}}{\partial C} &= \frac{\partial}{\partial C}\left[\sum_{i=1}^{N} \varepsilon_i^2 + \sum_{j=4}^{M} \lambda_j (1 - \|u_j\|^2)\right] \\ &= -2\sum_{i=1}^{N}\left[\sum_{j=4}^{M} (\mathbf{x}_i^T u_j) u_j - C\right] \\ &= -2\sum_{i=1}^{N} \varepsilon_i = 0. \end{aligned} \tag{9.20}$$

This, in turn, yields

$$\sum_{i=1}^{N} C = \sum_{i=1}^{N}\sum_{j=4}^{M}(\mathbf{x}_i^T u_j)u_j. \tag{9.21}$$

After the simplification of (9.21), we obtain

$$\begin{aligned} C &= \sum_{j=1}^{M}\left[\frac{1}{N}(\sum_{i=1}^{N}\mathbf{x}_i^T)u_ju_j\right] \\ &= \sum_{j=4}^{M}\mathbf{m}_x^T u_j u_j, \end{aligned} \tag{9.22}$$

where $\mathbf{m}_x^T = \frac{1}{N}\sum_{i=1}^{N}\mathbf{x}_i^T$. Hence, substituting (9.22) into (9.17) yields

$$\begin{aligned} \varepsilon_i &= \sum_{j=4}^{M}\mathbf{x}_i^T u_j u_j - \sum_{j=4}^{M}\mathbf{m}_x^T u_j u_j \\ &= \sum_{j=4}^{M}\left[(\mathbf{x}_i - \mathbf{m}_x)^T u_j\right] u_j, \end{aligned} \tag{9.23}$$

and substituting (9.23) into (9.18) yields

$$\begin{aligned} \mathcal{E}_{SSE} &= \sum_{i=1}^{N}\varepsilon_i^2 \\ &= \sum_{j=4}^{M} u_j^T\left[\sum_{i=1}^{N}(\mathbf{x}_i - \mathbf{m}_x)(\mathbf{x}_i - \mathbf{m}_x)^T\right] u_j \\ &= (N-1)\sum_{j=4}^{M} u_j^T C_x u_j, \end{aligned} \tag{9.24}$$

where $C_x = \frac{1}{N-1}\left[\sum_{i=1}^{N}(\mathbf{x}_i - \mathbf{m}_x)(\mathbf{x}_i - \mathbf{m}_x)^T\right]$ is a covariance matrix estimated over the data set $\mathbf{X}$. Differentiating $\mathcal{E}_{SSE}$ in (9.19) with respect to vectors u_j and setting it equal to zero gives:

$$\frac{\partial \mathcal{E}}{\partial u_j} = \frac{\partial}{\partial u_j}\left[\mathcal{E}_{SSE} + \sum_{j=4}^{M}\lambda_j(1 - \|u_j\|^2)\right]$$

$$\begin{aligned}
&= \frac{\partial}{\partial u_j}\left[(N-1)\sum_{j=4}^{M} u_j^T C_x u_j + \sum_{j=4}^{M} \lambda_j (1 - \|u_j\|^2)\right] \\
&= 2(N-1)\sum_{j=4}^{M} C_x u_j - 2\sum_{j=4}^{M} \lambda_j u_j \\
&= 2\sum_{j=4}^{M}[(N-1)C_x - \lambda_j]u_j \\
&= 0. \qquad (9.25)
\end{aligned}$$

This leads to the conclusion that u_j and λ_j are eigenvectors and eigenvalues of the covariance matrix $(N-1)C_x$, respectively. Therefore, the total transformation error is

$$\mathcal{E}_{SSEopt} = \sum_{j=4}^{Q} \lambda_j, \qquad (9.26)$$

where λ_j served in (9.19) and (9.25) as *Lagrange* multipliers, but are actually the eigenvalues associated with eigenvectors u_j, for $j = 4, 5, ..., N$.

If the projection eigenvectors u_1, u_2, and u_3 are associated with the three largest eigenvalues of the covariance matrix $(N-1)C_x$, we conclude that the minimum error is reached based on (9.26). Thus, each sampled data point of the Karhunen-Loève transform is projected into the three-dimensional component space determined by the three largest eigenvalues of the covariance matrix $(N-1)C_x$.

Use of the Karhunen-Loève transform for feature extraction is shown in Figure 9.6. The ellipses in Figure 9.6(a) are defined by the equation $\mathbf{x}^T\mathbf{R}\mathbf{x} = C$, where C is a constant and $\mathbf{R}$ is the covariance matrix of the mixture density $f_x(\mathbf{x})$. The eigenvectors, $\mathbf{u}_1$ and $\mathbf{u}_2$, are the major and minor axes, and the eigenvalues λ_1 and λ_2, where $\lambda_1 \geq \lambda_2$, represent the variances in these directions. If we are allowed to choose one component feature vector only, we should choose eigenvector $\mathbf{u}_1$. However, one major disadvantage of the Karhunen-Loève transform is that it does not consider the discrimination ability for the pattern classes (also see Duda and Hart [35]). Figure 9.6(b) shows this point of view. The mixture of the two signal pattern classes, ω_1 and ω_2, has a covariance matrix depicted by Figure 9.6(a).

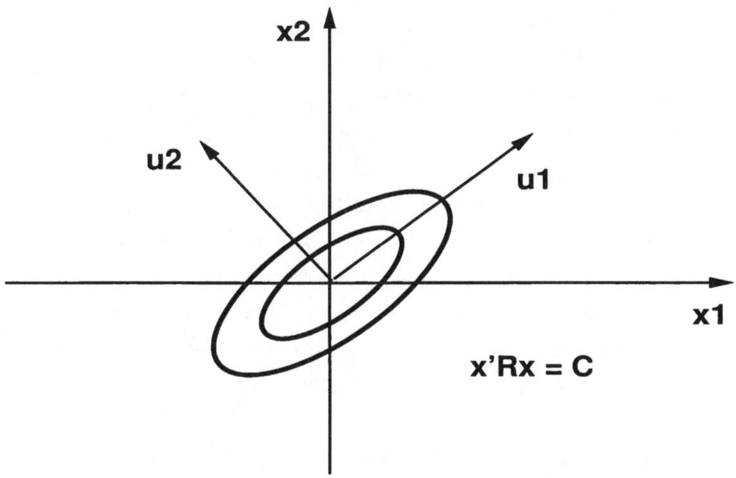

(a)

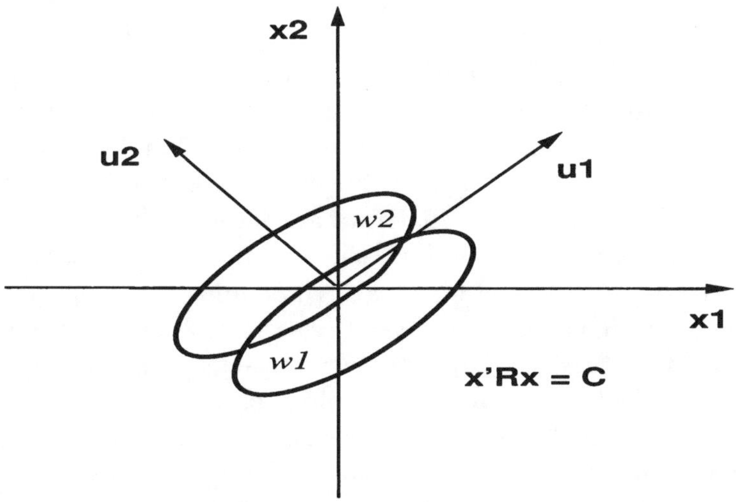

(b)

Figure 9.6 A scheme of feature extraction: (a) Feature extraction under normal signal pattern cases using the Karhunen-Loève transform. (b) Feature extraction under a pathological case using the Karhunen-Loève transform.

It should be obvious from the figure that for discrimination purposes one should choose the eigenvector $\mathbf{u}_2$ instead of the eigenvector $\mathbf{u}_1$. Thus, the Karhunen-Loève method is inadequate for this special case. We stress that this example is a contrived case. In practice, the Karhunen-Loève method often preserves discriminability.

In Chapter 10, we will describe a generalized principal components analysis based on the Karhunen-Loève transform. Also, we will introduce two relatively new transformations called *canonical discriminant analysis* and *optimal declustering analysis.* These two transformations give better performance of discrimination than principal components transformation under certain conditions. These facts are illustrated by the comprehensive results obtained from theoretical studies and practical applications.

9.4 Unsupervised Learning and Cluster Analysis

Unsupervised learning attempts to develop a pattern representation for the given training sample data without labeled information. In this case, the "natural clusters" are not apparent, but overlapping occurs as shown in Figure 9.7. Thus, in unsupervised learning, the objective is to partition the unlabeled training data into the clusters.

Unsupervised learning arises due to two possible situations as follows: (1) if no a priori label information is provided with the acquired digital sampled signals, we are faced with the problem of unsupervised statistical signal classification; (2) if the characteristic of the class-specific pattern-generating system is changing over time, a statistical signal classifier has to learn or adapt continuously. In both cases, learning happens without a teacher. Therefore, one technique called *cluster analysis* is often used to find a meaningful way of parametrically describing the signal pattern data. Then, these parameters provided by cluster analysis define the feature space.

Cluster analysis is a statistical tool that organizes the signal pattern vectors into groups or clusters. When successful, pattern vectors in the same cluster are very similar, whereas the pattern vectors in different clusters are distinct. In other words, cluster analysis places objects into groups or clusters such that objects in a given cluster are similar to each other in some sense, and objects in different clusters are dissimilar.

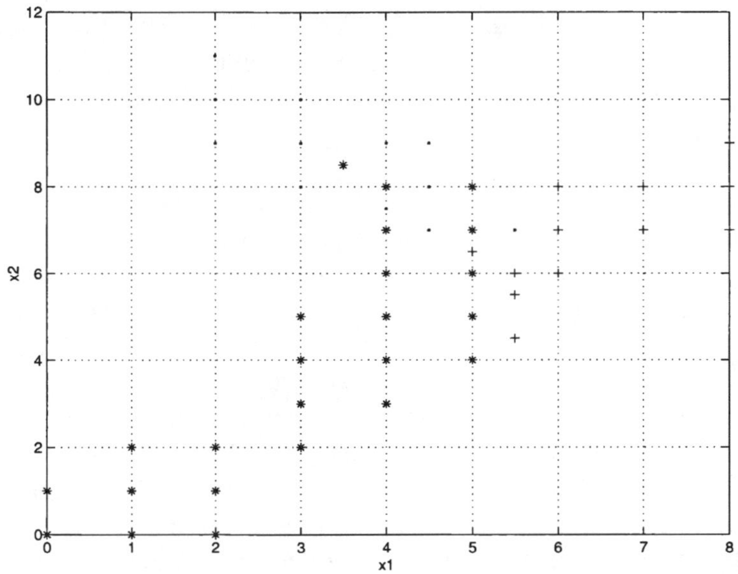

Figure 9.7 The "natural clusters" in the training data are not apparent and occur with overlapping, where "*," "·," and "+" indicate clusters ω_1, ω_2 and ω_3, respectively.

Generalization of cluster analysis is vague because most clustering methods have been developed in several different fields, with different definitions of clusters and similarity among objects. However, clustering methods can usually be divided into four groups:

1. *Partitioning* (or *disjoint*) clusters have each object in one and only one pattern cluster.
2. *Hierarchical* clusters are organized in such a way that one cluster may be entirely contained within another cluster. However, no other kind of overlap between clusters is allowed.
3. *Mode* (or *density*) seeking procedures assume that the pattern vectors will be allocated in a space in which there are several dense areas and regions. The areas between the dense areas and regions are very sparse. This method assumes the existence of original clusters.
4. *Clumping* is a probability or grade of membership for each object in each cluster. Clumping is different from

the above three techniques in that it permits the clusters to overlap. Clumping is also called *fuzzy* or *overlapping* clustering because this kind of cluster can be disjoint, hierarchical, or overlapping with any degree of overlap in cluster membership.

Partitioning clustering and hierarchical clustering belong to unsupervised clustering approaches. Those approaches employ general measures among pattern vectors and do not require a set of labeled training vectors. When the number of pattern vectors is large, hierarchical clustering may not be appropriate. In such cases, hierarchical methods have been shown to result in data partitions that are suboptimal in an agglomerate procedure.

Important introductions to cluster analyses can be found in Anderberg [36], Dubes and Jain [37], Everitt [38], Massart and Kaufman [39], and Milligan [40, 41]. Prospective researchers on statistical aspects of clustering methods should study the essential references by MacQueen [42], Scott and Symons [43], Hartigan [44, 45], Binder [46, 47], Wong and Schaack [48], and Wong and Lane [49]. The wide application value of cluster analysis in the field of statistical digital signal processing, featuring vector quantization, is described by Linde, Buzo, and Gray [50], Chang, Gray, and May [51], Gray [52], Rabiner and Juang [4], Miao [53], Abbas and Fahmy [54], and Barlaud et al. [55].

9.4.1 Hierarchical Approaches

In *hierarchical* approaches, the procedure proceeds sequentially so that only one pattern vector or group of vectors changes group membership during each step. The groups are nested according to the previous grouping at each step. Groups joined are never later separated in the clustering process. This hierarchical process is called *agglomerative* because the sizes of the clusters increase and the number of clusters decreases. Figure 9.8 shows a tree diagram for hierarchical clustering. The agglomerative process in hierarchical clustering begins to shift from the left branches to the right branches on the tree.

The value of a clustering criterion must be calculated to decide which groups are to be joined at each step in the hierarchical process.

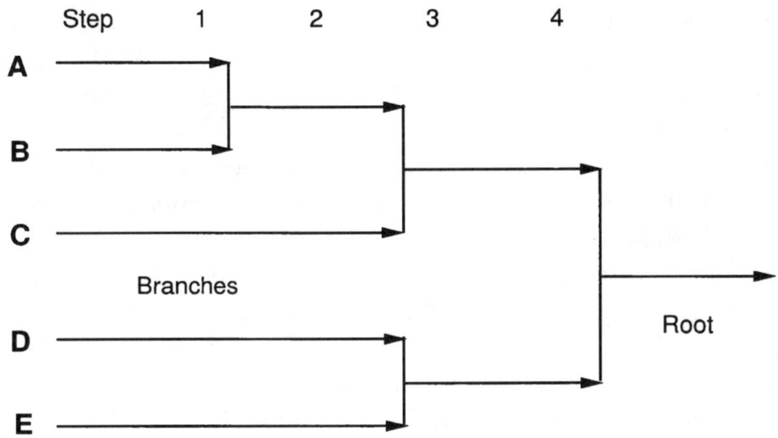

Figure 9.8 A general tree diagram for hierarchical clustering.

The clustering criterion is usually based on a measure of proximity between the groups. Such methods as *single linkage*, *complete linkage*, *centroid linkage*, and *average linkage* for proximity between groups can be used to determine which groups are to be joined.

Single Linkage

Let $\mathbf{x}_i$ be an ith pattern vector and D_{MN} be the distance or dissimilarity measure between clusters C_M and C_N:

$$\begin{aligned} D_{MN} &= min_{i \in C_M} min_{j \in C_N} \|\mathbf{x}_i - \mathbf{x}_j\|^2 \\ &= min_{i \in C_M} min_{j \in C_N} \left[(\mathbf{x}_i - \mathbf{x}_j)^T (\mathbf{x}_i - \mathbf{x}_j) \right] \\ &= min_{i \in C_M} min_{j \in C_N} \left(\mathbf{x}_i^T \mathbf{x}_i - 2\mathbf{x}_j^T \mathbf{x}_i + \mathbf{x}_j^T \mathbf{x}_j \right). \quad (9.27) \end{aligned}$$

It is also assumed that clusters C_M and C_N are joined to form a new signal pattern cluster C_S. Then the combinatorial formula gives the distance between the new cluster C_S and any other cluster C_T for single linkage as follows:

$$D_{ST} = min(D_{MT}, D_{NT}). \quad (9.28)$$

In the method of single linkage, the distance between two clusters is the minimum distance between a signal vector in one cluster

C_M and a vector in the other cluster C_N. One problem with the single linkage method is that it sacrifices performance in the recovery of compact clusters without imposing constraints on the shape of clusters. Another problem with the single linkage method is that it tends to chop off the tails of distributions before separating the main clusters. However, the single linkage method is able to detect elongated and irregular clusters during the agglomerative procedure in the hierarchical clustering.

The single linkage method was invented by Florek et al. [56], McQuity [57], and Sneath [58]. Later, many desirable theoretical properties of single linkage method were proved by Jardine and Sibson [59], Fisher and Ness [60], and Hartigan [45]. In the 1980s, the problems of applying the single linkage method in cluster studies were identified by Milligan [61].

Complete Linkage

Let $\mathbf{x}_i$ be an ith pattern vector and D_{MN} be the distance or dissimilarity measure between signal pattern cluster C_M and C_N:

$$\begin{aligned} D_{MN} &= max_{i \in C_M} max_{j \in C_N} \|\mathbf{x}_i - \mathbf{x}_j\|^2 \\ &= max_{i \in C_M} max_{j \in C_N} \left[(\mathbf{x}_i - \mathbf{x}_j)^T (\mathbf{x}_i - \mathbf{x}_j)\right] \\ &= max_{i \in C_M} max_{j \in C_N} \left(\mathbf{x}_i^T \mathbf{x}_i - 2\mathbf{x}_j^T \mathbf{x}_i + \mathbf{x}_j^T \mathbf{x}_j\right). \end{aligned} \tag{9.29}$$

It is also assumed that the clusters C_M and C_N are joined to form a new cluster C_S. Then the combinatorial formula gives the distance between the new cluster C_S and any other cluster C_T for complete linkage in the following way:

$$D_{ST} = max(D_{MT}, D_{NT}). \tag{9.30}$$

In the method of complete linkage, the distance between two clusters is the maximum distance between a vector in one cluster C_M and a vector in the other cluster C_N. One problem of the complete linkage method is that it is strongly biased toward producing clusters with roughly equal diameters. Another problem of the complete linkage method is that it severely distorts the clusters if they have outliers, possibly introduced by noise.

Centroid Linkage

Let $\mathbf{x}_i$ be the ith pattern vector and D_{MN} be the distance or dissimilarity measure between clusters C_M and C_N:

$$\begin{aligned} D_{MN} &= \|\overline{\mathbf{x}}_M - \overline{\mathbf{x}}_N\|^2 \\ &= (\overline{\mathbf{x}}_M - \overline{\mathbf{x}}_N)^T (\overline{\mathbf{x}}_M - \overline{\mathbf{x}}_N), \end{aligned} \tag{9.31}$$

where,

$$\overline{\mathbf{x}}_M = \frac{1}{N_M} \sum_{i \in C_M} \mathbf{x}_i \tag{9.32}$$

$$\overline{\mathbf{x}}_N = \frac{1}{N_N} \sum_{j \in C_N} \mathbf{x}_j. \tag{9.33}$$

It is also assumed that clusters C_M and C_N are joined to form a new cluster C_S. Then the combinatorial formula below gives the distance between the new cluster C_S and any other cluster C_T for centroid linkage:

$$D_{ST} = \frac{(N_M D_{MT} + N_N D_{NT})}{N_T} - \frac{N_M N_N D_{MN}}{N_T^2}. \tag{9.34}$$

In the method of centroid linkage, the distance between two clusters is *Euclidean distance* between the mean of a cluster C_M and the mean of another cluster C_N. The centroid linkage is more robust in overcoming the outlier (noise) problems than most other hierarchical methods mentioned above.

Average Linkage

Let $\mathbf{x}_i$ be the ith pattern vector and D_{MN} be the distance or dissimilarity measure between clusters, C_M and C_N:

$$\begin{aligned} D_{MN} &= \frac{1}{N_M N_N} \sum_{i \in C_M} \sum_{j \in C_N} \|\mathbf{x}_i - \mathbf{x}_j\|^2 \\ &= \frac{1}{N_M N_N} \sum_{i \in C_M} \sum_{j \in C_N} \left[(\mathbf{x}_i - \mathbf{x}_j)^T (\mathbf{x}_i - \mathbf{x}_j)\right] \\ &= \frac{1}{N_M N_N} \sum_{i \in C_M} \sum_{j \in C_N} \left(\mathbf{x}_i^T \mathbf{x}_i - 2\mathbf{x}_j^T \mathbf{x}_i + \mathbf{x}_j^T \mathbf{x}_j\right) \end{aligned}$$

$$= \frac{1}{N_M} \sum_{i \in C_M} \mathbf{x}_i^T \mathbf{x}_i - 2\overline{\mathbf{x}}_N^T \overline{\mathbf{x}}_M + \frac{1}{N_N} \sum_{j \in C_N} \mathbf{x}_j^T \mathbf{x}_j. \quad (9.35)$$

Adding the terms $\pm 2\overline{\mathbf{x}}_M^T \overline{\mathbf{x}}_M$ and $\pm 2\overline{\mathbf{x}}_N^T \overline{\mathbf{x}}_N$ on the right of (9.35) gives,

$$\begin{aligned} D_{MN} &= \left(\overline{\mathbf{x}}_M^T \overline{\mathbf{x}}_M - 2\overline{\mathbf{x}}_N^T \overline{\mathbf{x}}_M + \overline{\mathbf{x}}_N^T \overline{\mathbf{x}}_N\right) \\ &\quad + \left(\frac{1}{N_M} \sum_{i \in C_M} \mathbf{x}_i^T \mathbf{x}_i - 2\overline{\mathbf{x}}_M^T \overline{\mathbf{x}}_M + \overline{\mathbf{x}}_M^T \overline{\mathbf{x}}_M\right) \\ &\quad + \left(\frac{1}{N_N} \sum_{j \in C_N} \mathbf{x}_j^T \mathbf{x}_j - 2\overline{\mathbf{x}}_N^T \overline{\mathbf{x}}_N + \overline{\mathbf{x}}_N^T \overline{\mathbf{x}}_N\right) \\ &= \|\overline{\mathbf{x}}_M - \overline{\mathbf{x}}_N\| + \frac{1}{N_M} \sum_{i \in C_M} \left(\mathbf{x}_i^T \mathbf{x}_i - 2\overline{\mathbf{x}}_i^T \overline{\mathbf{x}}_M + \overline{\mathbf{x}}_M^T \overline{\mathbf{x}}_M\right) \\ &\quad + \frac{1}{N_N} \sum_{j \in C_N} \left(\mathbf{x}_j^T \mathbf{x}_j - 2\overline{\mathbf{x}}_j^T \overline{\mathbf{x}}_N + \overline{\mathbf{x}}_N^T \overline{\mathbf{x}}_N\right) \\ &= \|\overline{\mathbf{x}}_M - \overline{\mathbf{x}}_N\| + \frac{1}{N_M} \sum_{i \in C_M} \|\mathbf{x}_i - \overline{\mathbf{x}}_N\| \\ &\quad + \frac{1}{N_N} \sum_{j \in C_N} \|\mathbf{x}_j - \overline{\mathbf{x}}_N\|. \end{aligned} \quad (9.36)$$

As we can see, (9.36) is a modified version of (9.31) in the method of centroid linkage. It is also assumed that clusters C_M and C_N are joined to form a new cluster C_S. Then the combinatorial formula gives the distance between the new cluster C_S and any other cluster C_T for the average linkage in the following way:

$$D_{ST} = \frac{(N_M D_{MT} + N_N D_{NT})}{N_T}. \quad (9.37)$$

In the method of average linkage, the distance between two clusters is the average distance between pairs of vectors, one in each of the clusters C_M and C_N. The average linkage joins clusters with small variances and is slightly biased toward producing clusters with the same variance. However, the performance of the average linkage method is better than the performance of centroid linkage.

9.4.2 Partitioning Approaches

Partitioning methods begin with a given number of clusters. An initial allocation of the vectors into M groups is fulfilled by determining the proximity between each signal pattern vector and each group. Once the vectors change groups, the proximity is recalculated. The process of redistributing vectors to new groups continues until all vectors belong to the closest group. Unlike the hierarchical method, the partitioning method allows a vector to change group membership throughout the process of cluster formation.

K-Means Algorithm

A common algorithm of the partition method, which evaluates the proximity between groups using Euclidean distance between group "centroids," is the *k-means algorithm* and its variants. The k-means algorithm is based on the minimization of a performance index and requires computation cluster centers. This algorithm iteratively subdivides M training vectors into L clusters ($M \gg L$) and is guaranteed to converge to the locally optimal clustering. The k-means algorithm has been extensively addressed in the literature. Figure 9.9 shows a typical executive flowchart of the k-means algorithm. As we can see, each vector $\mathbf{x}_m$ is classified (or quantized) into the output of the cluster vector $\mathbf{x}$ that is nearest to it according to Euclidean distance (or whatever distortion measurement is adopted).

A typical procedure of the basic k-means algorithm can be described in the following steps:

1. Initialize by setting the iteration number $i = 0$ and choose a relatively small value for the vector ε. Choose a set of output vectors $\mathbf{Z}_k(0)$, $1 \leq k \leq L$, where k is the number of cluster centers.
2. Classify the vectors $\{\mathbf{x}_m, 1 \leq m \leq M\}$ into the clusters C_k by

$$\mathbf{x}_m \in C_k(i) \quad \text{iff} \parallel \mathbf{x}_m - \mathbf{Z}_k(i) \parallel \leq \parallel \mathbf{x}_m - \mathbf{Z}_j(i) \parallel, \tag{9.38}$$

for all $k \neq j$, and $D(i) = \parallel \mathbf{x}_m - \mathbf{Z}_k(i) \parallel$ is Euclidean distance (also called the distortion measurement).

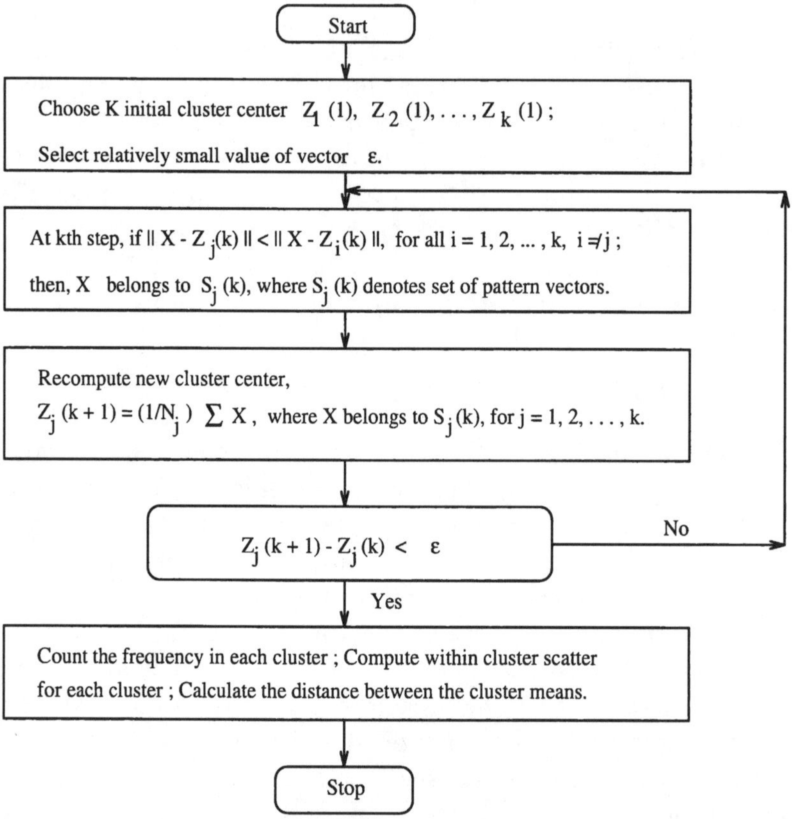

Figure 9.9 An executive flowchart of the k-means algorithm.

3. Recalculate the output vectors of every cluster by computing the centroid of the vectors that fall in each cluster C_k,

$$\mathbf{Z}_k(i) = \frac{1}{M_k} \sum_{m \in C_k} \mathbf{x}_m, \quad 1 \leq k \leq L. \tag{9.39}$$

Also compute Euclidean distance (or resulting distortion

measurement) $D(i)$ at the ith iteration,[3]

$$D(i) = \| \mathbf{x}_m - \mathbf{Z}_k(i) \| \quad 1 \le k \le L \text{ and } \mathbf{x}_m \in C_k(i). \quad (9.40)$$

4. If the change $\mathbf{Z}_k(i-1) - \mathbf{Z}_k(i) < \varepsilon$ (or $D(i-1) - D(i) < \varepsilon$) in the average distortion is relatively small, then terminate. Otherwise, go to step 2.

Note that an optimal partitioning algorithm solution, by using the k-means algorithm, does not essentially suggest that the solution will be completely admissible. In other words, the correct classes may not be found even though the correct number of clusters has been selected and the algorithm has converged. Therefore, we suggest that additional visual tools are needed to verify the cluster performance.

Example 9.3 Cluster Analysis Using k-Means Algorithm

In order to illustrate the procedure of the k-means algorithm, we consider the pattern vectors as shown in Figure 9.10. Using the k-means algorithm to find the cluster is described as follows:

1. Step 1. Choose the number of clusters, $K = 3$, corresponding to classes, C_1, C_2, and C_3. Initial the cluster centers, $\mathbf{z}_1(1) = \mathbf{x}_1 = (0,0)^T$, $\mathbf{z}_2(1) = \mathbf{x}_8 = (9,6)^T$, and $\mathbf{z}_3(1) = \mathbf{x}_{12} = (7,8)^T$.
2. Step 2. With the above initializations, we obtain

$$\begin{cases} \| \mathbf{x}_l - \mathbf{z}_1(1) \| < \| \mathbf{x}_l - \mathbf{z}_2(1) \| & \text{for } l = 1, 2, ..., 7 \\ \| \mathbf{x}_l - \mathbf{z}_1(1) \| < \| \mathbf{x}_l - \mathbf{z}_3(1) \| & \text{for } l = 1, 2, ..., 7 \end{cases} \quad (9.41)$$

and

$$\begin{cases} \| \mathbf{x}_k - \mathbf{z}_2(1) \| < \| \mathbf{x}_k - \mathbf{z}_1(1) \| & \text{for } k = 8, 9 \\ \| \mathbf{x}_k - \mathbf{z}_2(1) \| < \| \mathbf{x}_k - \mathbf{z}_3(1) \| & \text{for } k = 8, 9. \end{cases} \quad (9.42)$$

Therefore, $C_1(1) = \{\mathbf{x}_1, \mathbf{x}_2, ..., \mathbf{x}_7\}$ and $C_2(1) = \{\mathbf{x}_8, \mathbf{x}_9\}$. Similarly, the remaining pattern vectors are close to $\mathbf{z}_3(1)$, so $C_3(1) = \{\mathbf{x}_{10}, \mathbf{x}_{11}, ..., \mathbf{x}_{21}\}$.

---

3. Notice that if we set the number of cluster centers, $k = 2^B$, where B is the number of bits, and the distortion measurement, $D(i)$, is used to determine the convergence of the locally optimal clustering, then the k-means algorithm is the well-known method of *vector quantization*.

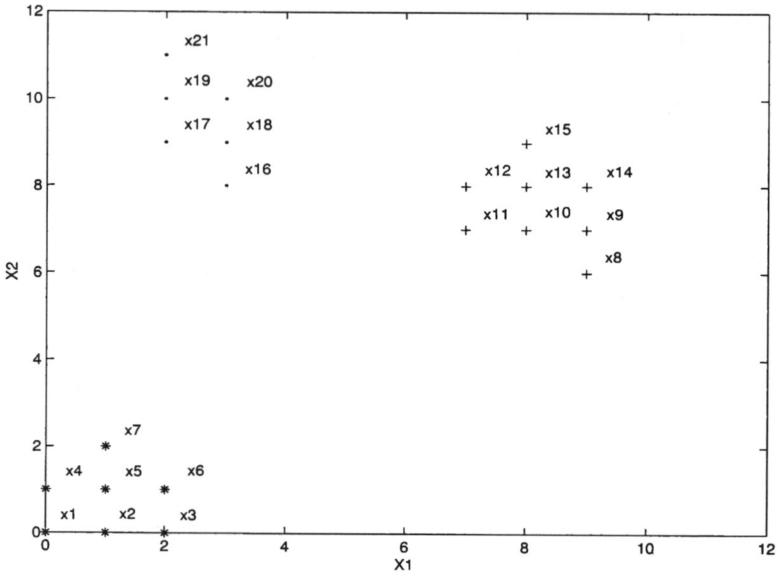

Figure 9.10 Sampled signal pattern vectors used to illustrate the procedure of the k-means algorithm to identify the signal pattern cluster classes.

3. Step 3. Updating the cluster centers, we obtain

$$\begin{aligned}
\mathbf{z}_1(2) &= \frac{1}{N_1} \sum_{\mathbf{x} \in C_1(1)} \mathbf{x} \\
&= \frac{1}{7}(\mathbf{x}_1 + \mathbf{x}_2 + \mathbf{x}_3 + \mathbf{x}_4 + \mathbf{x}_5 + \mathbf{x}_6 + \mathbf{x}_7) \\
&= \begin{pmatrix} 1.00 \\ 0.71 \end{pmatrix} \qquad (9.43)
\end{aligned}$$

$$\begin{aligned}
\mathbf{z}_2(2) &= \frac{1}{N_2} \sum_{\mathbf{x} \in C_2(1)} \mathbf{x} \\
&= \frac{1}{2}(\mathbf{x}_8 + \mathbf{x}_9) \\
&= \begin{pmatrix} 9.0 \\ 6.5 \end{pmatrix} \qquad (9.44)
\end{aligned}$$

$$\begin{aligned}\mathbf{z}_3(2) &= \frac{1}{N_3}\sum_{\mathbf{x}\in C_3(1)} \mathbf{x} \\ &= \frac{1}{12}(\mathbf{x}_{10}+\mathbf{x}_{11}+\mathbf{x}_{12}+...+\mathbf{x}_{19}+\mathbf{x}_{20}+\mathbf{x}_{21}) \\ &= \begin{pmatrix} 5.17 \\ 8.67 \end{pmatrix} \end{aligned} \quad (9.45)$$

4. Step 4. Since $\mathbf{z}_j(2) \neq \mathbf{z}_j(1)$, for $j = 1,2,3$, we return to step 2.
5. Step 2 (repeated). With the new digital signal pattern cluster centers, we obtain

$$\begin{cases} \| \mathbf{x}_l - \mathbf{z}_1(2) \| < \| \mathbf{x}_l - \mathbf{z}_2(2) \| & \text{for } l = 1,2,...,7 \\ \| \mathbf{x}_l - \mathbf{z}_1(2) \| < \| \mathbf{x}_l - \mathbf{z}_3(2) \| & \text{for } l = 1,2,...,7 \end{cases} \quad (9.46)$$

and

$$\begin{cases} \| \mathbf{x}_k - \mathbf{z}_2(2) \| < \| \mathbf{x}_k - \mathbf{z}_1(2) \| & \text{for } k = 8,...,11,13,...,15 \\ \| \mathbf{x}_k - \mathbf{z}_2(2) \| < \| \mathbf{x}_k - \mathbf{z}_3(2) \| & \text{for } k = 8,...,11,13,...,15 \end{cases} \quad (9.47)$$

Therefore, the class $C_1(2) = \{\mathbf{x}_1, \mathbf{x}_2, ..., \mathbf{x}_7\}$ and the class $C_2(2) = \{\mathbf{x}_8, ..., \mathbf{x}_{11}, \mathbf{x}_{13}, \mathbf{x}_{14}, \mathbf{x}_{15}\}$. Similarly, the remaining digital signal pattern vectors are close to $\mathbf{z}_3(2)$, so $C_3(2) = \{\mathbf{x}_{12}, \mathbf{x}_{16}, ..., \mathbf{x}_{21}\}$.

6. Step 3 (repeated). Update cluster centers again:

$$\begin{aligned}\mathbf{z}_1(3) &= \frac{1}{N_1}\sum_{\mathbf{x}\in C_1(2)} \mathbf{x} \\ &= \frac{1}{7}(\mathbf{x}_1+\mathbf{x}_2+\mathbf{x}_3+\mathbf{x}_4+\mathbf{x}_5+\mathbf{x}_6+\mathbf{x}_7) \\ &= \begin{pmatrix} 1.00 \\ 0.71 \end{pmatrix} \end{aligned} \quad (9.48)$$

$$\begin{aligned}\mathbf{z}_2(3) &= \frac{1}{N_2}\sum_{\mathbf{x}\in C_2(2)} \mathbf{x} \\ &= \frac{1}{7}(\mathbf{x}_8+\mathbf{x}_9+\mathbf{x}_{10}+\mathbf{x}_{11}+\mathbf{x}_{13}+\mathbf{x}_{14}+\mathbf{x}_{15}) \\ &= \begin{pmatrix} 8.29 \\ 7.43 \end{pmatrix} \end{aligned} \quad (9.49)$$

$$\begin{aligned} \mathbf{z}_3(3) &= \frac{1}{N_3} \sum_{\mathbf{x} \in C_3(2)} \mathbf{x} \\ &= \frac{1}{7}(\mathbf{x}_{12} + \mathbf{x}_{16} + \mathbf{x}_{17} + \mathbf{x}_{18} + \mathbf{x}_{19} + \mathbf{x}_{20} + \mathbf{x}_{21}) \\ &= \begin{pmatrix} 3.14 \\ 9.29 \end{pmatrix} \end{aligned} \tag{9.50}$$

7. Step 4 (repeated). Since $\mathbf{z}_j(3) \neq \mathbf{z}_j(2)$, for $j = 2, 3$, we return to step 2 once more.
8. Step 2 (repeated). With the new cluster centers, we have

$$\begin{cases} \| \mathbf{x}_l - \mathbf{z}_1(3) \| < \| \mathbf{x}_l - \mathbf{z}_2(3) \| & \text{for } l = 1, 2, ..., 7 \\ \| \mathbf{x}_l - \mathbf{z}_1(3) \| < \| \mathbf{x}_l - \mathbf{z}_3(3) \| & \text{for } l = 1, 2, ..., 7 \end{cases} \tag{9.51}$$

and

$$\begin{cases} \| \mathbf{x}_k - \mathbf{z}_2(3) \| < \| \mathbf{x}_k - \mathbf{z}_1(3) \| & \text{for } k = 8, 9, ..., 15 \\ \| \mathbf{x}_k - \mathbf{z}_2(3) \| < \| \mathbf{x}_k - \mathbf{z}_3(3) \| & \text{for } k = 8, 9, ..., 15. \end{cases} \tag{9.52}$$

Therefore, the class $C_1(3) = \{\mathbf{x}_1, \mathbf{x}_2, ..., \mathbf{x}_7\}$ and the class $C_2(3) = \{\mathbf{x}_8, \mathbf{x}_{11}, ..., \mathbf{x}_{15}\}$. Similarly, the remaining digital signal pattern vectors are close to $\mathbf{z}_3(3)$ so that the class $C_3(3) = \{\mathbf{x}_{16}, \mathbf{x}_{17}, ..., \mathbf{x}_{21}\}$.
9. Step 3 (repeated). Once again, we update the cluster centers:

$$\begin{aligned} \mathbf{z}_1(4) &= \frac{1}{N_1} \sum_{\mathbf{x} \in C_1(3)} \mathbf{x} \\ &= \frac{1}{7}(\mathbf{x}_1 + \mathbf{x}_2 + \mathbf{x}_3 + \mathbf{x}_4 + \mathbf{x}_5 + \mathbf{x}_6 + \mathbf{x}_7) \\ &= \begin{pmatrix} 1.00 \\ 0.71 \end{pmatrix} \end{aligned} \tag{9.53}$$

$$\begin{aligned} \mathbf{z}_2(4) &= \frac{1}{N_2} \sum_{\mathbf{x} \in C_2(3)} \mathbf{x} \\ &= \frac{1}{8}(\mathbf{x}_8 + \mathbf{x}_9 + \mathbf{x}_{10} + \mathbf{x}_{11} + \mathbf{x}_{12} + \mathbf{x}_{13} + \mathbf{x}_{14} + \mathbf{x}_{15}) \\ &= \begin{pmatrix} 8.13 \\ 7.50 \end{pmatrix} \end{aligned} \tag{9.54}$$

$$\begin{aligned} \mathbf{z}_3(4) &= \frac{1}{N_3} \sum_{\mathbf{x} \in C_3(3)} \mathbf{x} \\ &= \frac{1}{6}(\mathbf{x}_{16} + \mathbf{x}_{17} + \mathbf{x}_{18} + \mathbf{x}_{19} + \mathbf{x}_{20} + \mathbf{x}_{21}) \\ &= \begin{pmatrix} 2.50 \\ 9.50 \end{pmatrix} \end{aligned} \tag{9.55}$$

10. Step 4 (repeated). Since $\mathbf{z}_j(4) \neq \mathbf{z}_j(3)$, for $j = 2, 3$, we go to step 2 again.
11. Step 2 (repeated). This yields the same results as in the previous iteration. Therefore, $C_1(4) = \{\mathbf{x}_1, \mathbf{x}_2, ..., \mathbf{x}_7\}$, $C_2(4) = \{\mathbf{x}_8, \mathbf{x}_{11}, ..., \mathbf{x}_{15}\}$, and $C_3(4) = \{\mathbf{x}_{16}, \mathbf{x}_{17}, ..., \mathbf{x}_{21}\}$.
12. Step 3 (repeated). This also yields the same results after updating the cluster center,

$$\begin{aligned} \mathbf{z}_1(5) &= \begin{pmatrix} 1.00 \\ 0.71 \end{pmatrix}, \\ \mathbf{z}_2(5) &= \begin{pmatrix} 8.13 \\ 7.50 \end{pmatrix}, \\ \mathbf{z}_3(5) &= \begin{pmatrix} 2.50 \\ 9.50 \end{pmatrix}. \end{aligned} \tag{9.56}$$

13. Step 4 (repeated). Because of $\mathbf{z}_j(5) = \mathbf{z}_j(4)$, for $j = 1, 2, 3$, the algorithm has converged yielding these cluster centers: $\mathbf{z}_1 = \begin{pmatrix} 1.00 \\ 0.71 \end{pmatrix}$, $\mathbf{z}_2 = \begin{pmatrix} 8.13 \\ 7.50 \end{pmatrix}$, and $\mathbf{z}_3 = \begin{pmatrix} 2.50 \\ 9.50 \end{pmatrix}$.

As we can see, these results agree with the visual results of what we would expect from inspection of the given signal pattern sampled data as shown in Figure 9.10. In a similar way, the k-means algorithm can also be used to partition the unlabeled training data with overlapping in Figure 9.7 into three clusters.

9.4.3 Characteristic Methods for the Clustering Evaluation

To evaluate clustering methods, we must understand that most clustering methods are biased toward finding clusters possessing certain characteristics of size, shape, or dispersion. Methods based on the least-squares criterion [62], including the k-means algorithm, tend to find clusters with roughly the same number of points in each cluster. Such methods are not able to detect clusters with highly elongated or irregular shapes. The method of complete linkage is strongly biased toward producing clusters with roughly equal diameters and can be severely distorted if noise is present. The centroid linkage and average linkage are biased toward finding clusters of the same variance. The method of single linkage sacrifices performance in the recovery of compact clusters without imposing constraints on the shape of the clusters. However, the method of single linkage is able to detect elongated and irregular clusters during the agglomerative procedure in the hierarchical clustering. If we do not know what kinds of clusters to expect, we should use at least one of the relatively unbiased methods, such as single linkage, in clustering analysis.

If a feature space of the signal pattern vectors consists of a set of clusters with background noise, it may have difficulty in extracting the pattern clusters correctly. In particular, when we use a partitioning method including the k-means algorithm, the points of background noise are forced to join the clusters, thus biasing their statistics. Jolion and Rosenfeld [63] present a preprocessing technique in which points belonging to a cluster get relatively high weights, while background noise points get relatively low weights. When this technique is used before the k-means algorithm, it converges faster and produces more accurate estimates of the cluster statistics.

In unsupervised classification, one of most difficult problems is how to determine the number of clusters in the data. Unfortunately, there are no satisfactory methods for determining the number of population clusters for any type of cluster analysis (see Everitt [38, 64]). In 1985, Milligan and Cooper [61] evaluated the performance of 30 cluster choice criteria using a total of 432 sets of hierarchical solutions. The pseudo-F statistic showed the best results by correctly identifying the number of clusters in 390 out of 432 (90.3%) test cases.

It is also good to look at the pattern data graphically. We can make two-dimensional (2-D) scatter-plots identifying the clusters and assess the performance of the cluster analysis, such as the k-means algorithm, if there are only two or three measurement features. With more measurement features, we can compute component features for scatter-plots using such feature extraction methods as principal components analysis (Karhunen-Loève transformation), canonical discriminant analysis, optimal discriminant analysis, and optimal declustering analysis. These transform methods will be introduced in Chapter 10.

9.5 Signal Classification

Signal classification assigns input signal pattern vectors to one or more of L predefined classes based on extraction of significant measurement features or attributes. Processing analysis is tied to these attributes. Resorting to probability models in statistical signal classification is common. In this section, we introduce two approaches for statistical signal classification: *nonparametric signal classification* and *parametric signal classification.*

9.5.1 Nonparametric Signal Classification

We begin the study of statistical signal classifiers by considering one of the simplest and most intuitive approaches to nonparametric classification. Signal classification is based on "distance functions." The motivation of using distance functions follows naturally from the fact that the most obvious way of establishing a measure of similarity between signal pattern vectors is by determining their proximity. The measure of similarity can be considered as the distance between points in a *Euclidean space.* This simple technique of nonparametric signal classification is an effective tool for the solution of many problems in which the signal pattern classes exhibit a reasonably limited degree of variability.

Consider M signal pattern classes and assume that these classes are represented by their estimated mean vectors $\widehat{\mathbf{m}}_i$, where $i = 1, 2, ..., M$. Then, the Euclidean distance between an arbitrary signal pattern vector $\mathbf{x}$ and the ith signal pattern class ω_i is given by

$$\begin{aligned} D_i^2 &= \| \mathbf{x} - \widehat{\mathbf{m}}_i \|^2 \\ &= (\mathbf{x} - \widehat{\mathbf{m}}_i)^T (\mathbf{x} - \widehat{\mathbf{m}}_i) \\ &= \mathbf{x}^T \mathbf{x} - 2(\mathbf{x}^T \widehat{\mathbf{m}}_i - \frac{1}{2} \widehat{\mathbf{m}}_i{}^T \widehat{\mathbf{m}}_i), \end{aligned} \tag{9.57}$$

where

$$\widehat{\mathbf{m}}_i = \frac{1}{N_i} \sum_{k \in \omega_i} \mathbf{x}_k, \quad i = 1, 2, ..., M. \tag{9.58}$$

Note that choosing the minimum value of D_i^2 is equivalent to choosing the minimum D_i, since all distances are positive. From (9.57), we also see that the term $\mathbf{x}^T\mathbf{x}$ is independent of i in all D_i^2, $i = 1, 2, ..., M$. Therefore, choosing the minimum D_i^2 is equivalent to choosing the maximum $(\mathbf{x}^T\widehat{\mathbf{m}}_i - \frac{1}{2}\widehat{\mathbf{m}}_i{}^T\widehat{\mathbf{m}}_i)$. Consequently, we may define the decision functions

$$\widehat{d}_i(\mathbf{x}) = \mathbf{x}^T\widehat{\mathbf{m}}_i - \frac{1}{2}\widehat{\mathbf{m}}_i{}^T\widehat{\mathbf{m}}_i. \quad i = 1, 2, ..., M. \tag{9.59}$$

Generally, (9.59) defines a set of linear functions that assign signal pattern vectors in the signal sampled data set $\mathbf{X}$ to the correct class, ω_i. Then, the implementation of the signal classification rule is:

$$\text{choose } \omega_c \qquad \text{if } \widehat{d}_c(\mathbf{x}) = max_i[\widehat{d}_i(\mathbf{x})], \quad i = 1, 2, ..., M. \tag{9.60}$$

Equation (9.60) can be referred to as the *maximum signal classifier.* The maximum statistical signal classifier categorizes a signal pattern vector $\mathbf{x}$ based on the closest match between the unknown pattern vector and the respective pattern class, ω_i. This approach can also be called the *maximum correlation signal classifier.*

Example 9.4 Maximum Statistical Signal Classifier

In this example, we illustrate a maximum statistical signal classifier and boundary decision surface by reusing the data set as shown in Figure 9.10. The estimated means of the cluster centers are:

$$\widehat{\mathbf{m}}_1 = \begin{pmatrix} 1.00 \\ 0.71 \end{pmatrix}, \; \widehat{\mathbf{m}}_2 = \begin{pmatrix} 8.13 \\ 7.50 \end{pmatrix}, \text{ and } \widehat{\mathbf{m}}_3 = \begin{pmatrix} 2.50 \\ 9.50 \end{pmatrix}.$$

Considering (9.59), we obtain three decision functions

$$d_1(\mathbf{x}) = -0.75 + 1.00x_1 + 0.71x_2, \tag{9.61}$$
$$d_2(\mathbf{x}) = -61.17 + 8.13x_1 + 7.50x_2, \tag{9.62}$$
$$d_3(\mathbf{x}) = -48.25 + 2.50x_1 + 9.50x_2. \tag{9.63}$$

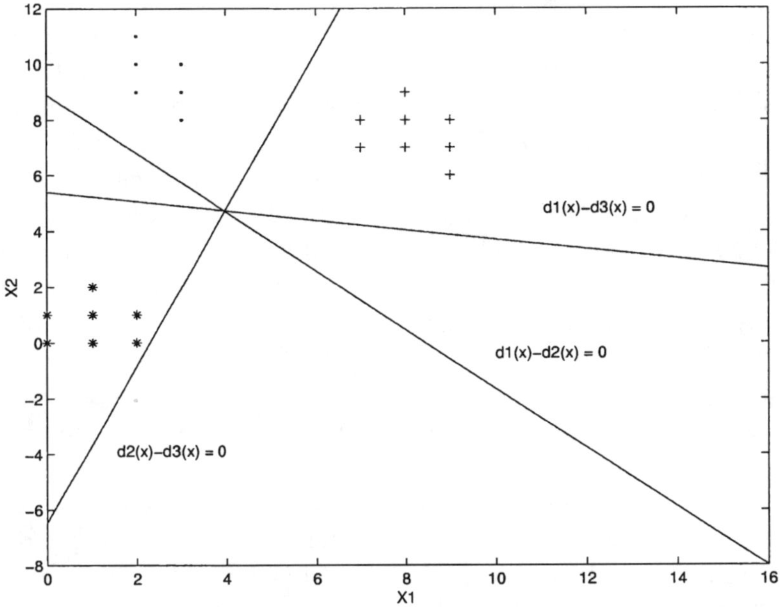

Figure 9.11 A set of decision surfaces is based on the maximum statistical signal classifier to classify the pattern vectors.

Therefore, the maximum statistical signal classifier assigns an unknown pattern vector $\mathbf{x}$ to pattern class ω_1 when $d_1(\mathbf{x}) > d_2(\mathbf{x})$ and $d_1(\mathbf{x}) > d_3(\mathbf{x})$. A pattern vector $\mathbf{x}$ is assigned to pattern class ω_2 when $d_2(\mathbf{x}) > d_1(\mathbf{x})$ and $d_2(\mathbf{x}) > d_3(\mathbf{x})$. Otherwise, a pattern vector $\mathbf{x}$ is assigned to pattern class ω_3.

The boundary of the decision surface is given by the equations

$$d_1(\mathbf{x}) - d_2(\mathbf{x}) = 60.42 - 7.13x_1 - 6.79x_2 = 0, \quad (9.64)$$
$$d_1(\mathbf{x}) - d_3(\mathbf{x}) = 47.50 - 1.50x_1 - 8.79x_2 = 0, \quad (9.65)$$
$$d_2(\mathbf{x}) - d_3(\mathbf{x}) = -12.93 + 5.63x_1 - 2.00x_2 = 0. \quad (9.66)$$

A set of decision surfaces is shown in Figure 9.11. As can be seen, classes are effectively dichotomized by the decision surfaces.

9.5.2 Parametric Signal Classification

The fundamental assumption of supervised or unsupervised statistical signal classification is that there is a multivariate probability distribution for each pattern class. Members of a pattern class are considered as signal pattern vectors sampled from a population. The population is distributed in a Q dimensional feature space according to the distribution related to that population.

When the conditional probability density functions (or class density functions) of the feature vectors for each class are given, the Bayes' classification function that minimizes the probability of misclassification error can be derived. In practice, the class density functions can be estimated from the pattern vectors, and then an optimal classification rule can be obtained when a large number of pattern vectors are available.

The probability is written as $p(\omega_i|\mathbf{x})$ when a pattern vector $\mathbf{x}$ comes from pattern class ω_i. If the pattern vector $\mathbf{x}$ is classified into pattern class ω_j when it really is a member of pattern class ω_i, the pattern classifier receives a loss equal to R_{ij}. Suppose that a pattern vector $\mathbf{x}$ belongs to any of the M pattern classes. Then the expected loss incurred in assigning pattern vector $\mathbf{x}$ to the pattern class ω_j is given by

$$r_j = \sum_{i=1}^{M} R_{ij} p(\omega_i|\mathbf{x}). \tag{9.67}$$

The optimal decision rule that minimizes the total expected loss is called the Bayes' minimum risk classifier. This Bayes' classifier represents the optimal measure of performance from a statistical point of view.

Using Bayes' formula (see Papoulis [65]),

$$p(\omega_i|\mathbf{x}) = \frac{p(\omega_i)p(\mathbf{x}|\omega_i)}{p(\mathbf{x})}, \tag{9.68}$$

equation (9.67) can be expressed in the form

$$r_j = \frac{1}{p(\mathbf{x})} \sum_{i=1}^{M} R_{ij} p(\mathbf{x}|\omega_i) p(\omega_i), \tag{9.69}$$

where $p(\mathbf{x}|\omega_i)$ is called the *maximum likelihood function* of pattern class ω_i. Since $1/p(\mathbf{x})$ does not depend on i or j, it can be eliminated from (9.69). Then, the expression for the expected loss reduces to

$$r_j = \sum_{i=1}^{M} R_{ij} p(\mathbf{x}|\omega_i) p(\omega_i), \tag{9.70}$$

from assigning pattern vector $\mathbf{x}$ to the pattern class ω_j.

Consider the $\{0, 1\}$ *loss* function,

$$R_{ij} = 1 - \delta_{ij} = \begin{cases} 0 & i = j \\ 1 & i \neq j. \end{cases} \tag{9.71}$$

Equation (9.71) indicates a normalized loss of unity for misclassification error and no loss for correct classification of a pattern vector. Substituting R_{ij} in (9.71) into (9.70) yields

$$\begin{aligned} r_j(\mathbf{x}) &= \sum_{i=1}^{M} (1 - \delta_{ij}) p(\mathbf{x}|\omega_i) p(\omega_i) \\ &= p(\mathbf{x}) - p(\mathbf{x}|\omega_j) p(\omega_j). \end{aligned} \tag{9.72}$$

The Bayes' classifier assigns a particular pattern vector $\mathbf{x}$ to the pattern class ω_i if

$$r_i(\mathbf{x}) < r_j(\mathbf{x}), \tag{9.73}$$

or

$$p(\mathbf{x}|\omega_i) p(\omega_i) > p(\mathbf{x}|\omega_j) p(\omega_j) \quad j = 1, 2, ..., M; \quad j \neq i. \tag{9.74}$$

It should be noted from (9.74) that the implementations of the decision functions are

$$d_i(\mathbf{x}) = p(\mathbf{x}|\omega_i) p(\omega_i), \quad i = 1, 2, ..., M, \tag{9.75}$$

where a pattern vector $\mathbf{x}$ is classified into the pattern class ω_i if $d_i(\mathbf{x}) > d_j(\mathbf{x})$ for all $j \neq i$. We usually refer to (9.75) as the *maximum likelihood signal classifier.*

9.5.3 Signal Classifiers for Gaussian Models

When a large number of pattern vectors are available, assuming that the class density function of the signal feature vectors for each class is a multivariate Gaussian density is often reasonable. Then the density function of each class can be estimated from the pattern vectors. The Bayes' classifier that minimizes the probability of misclassification error can be derived.

Let us consider M pattern classes with P measurement features described by the multivariate *Gaussian* density functions

$$p(\mathbf{x}|\omega_i) = \frac{1}{(2\pi)^{n/2}|\mathbf{C}_i|^{1/2}} exp[-\frac{1}{2}(\mathbf{x}-\mathbf{m}_i)^T\mathbf{C}_i^{-1}(\mathbf{x}-\mathbf{m}_i)], \qquad (9.76)$$

where $\mathbf{m}_i$ is the mean vector and $\mathbf{C}_i$ is the covariance matrix for the patterns of class ω_i, $i = 1, 2, ..., M$. The quadratic classifier assigns a pattern vector $\mathbf{x}$ to the pattern class ω_i associated with the maximum likelihood. Thus (9.76) defines a set of discriminant functions that optimally assign inputs to the correct distribution. The Gaussian density function is usually denoted by $N_p(\mathbf{m}_i, \mathbf{C}_i)$. Unfortunately, only estimates of $\mathbf{m}_i$ and $\mathbf{C}_i$ can be obtained due to finite sample sizes as well as noise. The $\mathbf{m}_i$ and $\mathbf{C}_i$ can be unbiased estimated using the following formulas:

$$\widehat{\mathbf{m}}_i = E_i[\mathbf{x}] = \frac{1}{N_i}\sum_{k=1}^{N_i}\mathbf{x}_k, \qquad (9.77)$$

and

$$\begin{aligned}\widehat{\mathbf{C}}_i &= E_i[(\mathbf{x}-\mathbf{m}_i)(\mathbf{x}-\mathbf{m}_i)^T] \\ &= \frac{1}{N_i-1}\sum_{k=1}^{N_i}(\mathbf{x}_k-\widehat{\mathbf{m}}_i)(\mathbf{x}_k-\widehat{\mathbf{m}}_i)^T, \qquad (9.78)\end{aligned}$$

where N_i is the total number of pattern vectors in the pattern classes ω_i, $i = 1, 2, ..., M$.

It is more convenient to work with the natural logarithms of the decision functions in (9.75) because the Gaussian density functions have an exponential form. From (9.75), we can use the estimated expression of the form

$$\begin{aligned}\widehat{D}_i(\mathbf{x}) &= \ln[\widehat{p}(\mathbf{x}|\omega_i)\widehat{p}(\omega_i)] \\ &= \ln[\widehat{p}(\mathbf{x}|\omega_i)] + \ln[\widehat{p}(\omega_i)]. \qquad (9.79)\end{aligned}$$

Substituting (9.76) into (9.79) yields

$$\widehat{D}_i(\mathbf{x}) = \ln[\widehat{p}(\omega_i)] - \frac{n}{2}\ln(2\pi) - \frac{1}{2}\ln|\widehat{\mathbf{C}}_i| - \frac{1}{2}[(\mathbf{x}-\widehat{\mathbf{m}}_i)^T\widehat{\mathbf{C}}_i^{-1}(\mathbf{x}-\widehat{\mathbf{m}}_i)]. \quad (9.80)$$

Since $(n/2)\ln(2\pi)$ does not depend on i, it can be dropped from (9.80). Then, $\widehat{D}_i(\mathbf{x})$ becomes

$$\widehat{D}_i(\mathbf{x}) = \ln[\widehat{p}(\omega_i)] - \frac{1}{2}\ln|\widehat{\mathbf{C}}_i| - \frac{1}{2}[(\mathbf{x}-\widehat{\mathbf{m}}_i)^T\widehat{\mathbf{C}}_i^{-1}(\mathbf{x}-\widehat{\mathbf{m}}_i)], \quad (9.81)$$

where $i = 1, 2, ..., M$. Equation (9.81) represents decision functions for Gaussian patterns. Note that $(\mathbf{x}-\widehat{\mathbf{m}}_i)^T\widehat{\mathbf{C}}_i^{-1}(\mathbf{x}-\widehat{\mathbf{m}}_i)$ is a modified form of the *weight Euclidean distance* (also called Mahalanobis distance) from the pattern vector $\mathbf{x}$ to the mean vector associated with ith pattern class ω_i. The middle term $\frac{1}{2}\ln|\widehat{\mathbf{C}}_i|$ represents the difference between statistical correlation and likelihood partitions. This term contains information about the spread of each distribution in the pattern space.

When pattern vector $\mathbf{x}$ is mutually uncorrelated with equal variance ($\widehat{\mathbf{C}}_i = \widehat{\sigma}^2\mathbf{I}$) within pattern class i, the class independent bias $\frac{1}{2}\ln|\widehat{\mathbf{C}}_i|$ can be eliminated. Then, (9.81) becomes

$$\widehat{D}_i(\mathbf{x}) = \ln[\widehat{p}(\omega_i)] - \frac{1}{2\sigma^2}[(\mathbf{x}-\widehat{\mathbf{m}}_i)^T(\mathbf{x}-\widehat{\mathbf{m}}_i)], \quad i = 1, 2, ..., M. \quad (9.82)$$

Therefore, the term $\frac{1}{2\sigma^2}[(\mathbf{x}-\widehat{\mathbf{m}}_i)^T(\mathbf{x}-\widehat{\mathbf{m}}_i)]$ creates *hyperspheres*, with each centered at the mean $\widehat{\mathbf{m}}_i$ of the signal pattern class. Moreover, (9.82) produces a direct mechanism to incorporate a priori probabilities into the discriminant function.

When pattern vector $\mathbf{x}$ is mutually uncorrelated with unequal variances ($\widehat{\mathbf{C}}_i = \widehat{\sigma}_{ii}^2$),

$$\widehat{\mathbf{C}}_i = \begin{pmatrix} \sigma_{11}^2 & 0 & \dots & 0 \\ 0 & \sigma_{22}^2 & \dots & 0 \\ \vdots & \vdots & \ddots & \vdots \\ 0 & 0 & \dots & \sigma_{pp}^2. \end{pmatrix}, \quad (9.83)$$

the inverse covariance matrix $\widehat{\mathbf{C}}_i^{-1}$ is

$$\widehat{\mathbf{C}}_i^{-1} = \begin{pmatrix} \frac{1}{\sigma_{11}^2} & 0 & \dots & 0 \\ 0 & \frac{1}{\sigma_{22}^2} & \dots & 0 \\ \vdots & \vdots & \ddots & \vdots \\ 0 & 0 & \dots & \frac{1}{\sigma_{pp}^2}. \end{pmatrix}. \tag{9.84}$$

The decision rule in (9.81) yields a weighted distance classifier. Note that, if $\frac{1}{\sigma_{11}^2} > \frac{1}{\sigma_{22}^2}$, we focus on pattern vector x_1 over x_2 because the pattern samples scatter more in the x_2 direction than in the x_1 direction. Thus, we expect more deviation of x_2 from its mean than for x_1.

Generally, (9.81) defines a set of discriminant functions that optimally assign pattern vectors to the correct pattern class. The implementation of the classification rule results in,

$$\text{Choosing } \omega_c \text{ if } \widehat{D}_c(\mathbf{x}) = min_i[\widehat{D}_i(\mathbf{x})], \quad i = 1, 2, ..., M. \tag{9.85}$$

Equation (9.85) is called the *minimum distance classifier.*

It follows from (9.81) that the decision boundary between the pattern class ω_i and the pattern class ω_j is defined by the condition where $\widehat{D}_i(\mathbf{x}) - \widehat{D}_j(\mathbf{x})$, or

$$\begin{aligned} \widehat{D}(\mathbf{x}) &= \widehat{D}_i(\mathbf{x}) - \widehat{D}_j(\mathbf{x}) \\ &= \ln\left[\frac{\widehat{p}(\omega_i)}{\widehat{p}(\omega_j)}\left|\frac{\widehat{\mathbf{C}}_j}{\widehat{\mathbf{C}}_i}\right|^{1/2}\right] + \frac{1}{2}[(\mathbf{x} - \widehat{\mathbf{m}}_j)^T\widehat{\mathbf{C}}_j^{-1}(\mathbf{x} - \widehat{\mathbf{m}}_j)] \\ &\quad -\frac{1}{2}[(\mathbf{x} - \widehat{\mathbf{m}}_i)^T\widehat{\mathbf{C}}_i^{-1}(\mathbf{x} - \widehat{\mathbf{m}}_i)]. \end{aligned} \tag{9.86}$$

Equation (9.86) is a quadratic discriminant function describing a *hyperquadratic.*

If all the covariance matrices are equal (i.e., $\mathbf{C}_i = \mathbf{C}$, for $i = 1, 2, ..., M$), then the decision boundary is defined by

$$\begin{aligned} \widehat{D}(\mathbf{x}) &= \mathbf{x}^T\widehat{\mathbf{C}}^{-1}(\widehat{\mathbf{m}}_i - \widehat{\mathbf{m}}_j) \\ &\quad -\frac{1}{2}(\widehat{\mathbf{m}}_i + \widehat{\mathbf{m}}_j)^T\widehat{\mathbf{C}}^{-1}(\widehat{\mathbf{m}}_i - \widehat{\mathbf{m}}_j) + \ln\left[\frac{\widehat{p}(\omega_i)}{\widehat{p}(\omega_j)}\right] \end{aligned}$$

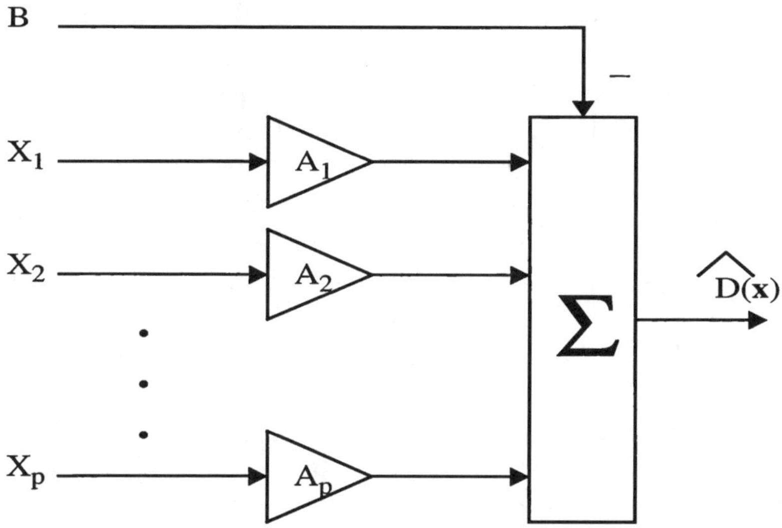

Figure 9.12 Signal flow-graph of the implementation of a Gaussian classifier when all the covariance matrices are equal.

$$
\begin{aligned}
&= \widehat{\mathbf{A}}^T \mathbf{x} - \widehat{B} \\
&= \sum_{i=1}^{p} A_i X_i - \widehat{B},
\end{aligned} \tag{9.87}
$$

where $\widehat{\mathbf{A}} = \mathbf{C}^{-1}(\widehat{\mathbf{m}}_i - \widehat{\mathbf{m}}_j)$ is a matrix, and $\widehat{B} = -\ln\left[\frac{\widehat{p}(\omega_i)}{\widehat{p}(\omega_j)}\right] + \frac{1}{2}(\widehat{\mathbf{m}}_i + \widehat{\mathbf{m}}_j)^T \widehat{\mathbf{C}}^{-1}(\widehat{\mathbf{m}}_i - \widehat{\mathbf{m}}_j)$ is a constant term. Equation (9.87) represents a linear decision function describing a *hyperplane*. This equation may thus be implemented by a linear combiner as shown in Figure 9.12.

Example 9.5 The Gaussian-Model-Based Classifier

In this example, we consider a minimum distance classifier based on the Gaussian model using the pattern vectors as shown in Figure 9.10. It has been shown that unbiased estimates of the mean vector $\widehat{\mathbf{m}}_i$ and the covariance matrix $\widehat{\mathbf{C}}_i$ can be obtained using (9.77) and (9.78). For this example, we obtain the estimate of mean vector $\widehat{\mathbf{m}}_i$ and covariance matrix $\widehat{\mathbf{C}}_i$,

$$\widehat{\mathbf{m}}_1 = \begin{pmatrix} 1.00 \\ 0.71 \end{pmatrix}, \widehat{\mathbf{m}}_2 = \begin{pmatrix} 8.13 \\ 7.50 \end{pmatrix}, \widehat{\mathbf{m}}_3 = \begin{pmatrix} 2.50 \\ 9.50 \end{pmatrix}$$

$$\widehat{\mathbf{C}}_1 = \begin{pmatrix} 0.67 & 0 \\ 0 & 0.57 \end{pmatrix}, \widehat{\mathbf{C}}_1^{-1} = \begin{pmatrix} 1.50 & 0.00 \\ 0.00 & 1.75 \end{pmatrix}$$

$$\widehat{\mathbf{C}}_2 = \begin{pmatrix} 0.70 & -0.21 \\ -0.21 & 0.86 \end{pmatrix}, \widehat{\mathbf{C}}_2^{-1} = \begin{pmatrix} 1.56 & 0.39 \\ 0.39 & 1.26 \end{pmatrix}$$

$$\widehat{\mathbf{C}}_3 = \begin{pmatrix} 0.30 & -0.30 \\ -0.30 & 1.10 \end{pmatrix}, \widehat{\mathbf{C}}_3^{-1} = \begin{pmatrix} 4.58 & 1.25 \\ 1.25 & 1.25 \end{pmatrix}.$$

Since the covariance matrices, C_i, $i = 1, 2, 3$, are not equal, the decision functions based on a Gaussian model are given by (9.81). If we further assume equal priori probabilities, $p(\omega_1) = p(\omega_2) = p(\omega_3) = \frac{1}{3}$, then the term $\ln[p(\omega_i)]$ drops out. As a result, we obtain

$$\widehat{D}_i(\mathbf{x}) = -\frac{1}{2}\ln|\widehat{\mathbf{C}}_i| - \frac{1}{2}[(\mathbf{x} - \widehat{\mathbf{m}}_i)^T \widehat{\mathbf{C}}_i^{-1}(\mathbf{x} - \widehat{\mathbf{m}}_i)], \quad i = 1, 2, 3. \quad (9.88)$$

Now, substituting the values of $\widehat{\mathbf{m}}_i$ and $\widehat{\mathbf{C}}_i$ into the above (9.88) yields

$$\begin{aligned}
\widehat{D}_1(\mathbf{x}) &= -0.75x_1^2 + 1.5x_1 + 1.24x_2 - 0.88x_2^2 - 0.71 \\
\widehat{D}_2(\mathbf{x}) &= -0.78x_1^2 + 15.59x_1 - 0.39x_1x_2 + 12.24x_2 \\
&\quad - 0.63x_2^2 - 110.48 \\
\widehat{D}_3(\mathbf{x}) &= -2.29x_1^2 + 23.33x_1 - 1.25x_1x_2 + 15x_2 - 0.63x_2^2 - 70.
\end{aligned}$$

Therefore, the minimum distance classifier assigns a pattern vector $\mathbf{x}$ to the pattern class ω_1 when $\widehat{D}_1(\mathbf{x}) < \widehat{D}_2(\mathbf{x})$ and $\widehat{D}_1(\mathbf{x}) < \widehat{D}_3(\mathbf{x})$. A pattern vector $\mathbf{x}$ is assigned to the pattern class ω_2 when $\widehat{D}_2(\mathbf{x}) < \widehat{D}_1(\mathbf{x})$ and $\widehat{D}_2(\mathbf{x}) < \widehat{D}_3(\mathbf{x})$. Otherwise, a pattern vector $\mathbf{x}$ is assigned to the pattern class ω_3.

9.5.4 Homogeneity Test of Covariance Matrices

One important question regarding the classification criterion based on a Gaussian model is whether or not the individual within-class covariance matrix $\widehat{\mathbf{C}}_i$ or the pooled covariance matrix $\widehat{\mathbf{C}}$ should be used in (9.81). The determination of whether the classification criterion should be based on the within-class covariance matrix $\widehat{\mathbf{C}}_i$ or the pooled covariance matrix $\widehat{\mathbf{C}}$, is known as the *homogeneity test* (see Anderson [66], Kendall and Stuart [67], and Morrison [68]).

The hypothesis of the homogeneity test for covariance matrices is as follows: the null hypothesis H_0: $\widehat{\mathbf{C}}_1 = \cdots = \widehat{\mathbf{C}}_K$, where K is the number of pattern classes. Let P be the number of pattern features, N be the total number of pattern sample observations, and N_i be the total number of pattern observations for pattern classes ω_i, $i = 1, 2, ..., K$. Also, let V be defined as the ratio of the product of the determinants of the within-class covariance matrices $\widehat{\mathbf{C}}_i$ to the pooled covariance matrix $\widehat{\mathbf{C}}$, all raised to a power depending on the number of samples:

$$V = \frac{\prod_i^K \left|\widehat{\mathbf{C}}_i\right|^{N_i/2}}{\left|\widehat{\mathbf{C}}\right|^{N/2}}. \tag{9.89}$$

In addition, defining of a weight scalar γ is

$$\gamma = 1 - \left[\frac{2P^2 + 3P - 1}{6(P+1)(K-1)}\right]\left[\sum_{i=1}^{K}\left(\frac{1}{N_i - 1}\right) - \frac{1}{N-K}\right] \tag{9.90}$$

with the degrees of freedom (DF), where

$$DF = \frac{P(P+1)(K-1)}{2}. \tag{9.91}$$

It has been shown that the above equations can be formed as the *chi-square* statistic,

$$\begin{aligned}
\mathcal{X}^2 &= -2(\gamma)\left[\ln\left(\frac{N^{PN/2}V}{\prod_i^K N_i^{PN_i/2}}\right)\right] \\
&= \left\{-2 + \left[\frac{2P^2 + 3P - 1}{3(P+1)(K-1)}\right]\left[\sum_{i=1}^{K}\left(\frac{1}{N_i - 1}\right) - \frac{1}{N-K}\right]\right\} \\
&\quad \cdot \left\{\ln\left[\left(\frac{N^{PN/2}}{\prod_i^K N_i^{PN_i/2}}\right)\left(\frac{\prod_i^K \left|\widehat{\mathbf{C}}_i\right|^{N_i/2}}{\left|\widehat{\mathbf{C}}\right|^{N/2}}\right)\right]\right\}.
\end{aligned} \tag{9.92}$$

Under the null hypothesis H_0, (9.92) is distributed approximately as asymptotic chi-square with $\frac{P(P+1)(K-1)}{2}$ degrees of freedom. If the chi-square value exceeds a critical value, it will be significant at the α level. If this is the case, the within-class covariance matrix $\widehat{\mathbf{C}}_i$ is used for the basis of the weighted Euclidean distance in (9.81). Otherwise, the pooled covariance matrix $\widehat{\mathbf{C}}$ is used for the discriminant function.

9.5.5 Implementation of Statistical Signal Classifiers

The implementation of a statistical signal classification rule based on the minimum distance classifier was introduced in Section 9.5.3. In this section, we present the implementation of the signal classification rule based on the maximum likelihood signal classifier.

A simplified block diagram for implementing the maximum likelihood classifier is shown in Figure 9.13, where $p(\mathbf{x}|\omega_i)$ is the maximum likelihood function and $p(\omega_i)$ is a priori probability of pattern class ω_i. Hence, a pattern vector $\mathbf{x}$ is classified into pattern class ω_c if

$$p(\omega_c)p(\mathbf{x}|\omega_c) = max_i\,[p(\omega_i)p(\mathbf{x}|\omega_i)] \quad i = 1, 2, ..., M. \tag{9.93}$$

Equation (9.93) is called the Bayes' maximum likelihood classifier.

Under most circumstances, the maximum likelihood signal classifier assigns the unknown vector $\mathbf{x}$ to the pattern class for which a posterior i membership probability is maximum. Such a posterior i probability of membership in each pattern class ω_i is given by

$$P(\omega_i|\mathbf{x}) = \frac{exp[\widehat{D}_i(\mathbf{x})]}{\sum_{k=1}^{M} exp[\widehat{D}_k(\mathbf{x})]}, \quad i = 1, 2, ..., M. \tag{9.94}$$

Equation (9.94) is known as the *posterior maximum likelihood signal classifier*. Therefore, an observation of pattern vector $\mathbf{x}$ is classified into the pattern class ω_c if setting $i = c$ produces the largest value of probability $P(\omega_i|\mathbf{x})$.

9.6 Estimated Probability of Misclassification

Without perfect measurements, all rules for statistical signal classification have some probability of misclassification. In most cases, even the optimal decision regions will contain the patterns

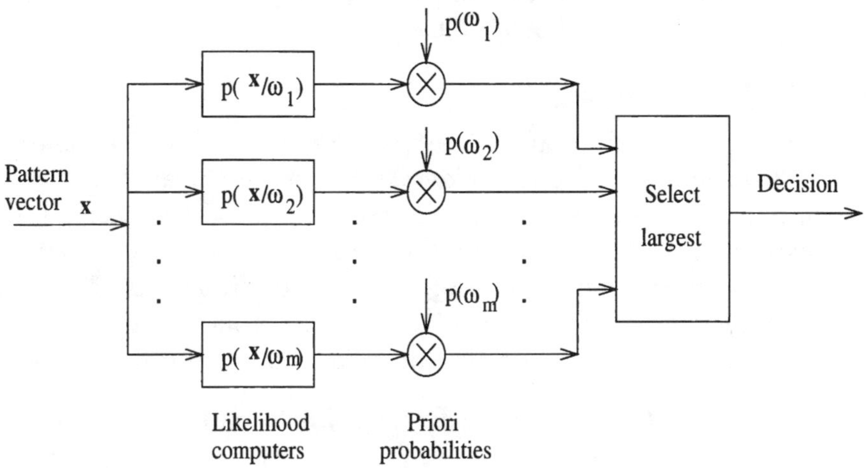

Figure 9.13 A general block diagram for implementing the maximum likelihood signal classifier.

out of class. This is especially true when decision rules are simplified to accommodate models such as Gaussian distributions.

The optimal Bayes's probability of misclassification error, given the unknown pattern $\mathbf{x}$, denoted by P_e^B, is given by

$$P_e^B = 1 - \sum_{i=1}^{M} max_i[p(\mathbf{x}|\omega_i)p(\omega_i)], \tag{9.95}$$

where $p(\mathbf{x}|\omega_i)$ is a class-dependent conditional probability density and $p(\omega_i)$ is a priori probability of the ith pattern class ω_i.

In most problems of statistical signal classification, we are interested in $\widehat{P}_e$ ($\widehat{P}_e \geq P_e^B$), where $\widehat{P}_e$ represents the probability of misclassification error in future performance. When the statistical signal classifier is trained on a set of sample data, there are two basic approaches in estimating the probability error $\widehat{P}_e$ as described below.

Parametric Approaches

In parametric approaches, it is assumed that the unknown pattern distributions belong to a parametric family. In many cases, it is also assumed that the pattern vectors follow a multivariate Gaussian distribution. Then, the probability of error is determined based on the *error function* (see Schalkoff [15], Fukunaga and Hayes [26], Tou and Gonzalez [32] and Postaire [69]).

Nonparametric Approaches

In nonparametric approaches, the statistical signal classifier is trained on one set of pattern data, and tested on another set of pattern data. Examples of nonparametric approaches include the *resubstitution method*, the *holdout method*, the *leave-one-out method*, and the *rotation method* (see Highleyman [70], Toussaint [71], Hills [72], Kanal and Chandrasekaran [73], Sorum [74], Mardia, Kent, and Bibby [75], Raudys and Jain [76], and Ney, Essen, and Kneser [77]).

9.6.1 Estimation of Probability Error Using Parametric Approaches

In the parametric approach, it is assumed that the unknown distribution belongs to a parametric family, often the Gaussian distribution. Consider two pattern classes, ω_i and ω_j, for which the pattern vectors $\mathbf{x}$ are characterized by multivariate Gaussian density functions, $p(\mathbf{x}|\omega_i) \sim N_d(\mathbf{m}_i, \mathbf{C})$ and $p(\mathbf{x}|\omega_j) \sim N_d(\mathbf{m}_j, \mathbf{C})$, with equal covariance matrices.

We define the probability of error P_e as $P_e = P(\text{choose } \omega_i$ where the pattern is actually from $\omega_j) + P(\text{choose } \omega_j$ where the pattern is actually from $\omega_i)$. Then, the probability of error is given by

$$\begin{aligned} P_e &= P(\omega_i)P(error|\omega_i) + P(\omega_j)P(error|\omega_j) \\ &= P(\omega_i)P(L_{ij} < \beta|\omega_i) + P(\omega_j)P(L_{ij} > \beta|\omega_j), \end{aligned} \tag{9.96}$$

where L_{ij} is a function of the logarithm of the *likelihood ratio*,

$$\begin{aligned} L_{ij} &= \ln\left[\frac{p(\mathbf{x}|\omega_i)}{p(\mathbf{x}|\omega_j)}\right] \\ &= -\frac{1}{2}(\mathbf{x}-\mathbf{m}_i)^T\mathbf{C}^{-1}(\mathbf{x}-\mathbf{m}_i) + \frac{1}{2}(\mathbf{x}-\mathbf{m}_j)^T\mathbf{C}^{-1}(\mathbf{x}-\mathbf{m}_j) \\ &= \mathbf{x}^T\mathbf{C}^{-1}(\mathbf{m}_i-\mathbf{m}_j) - \frac{1}{2}(\mathbf{m}_i+\mathbf{m}_j)^T\mathbf{C}^{-1}(\mathbf{m}_i-\mathbf{m}_j), \end{aligned} \tag{9.97}$$

and

$$\beta = \ln\left[\frac{p(\omega_i)}{p(\omega_j)}\right]. \tag{9.98}$$

Note that L_{ij} is a linear combination of the components of pattern vectors $\mathbf{x}$, which have Gaussian distribution. Then, L_{ij} is also characterized by a Gaussian distribution. The mean of L_{ij} is

$$\begin{aligned} E_i\{L_{ij}\} &= \mathbf{m}_i^T\mathbf{C}^{-1}(\mathbf{m}_i - \mathbf{m}_j) - \frac{1}{2}(\mathbf{m}_i + \mathbf{m}_j)^T\mathbf{C}^{-1}(\mathbf{m}_i - \mathbf{m}_j) \\ &= \frac{1}{2}(\mathbf{m}_i - \mathbf{m}_j)^T\mathbf{C}^{-1}(\mathbf{m}_i - \mathbf{m}_j) \\ &= \frac{1}{2}MD_{ij}, \end{aligned} \tag{9.99}$$

where MD_{ij} is referred to as the *squared Mahalanobis distance* between two density functions $p(\mathbf{x}|\omega_i)$ and $p(\mathbf{x}|\omega_j)$.

The variance of L_{ij} is defined by

$$\begin{aligned} Var_i\{L_{ij}\} &= E_i\{(L_{ij} - E_i\{L_{ij}\})^2\} \\ &= E_i\{(\mathbf{m}_i - \mathbf{m}_j)^T\mathbf{C}^{-1}(\mathbf{x} - \mathbf{m}_i) \\ &\qquad \cdot (\mathbf{x} - \mathbf{m}_i)^T\mathbf{C}^{-1}(\mathbf{m}_i - \mathbf{m}_j)\} \\ &= MD_{ij}. \end{aligned} \tag{9.100}$$

Therefore, L_{ij} is Gaussian distribution described as follows:

$$L_{ij} \Longleftrightarrow \begin{cases} N_d(\frac{1}{2}MD_{ij}, MD_{ij}) & \mathbf{x} \in \omega_i \\ N_d(-\frac{1}{2}MD_{ij}, MD_{ij}) & \mathbf{x} \in \omega_j. \end{cases} \tag{9.101}$$

As a result, the probability of error based on (9.96) is given by

$$\begin{aligned} P_e &= P(\omega_i)\int_{-\infty}^{\beta} \frac{1}{\sqrt{2\pi MD_{ij}}} exp\left[-\frac{(L_{ij} - \frac{1}{2}MD_{ij})^2}{2MD_{ij}}\right] dL_{ij} \\ &\quad + P(\omega_j)\int_{\beta}^{\infty} \frac{1}{\sqrt{2\pi MD_{ij}}} exp\left[-\frac{(L_{ij} + \frac{1}{2}MD_{ij})^2}{2MD_{ij}}\right] dL_{ij}. \end{aligned} \tag{9.102}$$

Assuming the equal priori probabilities, $P(\omega_i) = P(\omega_j)$, yields $\beta = 0$ and a probability of error

$$P_e = \frac{1}{2}\int_{-\frac{1}{2}\sqrt{MD_{ij}}}^{\infty} \frac{1}{\sqrt{2\pi}} exp\left(-\frac{z^2}{2}\right) dz$$

$$+\frac{1}{2}\left[1-\int_{-\infty}^{\frac{1}{2}\sqrt{MD_{ij}}}\frac{1}{\sqrt{2\pi}}exp\left(-\frac{z^2}{2}\right)\right]dz$$

$$= \frac{1}{2}\int_{\frac{1}{2}\sqrt{MD_{ij}}}^{\infty}\frac{1}{\sqrt{2\pi}}exp\left(-\frac{z^2}{2}\right)dz. \quad (9.103)$$

As can be seen from (9.103), the function that relates the squared Mahalanobis distance to the probability of error is the integrated Gaussian density with zero mean and unit variance.

Now, let $y = \frac{z}{\sqrt{2}}$, and $dy = \frac{1}{\sqrt{2}}dz$. If we substitute z and dz into (9.103), it can be rewritten in terms of the *error function* and the *complementary error function* as the following:

$$\begin{aligned} P_e &= \frac{1}{2}\int_{\frac{1}{2}\sqrt{MD_{ij}/2}}^{\infty}\frac{1}{\sqrt{\pi}}exp(-y^2)dy \\ &= \frac{1}{4}erfc\left(\frac{1}{2}\sqrt{\frac{MD_{ij}}{2}}\right) \\ &= \frac{1}{4}\left[1-erf\left(\frac{1}{2}\sqrt{\frac{MD_{ij}}{2}}\right)\right]. \end{aligned} \quad (9.104)$$

The error function is defined by

$$erf(x) = \int_0^x \frac{2}{\sqrt{\pi}}exp(-x^2)dx, \quad (9.105)$$

and the complementary error function is

$$\begin{aligned} erfc(x) &= \int_x^{\infty}\frac{2}{\sqrt{\pi}}exp(-x^2)dx \\ &= 1-erf(x). \end{aligned} \quad (9.106)$$

The relationship between the probability of error, P_e, and the squared Mahalanobis distance, MD_{ij}, is shown in Figure 9.14. It is observed that P_e is a monotonic decreasing function of MD_{ij}. As we can see, the probability of error is less than 2.5% when $MD_{ij} = 16$ (Mahalanobis distance = 4).

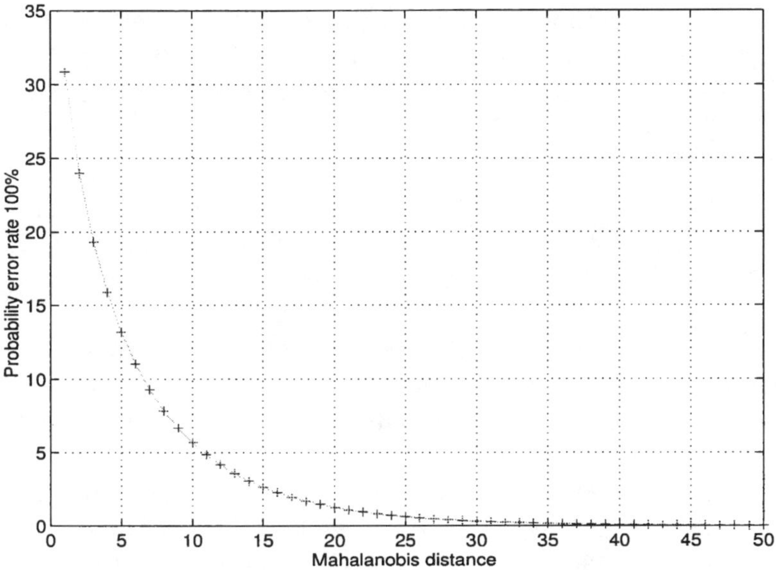

Figure 9.14 The relationship between the probability of misclassification error P_e and the squared Mahalanobis distance MD_{ij}.

9.6.2 Estimation of Probability Error Using Nonparametric Approaches

Let $\{\mathbf{X}, \theta\} = \{\mathbf{x}_1, \theta_1; \mathbf{x}_2, \theta_2 ...; \mathbf{x}_N, \theta_N\}$ be a set of N pattern vectors, where $\mathbf{x}_i$ and θ_i denote the information of statistical signal classification and the label of the ith pattern vector, respectively. It is also assumed that each θ_i associated with a pattern vector $\mathbf{x}_i$ is the correct label. In other words, the pattern vectors have been correctly predefined.

A summary of four nonparametric approaches for estimating the probability of error is listed below.

The Resubstitution Method

The *resubstitution method* is also called the R method. The statistical signal classifier is trained on a pattern data $\{\mathbf{X}, \theta\}$ and is tested on the same pattern data $\{\mathbf{X}, \theta\}$. The advantage of this estimation method is that it uses the pattern data very efficiently. However, this method results in optimistically biased estimates of asymptotic probability of error. Therefore, this R method should be used only when the sample size of pattern vectors is sufficiently large.

The Holdout Method

The *holdout method* is also known as the H method. The pattern data $\{\mathbf{X}, \theta\}$ are partitioned into two mutually exclusive data sets $\{\mathbf{X}, \theta\}_1$ and $\{\mathbf{X}, \theta\}_2$. We train the statistical signal classifier on data $\{\mathbf{X}, \theta\}_1$ and test the statistical signal classifier on data $\{\mathbf{X}, \theta\}_2$. The advantage of this estimation method is that it results in an unbiased estimate of expected probability of error when the sample size of pattern vectors is large. The disadvantage of this method is that it uses the data inefficiently. This method pessimistically estimates the performance of statistical signal classifiers. In addition, this method is unreliable sometimes when the sample size of pattern vectors is relatively small. Therefore, the H method should only be used when the sample size of pattern vectors is large.

The Leave-One-Out Method

The *leave-one-out method* is described as the U method in the literature. We take one pattern vector $\{\mathbf{x}_i, \theta_i\}$, where $i = 1, 2, ...N$, out of the entire set of data $\{\mathbf{X}, \theta\}$, and train the statistical signal classifier on the pattern vectors $\{\mathbf{X}, \theta\}$ excluding the pattern vector $\{\mathbf{x}_i, \theta_i\}$. Then, the statistical signal classifier is tested on this vector $\{\mathbf{X}_i, \theta_i\}$. The detailed procedure of this method is provided in the following steps:

1. Take a pattern vector $\{\mathbf{x}_i, \theta_i\}$ out of the entire data set $\{\mathbf{X}, \theta\}$,

$$\{\mathbf{X}, \theta\}_i = \{\mathbf{X}_i, \theta_i\}, \quad i = 1, 2, ..., i-1; \; i+1, ..., N. \tag{9.107}$$

2. Train the statistical signal classifiers on the remaining set of pattern vectors $\{\mathbf{X}, \theta\}_i$.
3. Test the statistical signal classifiers on the single left-out point. If the test is correct, then the error e_i is zero; Otherwise, the error e_i equals one.

4. Repeat steps 1, 2, and 3, for $i = 1, 2, ..., N$.
5. Estimate the probability of error,

$$\widehat{P}_e[U] = \frac{1}{N}\sum_{i=1}^{N} e_i, \quad i = 1, 2, ..., N. \tag{9.108}$$

The advantage of this method is that it results in an unbiased estimate of expected probability of error. The disadvantage of the method is that it yields a larger variance. In addition, for some types of classification algorithms, it requires substantial computation time compared with the *holdout method.* Therefore, the U method should only be used when the sample size of vectors is small or when the number of pattern vectors is less than the number of pattern classes.

The Rotation Method

The *rotation method* is referred to as the Π method. In this method, we take a subset of pattern vectors $\{\mathbf{X}, \theta\}_{TS}$ out of data $\{\mathbf{X}, \theta\}$. Usually, the subset of pattern vectors is less than half of the total number of pattern vectors. We train the statistical signal classifier on the remaining pattern data $\{\mathbf{X}, \theta\}$ without containing the subset of pattern vectors $\{\mathbf{X}, \theta\}_{TS}$. Then, we test the statistical signal classifier on the subset of data $\{\mathbf{X}, \theta\}_{TS}$. The detailed procedure of this method is described in the following steps:

1. Take a subset of pattern vectors

$$\begin{aligned} \{\mathbf{X}, \theta\}_{i(TS)} &= \{\mathbf{X}_i, \theta_i\}, \quad i = 1, 2, ..., P, \\ \{\mathbf{X}, \theta\}_{i(TR)} &= \{\mathbf{X}_i, \theta_i\}, \quad i = P+1, ..., N, \end{aligned} \tag{9.109}$$

where $1 \le P \le \frac{N}{2}$ and $\frac{N}{P}$ is integer.

2. Train the statistical signal classifiers on the remaining data $\{\mathbf{X}, \theta\}_{i(TR)}$.
3. Test the statistical signal classifiers on the subset of data $\{\mathbf{X}, \theta\}_{i(TS)}$ to obtain the estimated probability of misclassification error $\widehat{P}_e[\Pi]_i$.

4. Repeat steps 1, 2, and 3 in such way that $\{\mathbf{X}, \theta\}_{i(TS)}$ and $\{\mathbf{X}, \theta\}_{i(TR)}$ are disjoint for $i \neq j$, $i = 1, 2, ..., \frac{N}{P}$, and $j = 1, 2, ..., \frac{N}{P}$.
5. Estimate the probability of error,

$$\widehat{P}_e[\Pi] = \frac{P}{N} \sum_{i=1}^{N/P} \widehat{P}_e[\Pi]_i, \quad i = 1, 2, ..., \tfrac{N}{P}. \tag{9.110}$$

The advantage of this method is that it has less bias than the holdout method and less computation than the leave-one-out method. However, this method still requires complex implementation. Therefore, the Π method should be used when the sample size of the data set is neither large nor small.

Note that the rotation method converges to the leave-one-out method when $P = 1$. The rotation method tends to the holdout method when $P = \frac{N}{2}$. In other words, the holdout method and the leave-one-out method are special cases of the rotation method.

A detailed theoretical development of four nonparametric approaches for estimating probability of error $\widehat{P}_e$ is omitted. Readers are encouraged to study the papers mentioned in the references including Highleyman [70], Hills [72], Kanal and Chandrasekaran [73], Sorum [74], Toussaint [71], Mardia, Kent, and Bibby [75], Raudys and Jain [76], and Ney, Essen, and Kneser [77].

9.7 Summary

In this chapter, we first presented an introduction to statistical signal classification and its applications. Next, we introduced the statistical pattern representation and defined the definition of terminologies. The digital signal preprocessor was also addressed. Then we discussed the technical methods of feature extraction theory, with emphasis on selection of measurement features including the $\mathcal{F}$-test and the stepwise discriminant analysis, and dimensionality reduction. Unsupervised learning and cluster analysis were discussed, with a focus on partitioning and hierarchical approaches. Subsequently, we presented statistical signal classifications including nonparametric and parametric approaches, signal classifiers based on Gaussian models, implementation methods. The estimated probability of misclassification was discussed by using parametric and nonparametric approaches. Moreover, we provided mathematical treatments for the above techniques and described the advantages and the disadvantages of various schemes. In addition, we gave many examples illustrating the basic ideas so that the mathematical treatments could be easily understood.

Given a training pattern data set, it has been shown that there is a "peaking" phenomenon in the finite-sample-size case. The optimal number of measurement features can be found when the probability of misclassification error is minimumal. This usually depends on the training sample size, the type of classification function, and the distributions of the pattern vectors. However, the optimal tools do not exist for finding the optimal number of measurement features in the statistical signal classification. The $\mathcal{F}$-test method and stepwise discriminant analysis introduced in this chapter are valuable tools to determine a good subset of discriminatory measurement features for the statistical signal classification along with cross-validation.

There is little doubt that unsupervised learning and cluster analysis using only unlabeled pattern data are challenging. Partitioning and hierarchical approaches that belong to unsupervised learning are often used for cluster analysis. These approaches utilize general measures among pattern vectors and do not require a set of labeled training vectors. However, hierarchical approaches may not be appropriate if the number of measurement vectors is large. This is because the hierarchical methods result in data partitions with suboptimal solution in an agglomerate procedure. Therefore, in such cases, partitioning approaches such as the k-means algorithm are useful to find the cluster information.

So far, we have seen that a variety of approaches are possible in the field of statistical signal classification. We do not attempt to cover all the classification schemes proposed in the literature, but have chosen instead to emphasize the fundamental principles and practical application issues.

In the next chapter, we focus on the introduction of transform methods for reducing highly multidimensional features into lower component features without degrading the information lost.

References

[1] Sastry, R., N. Ranganathan, and K. Remedios, "CASM: A VLSI Chip for Approximate String Matching," *IEEE Trans. on Pattern Analysis and Machine Intelligence*, Vol. 17, No. 8, pp. 824-830, August 1995.

[2] Cheng, H. D., and D. C. Xia, "A Novel Parallel Approach to Character Recognition and Its VLSI Implementation," *Pattern Recognition*, Vol. 29, No. 1, pp. 97-119, January 1996.

[3] Gong, L., and C. A. Kulikowski, "Composition of Image Analysis Processes Through Object-Centered Hierarchical Planning," *IEEE Trans. on Pattern Analysis and Machine Intelligence*, Vol. 17, No. 10, pp. 997-1009, October 1995.

[4] Rabiner, L., and B. Juang, *Fundamentals of Speech Recognition*, Prentice-Hall, Englewood Cliffs, New Jersey, 1993.

[5] Blue, J. L., G. T. Candela, P. J. Grother, R. Chellappa, and C. L. Wilson, "Evaluation of Pattern Classifiers for Fingerprint and OCR Applications," *Pattern Recognition*, Vol. 27, No. 4, pp. 485-501, April 1994.

[6] Lu, Y., and M. Shridhar, "Character Segmentation in Hardwritten Words: an Overview," *Pattern Recognition*, Vol. 29, No. 1, pp. 77-96, January 1996.

[7] Olson, C. F., "Probabilistic Indexing for Object Recognition," *IEEE Trans. on Pattern Analysis and Machine Intelligence*, Vol. 17, No. 5, pp. 518-522, May 1995.

[8] Jia, X., and M. S. Nixon, "Extending the Feature Vector for Automatic Face Recognition," *IEEE Trans. on Pattern Analysis and Machine Intelligence*, Vol. 17, No. 12, pp. 1167-1176, December 1995.

[9] Chen, C. H. (eds.), *Digital Waveform Processing and Recognition*, CRC Press, Boca Raton, Florida, 1982.

[10] Miao, J., and X. Yu, "Detection and Handling System of Pulse Waveforms in Human Finger Using Minicomputer," *Journal of Electric Automation*, No. 4, pp. 34-36, 1985.

[11] Miao, J., and M. A. Clements, "Unsupervised Pattern Recognition for Digital Waveform Classification from Radiation Detectors," *Proc. IEEE Int. Conf. Acoustics, Speech and Signal Processing*, Vol. 4, pp. 2751-2754, May 1995.

[12] Miao, J., S. Sun, and X. Yu, "Synchronous Controller for Cardiac Assistance Pumps Using Noninvasive Measurement of the Digital Waveforms of Blood Pressure," *Journal of Medical Instrumentation*, Vol. 9, No. 3, pp. 34-37, May 1985.

[13] Miao, J., P. J. Benkeser, and F. T. Nichols, "A Computer-Based Statistical Pattern Recognition for Doppler Spectral Waveforms of Intracranial Blood Flow," *Computer in Biology and Medicine: An International Journal*, Vol. 26, No. 1, pp. 53-63, February 1996.

[14] Novak, L. M., G. J. Owirk, and C. M. Netishen, "Radar Target Identification Using Spatial Matched Filters," *Pattern Recognition*, Vol. 27, No. 4, pp. 607-617, April 1994.

[15] Schalkoff, R., *Pattern Recognition: Statistical, Structural and Neural Approaches*, John Wiley & Sons, Inc., New York, 1992.

[16] Echard, J. D., M. Burrow, J. Miao, and D. Humphrey, "Pattern Recognition Techniques for Cortical Control of a Robotic Arm," *Proc. 13th IEEE Southern Biomedical Engineering Conf.*, Washington, D.C., April 16-17, 1994.

[17] Patrick, E. A., and J. C. Hancock, "Nonsupervised Sequential Classification and Recognition of Patterns," *IEEE Trans. on Information Theory*, Vol. IT-12, No. 3, pp. 362-372, July 1966.

[18] Leboucher, G., and G. E. Lowitz, "What a Histogram Can Really Tell the Classifier," *Pattern Recognition*, Vol. 10, pp. 351-357, 1978.

[19] Yarman-Vural, F., and E. Ataman, "Noise, Histogram and Cluster Validity for Gaussian-Mixtured Data," *Pattern Recognition*, Vol. 20, No. 4, pp. 385-401, 1987.

[20] Miao, J., "Component Feature-Based Digital Waveform Analysis and Classification," Ph.D. Dissertation, Georgia Institute of Technology, Atlanta, Georgia, April 1996.

[21] Dudgeon, D. E., and R. M. Mersereau, *Multidimensional Digital Signal Processing*, Prentice-Hall, Englewood Cliffs, New Jersey, 1984.

[22] Quackenbush, S. R., T. B. Barnwell, and M. A. Clements, *Objective Measures of Speech Quality*, Englewood Cliffs, Prentice Hall, New Jersey, 1988.

[23] Oppenheim, A. V., and R. W. Schafer, *Discrete-Time Signal Processing*, Prentice-Hall, Englewood Cliffs, New Jersey, 1989.

[24] Proakis, J. G., and D. G. Manolakis, *Digital Signal Processing: Principles, Algorithms, and Applications*, 3d ed., Englewood Cliffs, Prentice Hall, New Jersey, 1996.

[25] Aeberhard, S., D. Coomans, and O. D. Vel, "Comparative Analysis of Statistical Pattern Recognition Methods in High Dimensional Settings," *Pattern Recognition*, Vol. 27, No. 8, pp. 1065-1077, 1994.

[26] Fukunaga, F., and R. R. Hayes, "Effects of Sample Size in Classifier Design," *IEEE Trans. on Pattern Analysis and Machine Intelligence*, Vol. 11, No. 8, pp. 873-885, August 1989.

[27] Costanza, M. C., and A. A. Afifi, "Comparison of Stopping Rules in Forward Stepwise Discriminant Analysis," *Journal of the American Statistical Association*, Vol. 74, pp. 77-785, 1979.

[28] Jennrich, R. I., "Stepwise Discriminant Analysis," *Statistical Methods for Digital Computer*, K. Enslein, A. Ralston, and H. Wilf (eds.), John Wiley & Sons, New York, 1977.

[29] Wilks, S. S., *Mathematical Statistics*, John Wiley & Sons, New York, 1962.

[30] Rao, C. R., *Linear Statistical Interference and Its Applications*, John Wiley & Sons, New York, 1965.

[31] Young, T. Y., and T. W. Calvert, *Classification, Estimation and Pattern Recognition*, American Elsevier Publishing Company, Inc., New York, 1974.

[32] Tou, J. T., and R. C. Gonzalez, *Pattern Recognition Principles*, 4th printing, Addison-Wesley Publishing Company, Reading, Massachusetts, 1981.

[33] Fu, K. S., "Recent Developments in Pattern Recognition," *IEEE Trans. on Computers*, Vol. C-29, No. 10, pp. 845-854, October, 1980.

[34] Siedlecki, W., K. Siedlecka, and J. Sklansky, "An Overview of Mapping Techniques for Exploratory Pattern Analysis," *Pattern Recognition*, Vol. 21, No. 5, pp. 411-429, 1988.

[35] Duda, R. O., and P. E. Hart, *Pattern Classification and Scene Analysis*, John Wiley & Sons, New York, 1973.

[36] Anderberg, M. R., *Cluster Analysis for Application*, Academic Press, New York, 1973.

[37] Dubes, R., and A. K. Jain, "Clustering Methodologies in Exploratory Data Analysis," *Advances in Computers*, M. C. Yovits (eds.), Academic Press, pp. 113-228, New York, 1981.

[38] Everitt, B. S., *Cluster Analysis*, 2d ed., London: Heineman Educational Books Ltd., 1980.

[39] Massart, D. L., and L. Kaufman, *The Interpretation of Analytical Chemical Data by the Use of Cluster Analysis*, John Wiley & Sons, New York, 1983.

[40] Milligan, G. W., "An Examination of the Effect of Six Types of Error Perturbation on Fifteen Clustering Algorithms," *Psychometrika*, Vol. 45, pp. 325-342, 1980.

[41] Milligan, G. W., "A Review of Monte Carlo Test of Cluster Analysis," *Multivariate Behavioral Research*, Vol. 16, pp. 379-407, 1981.

[42] MacQueen, J. B., "Some Methods for Classification and Analysis of Multivariate Observations," *Proc. 5th Berkeley Symp. Mathematical Statistics and Probability,* Vol. 1, pp. 281-297, 1967.

[43] Scott, A. J., and M. J. Symons, "Clustering Methods Based on Likelihood Ratio Criteria," *Biometrics*, Vol. 27, pp. 387-397, 1971.

[44] Hartigan, J. A., *Clustering Algorithms*, John Wiley & Sons, New York, 1975.

[45] Hartigan, J. A., "Consistency of Single Linkage for High-Density Clusters," *Journal of the American Statistical Association*, Vol. 76, pp. 388-394, 1981.

[46] Binder, D. A., "Bayesian Cluster Analysis," *Biometrike*, Vol. 65, pp. 31-38, 1978.

[47] Binder, D. A., "Approximation to Bayesian Clustering Rules," *Biometrike*, Vol. 68, pp. 275-285, 1981.

[48] Wong, M. A., and C. Schaack, "Using the *k*th Nearest Neighbor Clustering Procedure to Determine the Number of Subpopulations," *Proc. of the Statistical Computing Section, American Statistical Association*, pp. 40-48, 1982.

[49] Wong, M. A., and T. Lane, "A *k*th Nearest Neighbor Clustering Procedure," *Journal of the Royal Statistical Society*, Series B, Vol. 45, pp. 362-368, 1983.

[50] Linde, Y., A. Buzo, and R. M. Gray, "An Algorithm for Vector Quantizer Design," *IEEE Trans. on Communication*, Vol. COM-28, pp. 84-95, January

1980.

[51] Chang, P., R. M. Gray, and J. May, "Fourier Transform Vector Quantization for Speech Coding," *IEEE Trans. on Communications*, Vol. COM-35, No. 10, October 1987.

[52] Gray, R. M., "Vector Quantization," *Readings in Speech Recognition*, A. Waibel and K. Lee (eds.), Morgan Kaufmann Publishers, Inc., San Mateo, California, 1990.

[53] Miao, J., "Vector Quantization of DCT Components for Speech Coding," *Proc. Australian Int. Conf. of Speech Science and Technology*, Roberto Togneri (eds.), Vol. 2, pp. 551-556, Perth, Australia, December 1994.

[54] Abbas, H. M., and M. M. Fahmy, "Classified Vector Quantization Using Variance Classifier and Maximum Likelihood Clustering," *Pattern Recognition Letters*, Vol. 15, pp. 49-55, 1994.

[55] Barlaud, M., P. Sole, T. Gaidon, M. Antonini, and P. Mathieu, "Pyramidal Lattice Vector Quantization for Multiscale Image Coding," *IEEE Trans. on Image Processing*, Vol. 3, No. 4, pp. 367-381, July 1994.

[56] Florek, K., J. Lukaszewicz, J. Perkal, and S. Zubrzycki, "Taksonomia Wroclawska," *Przeglad Antropol*, Vol. 17, pp. 193-211, 1951.

[57] McQuity, L. L., "Elementary Linkage Analysis for Isolating Orthogonal and Oblique Types and Typal Relevancies," *Educational and Psychological Measurement*, Vol. 17, pp. 207-229, 1957.

[58] Sneath, P. H. A., "The Application of Computers to Taxonomy," *Journal of General Microbiology*, Vol. 17, pp. 201-226, 1957.

[59] Jardine, N., and R. Sibson, *Mathematical Taxonomy*, John Wiley & Sons, New York, 1971.

[60] Fisher, L., and J. W. Ness, "Admissible Clustering Procedures," *Biometrika*, Vol. 58, pp. 91-104, 1971.

[61] Milligan, G. W., and M. C. Cooper, "An Examination of Procedures for Determining the Number of Clusters in a Data Set," *Psychometrika*, Vol. 50, pp. 159-179, 1985.

[62] Sarle, W. S., "Cluster Analysis by Least Squares," *Proc. 7th Annual SAS User Group Int. Conf.*, pp. 651-653, 1982.

[63] Jolion, J. M., and A. Rosenfeld, "Cluster Detection in Background Noise," *Pattern Recognition*, Vol. 22, No. 5, pp. 603-607, 1989.

[64] Everitt, B. S., "Unresolved Problems in Cluster Analysis," *Biometrics*, Vol. 35, pp. 169-181, 1979.

[65] Papoulis, A., *Probability, Random Variables, and Stochastic Processes*, 3d ed., McGraw-Hill, Inc., New York, 1991.

[66] Anderson, T. W., *An Introduction to Multivariate Statistical Analysis*, John Wiley & Sons, New York, 1958.

[67] Kendall, M. G., and A. Stuart, *The Advanced Theory of Statistics*, Vol. 3, Charles Griffin and Company, Ltd., London, 1961.

[68] Morrison, D. F., *Multivariate Statistical Methods*, 2d ed., McGraw-Hill, New

York, 1976.

[69] Postaire, J. G., "An Unsupervised Bayes Classifier for Normal Patterns Based on Marginal Densities Analysis," *Pattern Recognition*, Vol. 15, No. 2, pp. 103-111, 1982.

[70] Highleyman, W. H., "The Design and Analysis of Pattern Recognition Experiments," *Bell Syst. Technology Journal*, Vol. 41, pp. 723-744, March, 1962.

[71] Toussaint, G. T., "Bibliography on Estimation of Misclassification," *IEEE Trans. on Information Theory*, Vol. IT-20, No. 4, pp. 472-479, 1974.

[72] Hills, M., "Allocation Rules and Their Error Rates," *Journal Royal Statistical Society.*, Series B, No. 28, pp. 1-31, 1966.

[73] Kanal, L., and B. Chandrasekaran, "On Dimensionality and Sample Size in Statistical Pattern Recognition," *Proc. Natl. Electron Conf.*, Vol. 24, pp. 2-7, 1968.

[74] Sorum, M., "Estimating the Conditional Probability of Misclassification," *Technometrics*, Vol. 13, pp. 333-343, May, 1971.

[75] Mardia, K. V., J. T. Kent and J. M. Bibby, *Multivariate Analysis*, 3d ed., Academic Press Inc., New York, 1982.

[76] Raudys, S. J., and A. K. Jain, "Small Sample Size Effects in Statistical Pattern Recognition: Recommendations for Practitioners," *IEEE Tran. on Pattern Analysis and Machine Intelligence*, Vol. 13, No. 3, pp. 252-264, March 1991.

[77] Ney, H., U. Essen, and R. Kneser, "On the Estimation of Small Probabilities by Leaving-One-Out," *IEEE Trans. on Pattern Analysis and Machine Intelligence*, Vol. 17, No. 12, pp. 1202-1212, December 1995.

10

Transform-Based Statistical Signal Classification

10.1 Introduction

In Chapter 9, we introduced the theory and structure of statistical signal classification as well as its implementation. We addressed supervised and unsupervised statistical signal classification and discussed specific approaches, including feature extraction theory, cluster analysis, specification of the optimum decision function, and estimated probability of misclassification error. In general, a multidimensional statistical signal classification tends to achieve high accuracy in identifying desired targets, but it also has high computational complexity and large memory requirements.

In dealing with multidimensional feature observations, visualization and understanding are often aided by representing the observations in a lower-dimensional space. In particular, two-dimensional (2-D) scatter plots are helpful in exploring relationships between the cluster groups, in assessing the group-conditional distributions, and in identifying a typical feature observation.

The goal of this chapter is to present linear transformation methods and their operations; we include five major families of transformations: principal components, canonical correlation, canonical discriminant, optimal discriminant, and optimal declustering. These transforms belong to the class of multivariate techniques and are useful for examining relationships among the multidimensional measurement features. They are also used for summarizing data and for specifying linear combinations of the measurement features without losing significant information. One common property of these transforms is that they do not depend on any assumption of probability distribution. Only knowledge of the first two moments of the cluster group-conditional distribution is required. For plotting 2-D features, principal components, canonical correlation, canonical discriminant, optimal discriminant, and/or optimal declustering transforms are especially valuable. These also serve as approaches for exploring training data characteristics. Therefore, focusing on the design of the 2-D transform-based statistical signal classification system is a primary goal in this chapter.

Transform-based statistical signal classification systems have several important properties as follows:

- The operation of transform-based statistical signal classification allows for the substitution of very-high-dimensional measurement features by 2-D transform features while minimizing the information lost in the training data and test data.
- A 2-D transform-based statistical signal classification system can simplify the structure of the covariance matrix to make visual interpretation of cluster graphics better for assessing the results of cluster performance.
- The performance of 2-D transform-based statistical signal classification is often comparable to classification using multidimensional feature sets. In addition, the efficiency of computation and the storage requirements can be significantly improved.
- The 2-D transform-based statistical signal classification system is particularly important in real-time applications for analyzing signal sources with thousands of count rates per second, sampled with an extremely high-speed A/D converter.

Since the transformed features reduce the dimensionality of the pattern space, obviously some of the information contained in it is lost. However, we are usually willing to use transformed features in statistical signal classification, due to one of the following reasons:

- The transform feature space is physically more meaningful than the original measurement feature space. For example, it is well known in speech, audio, and image processing that the frequency spectrum is often more meaningful than the raw signals. If a discrete Fourier transform or discrete cosine transform is taken, the corresponding frequency components are often better features for classification than the time-domain features. Moreover, combining the frequency components can reduce the dimensionality.
- If the measurement and classification are performed at different locations, then it is necessary to transmit the data over a digital communication channel or wireless channel.
- There may be prior knowledge that the measurement features are redundant and the data is highly correlated. Then the dimensionality may be reduced with very little loss of information.
- The efficiency of computation and storage requirements in the hardware are significantly improved because of dimensionality reduction.

This chapter is organized as follows. Section 10.1 has explained why we need to develop transform-based statistical signal classification methods. Section 10.2 gives the high-level system architecture of transform-based statistical signal classification. Section 10.3 provides an overview of the theory of the generalized principal components transform. The quantity analysis and benefits of the principal components transform are also discussed. The canonical correlation transform is introduced in Section 10.4. Subsequently, Section 10.5 provides the theory of the canonical discriminant transform, which is derived from the concepts of canonical correlation and principal components transforms. We develop the canonical discriminant transform with two approaches, including the differentiation of discriminant function with constraints and without constraints. The optimal discriminant transform and the generalized optimal declustering transform are introduced in Sections 10.6 and 10.7, respectively. We show that the optimal discriminant transform is a special case in the generalized optimal declustering transform. The transform-based statistical signal classifiers are presented in Section 10.8. In Section 10.9, application and performance evaluations of the transform-based statistical signal classification are given, with emphasis on using the first two transform features in statistical signal classification. Finally, the chapter is summarized in Section 10.10.

10.2 System Architectures of Transform-Based Statistical Signal Classification

In this section, we introduce the high-level system architecture of 2-D transform-based supervised and unsupervised statistical signal classification systems. As shown in Figure 10.1(a), the training signal classification system contains five subsystems: a digital signal preprocessor, a selector of measurement features, a components analysis, a cluster analysis, and an optimal decision function. The implementation of 2-D transform-based statistical signal classification, as shown in Figure 10.1(b), consists of three subsystems: digital signal preprocessing, dimensionality reduction, and implementation of optimal decision functions. In particular, the integration of the transform-based features and optimal decision functions forms the central computational mechanism within the DSP system.

The functions at the upper- and lower-level block diagrams as shown in Figure 10.1 were introduced in Chapter 9. However, in the components analysis, we only presented the Karhunen-Loève transform. In this chapter, the methodology of components analysis is referred to as transform-based features analysis. Using the transform methods including principal components, canonical correlation, canonical discriminant, optimal discriminant, and optimal declustering, we can generate the 2-D transform-based features from highly dimensional measurement features for statistical signal classification. Note that the transformation theory is based on the theory of matrix algebra of linear transform, which plays a central role in transform-based statistical signal classification. The matrix algebra for linear transform is reviewed in the Appendix.

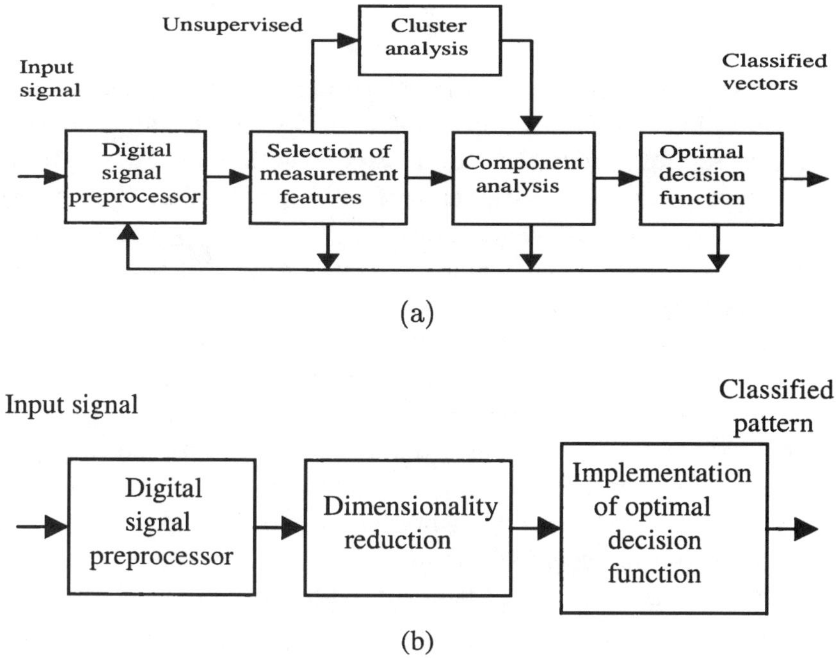

Figure 10.1 (a) A training block diagram of the transform-based supervised and unsupervised statistical signal classification system. (b) A block diagram for implementation of a classifier based on the transform features and optimal decision functions.

10.3 Generalized Principal Components Transform

Generally, the principal components transform looks for a few linear combinations that can be used to summarize the training data, losing in the process as little information as possible. This transform, which attempts to reduce dimensionality, can be described as "*parsimonious summarization*" of the observation data, according to the references by Mardia, Kent, and Bibby [1] and Miao [2].

If $\mathbf{x}$ is a random vector with mean $\mathbf{m}$ and covariance (or correlation) matrix $\mathbf{R}$, then the principal components transform is the linear transformation in the following:

$$\mathbf{x} \Longrightarrow \mathbf{y} = \mathbf{B}^T(\mathbf{x} - \mathbf{m}), \tag{10.1}$$

where $\mathbf{B}$ is an orthogonal matrix. Note that the covariance matrix and correlation matrix are all denoted by using the same notation $\mathbf{R}$ in this section, since a random vector with nonzero mean can be transformed into one with zero mean by translation. The matrix $\mathbf{B}^T\mathbf{R}\mathbf{B}$ is diagonal with eigenvalues $\lambda_1 \geq \lambda_2 \geq ... \geq \lambda_p \geq 0$. The strict positivity of the eigenvalues λ_i is guaranteed if the covariance (or correlation) matrix $\mathbf{R}$ is positive definite. This representation of covariance (or correlation) matrix $\mathbf{R}$ follows from the spectral decomposition theorem (see Section A.7 in the Appendix).

Given training data with p measurement features, p principal components features can be calculated. Each of the principal components features is a linear combination of the original p measurements, with coefficients equal to the eigenvectors of the covariance (or correlation) matrix $\mathbf{R}$. The eigenvectors are customarily taken with unit-norms. The principal components features are sorted in descending order of the corresponding eigenvalues, which are themselves equal to the variances of the extracted components. This method is also known as the Karhunen-Loève transformation (see Chien and Fu [3], Fukunage and Koontz [4], Young [5], and Duchene and Leclercq [6]).

Let a training data matrix $\mathbf{X}$ represent N observations of a p-dimensional random vector. We assume without loss of generality that $\mathbf{X}$ has zero mean, $E[\mathbf{X}] = 0$, because a random vector with nonzero mean can be translated into another random vector with zero mean. The covariance (or correlation) matrix $\mathbf{R}$ of the data vector is then defined by

$$\mathbf{R} = E[\mathbf{X}^T\mathbf{X}]. \tag{10.2}$$

We observe that the covariance (or correlation) matrix $\mathbf{R}$ is symmetric, which means that

$$\mathbf{R}^T = \mathbf{R}. \tag{10.3}$$

The next issue to be considered is that the eigenvalue and eigenvector problem can be solved in the following:

$$(\mathbf{R} - \Lambda\mathbf{I})\mathbf{U} = 0, \tag{10.4}$$

where Λ is a diagonal matrix defined by the eigenvalues of the covariance (or correlation) matrix $\mathbf{R}$,

$$\Lambda = diag[\lambda_1, \lambda_2, ..., \lambda_p], \tag{10.5}$$

and $\mathbf{U}$ is the matrix of associated eigenvectors

$$\mathbf{U} = [\mathbf{u}_1, \mathbf{u}_2, ..., \mathbf{u}_p]. \tag{10.6}$$

The matrix $\mathbf{U}$ is an orthonormal matrix in the sense that its column vectors satisfy the conditions

$$\mathbf{U}^T\mathbf{U} = \mathbf{I} \qquad and \qquad \mathbf{U}^T = \mathbf{U}^{-1}. \tag{10.7}$$

This means that we can write (10.4) in a form known as the *orthogonal similarity transformation*,

$$\mathbf{U}^T\mathbf{R}\mathbf{U} = \Lambda. \tag{10.8}$$

The principal components features in the matrix $\mathbf{Z}$ can be determined by

$$\mathbf{Z} = \mathbf{X}\mathbf{U}. \tag{10.9}$$

Since the eigenvectors in the matrix $\mathbf{U}$ are mutually orthogonal, the principal components features in the matrix $\mathbf{Z}$ are also mutually orthogonal. Therefore, the matrix $\mathbf{Z}^T\mathbf{Z}$ is the diagonal matrix of p eigenvalues λ_k $(k = 1, 2, ..., p)$ that decrease in magnitude. The eigenvalues λ_k in the matrix $\mathbf{Z}^T\mathbf{Z}$ are given by

$$\sum_{i=1}^{n} z_{ik}^2 = \lambda_k, \quad k = 1, 2, ..., p. \tag{10.10}$$

As can be seen, the sum of squares for each principal components feature is calculated by using the corresponding eigenvalue.

The most important properties of the principal components transform are as follows:

- The eigenvectors are mutually orthogonal, so the principal components features show jointly perpendicular directions through the space of the original measurement features.

- The principal components features are linear transformations and are not correlated with each other.
- The total variance of all the principal components features or the sum of all the eigenvalues, $\sum_{i=1}^{P} \lambda_i$, is the same as that of all the original measurement features.
- The first principal components feature has the largest variance and the last principal components feature has the smallest variance of any unit-length linear combination of the original measurement features.
- The principal components features of a random vector are *not scale-invariant.* Note that non-scale-variance is a disadvantage of principal components analysis.

10.3.1 The Quantity Analysis of Principal Component Features

In this section, we introduce the method of measuring the quantity of principal component features. The measurement method can help a reader selecting the right principal component features to represent the given original data, while minimizing the quantity error.

We have shown that the principal components features in (10.9) are given by

$$\mathbf{Z} = \mathbf{X}\mathbf{U}. \tag{10.11}$$

Thus, the training data matrix $\mathbf{X}$ can be estimated by using the matrix $\mathbf{Z}$ and $\mathbf{U}$. In other words, in matrix notation, the approximation training data matrix $\widehat{\mathbf{X}}$ is calculated by

$$\widehat{\mathbf{X}} = \mathbf{Z}\mathbf{U}^T. \tag{10.12}$$

Therefore, the quantity analysis to be minimized is the error $\mathcal{E}$

$$\begin{aligned} \mathcal{E} &= tr[(\mathbf{X} - \widehat{\mathbf{X}})^T(\mathbf{X} - \widehat{\mathbf{X}})] \\ &= tr(\mathbf{X}^T\mathbf{X}) - \sum_{k=1}^{f} \lambda_k \\ &= \sum_{j=1}^{p} \lambda_j - \sum_{k=1}^{f} \lambda_k, \end{aligned} \tag{10.13}$$

where p and f are the number of measurement features and the selected number of principal components features, respectively. If we set $f = p$, then the quantity analysis of error $\mathcal{E}$ in (10.13) is equal to

$$\mathcal{E} = \sum_{j=1}^{p} \lambda_j - \sum_{k=1}^{f} \lambda_k = 0. \tag{10.14}$$

If we select $f < p$, then the quantity analysis of error $\mathcal{E}$ in (10.13) yields

$$\begin{aligned} \mathcal{E} &= \sum_{j=1}^{p} \lambda_j - \sum_{k=1}^{f} \lambda_k \\ &= \sum_{j=f+1}^{p} \lambda_j. \end{aligned} \tag{10.15}$$

This means that the quantity analysis of error $\mathcal{E}$ is a sum of the eigenvalues λ_j, where $j = f + 1, ..., p$.

Each of the eigenvectors produce a portion of the total variation in matrix $\mathbf{X}$ as evaluated by $tr(\mathbf{X}^T\mathbf{X})$. Obviously, it is essential for a reader to correctly select f ($f < p$) principal components features to represent the given original information data, while minimizing the information lost. For this purpose, the contribution percentage of eigenvalues to the total variance is used to determine whether its associated eigenvector and the corresponding component features should be chosen. Such a contribution percentage L_k of eigenvalues λ_k is defined by

$$L_k = \frac{\lambda_k}{\sum_{i=1}^{p} \lambda_i} \times 100\%, \quad k = 1, 2, ..., p. \tag{10.16}$$

The number of principal component features actually used for the approximation of the matrix $\mathbf{X}$ can be guided by the measure of cumulative percentage C_f

$$C_f = \frac{\sum_{k=1}^{f} \lambda_k}{\sum_{k=1}^{p} \lambda_k} \times 100\%, \quad f \leq p. \tag{10.17}$$

In practice, this ratio can be close to 100%, although f is much less than p.

Thus, to achieve dimensionality reduction on some input data, we compute the eigenvalues and eigenvectors of the covariance (or correlation) matrix $\mathbf{R}$ of the input data vector, and then project the data orthogonally onto the subspace spanned by the eigenvectors belonging to the largest eigenvalues. This method of data representation is also called *subspace decomposition* (see Oja [7]).

The reduction in dimensions afforded by the principal components transform can be used graphically. Thus if the first two components features explain "most" of the variance, then a scatter-gram showing the distribution of the objects on these two dimensions will often give a fair indication of the overall distribution of the data. Such plots can augment the graphical methods for visual explanation of the original measurement features.

10.3.2 Benefits of Principal Components Transform

The advantage of the principal components transform is that it can transform a higher dimensional space to one of lower dimensional space before any clustering takes place. However, the principal components transform, if used before cluster analysis, must be carried out with extreme caution because of the weighting and scaling effects that the transform can introduce.

A disadvantage of the principal components transform is that it does not take discrimination between cluster groups into consideration (see Young and Calvert [8], and Miao [9]). In other words, this procedure does nothing either to conserve or enhance the separation between cluster groups (see Nagy [10] and Siedlecki, Siedlecka, and Sklansky [11]). Moreover, this procedure can easily be derailed so that the principal axes do not preserve the separation between the cluster groups if the clusters are bar-shaped or if irrelevant features (noise) are added (see Liu, Yammashita, and Ogana [12] and Shum, Ikeuchi, and Reddy [13]).

10.4 Canonical Correlation Transform

The *canonical correlation transform* involves partitioning a collection of variables into two-set random vectors, an $\mathbf{x}$-set vector and a $\mathbf{y}$-set vector. The objective of the canonical correlation transform is then to find two linear combinations $\eta = \mathbf{a}^T\mathbf{x}$ and $\psi = \mathbf{b}^T\mathbf{y}$ such that η and ψ have the largest possible correlation. Such linear combinations can give insight into the relationships between the two sets of random vectors.

Suppose that $\mathbf{x}$ is a q-dimensional random vector and $\mathbf{y}$ is a p-dimensional random vector. Suppose further that $\mathbf{x}$ and $\mathbf{y}$ have means $\mathbf{m}_x$ and $\mathbf{m}_y$, and covariance matrices

$$E\left[(\mathbf{x}-\mathbf{m}_x)(\mathbf{x}-\mathbf{m}_x)^T\right] = \mathbf{R}_x \quad (10.18)$$

$$E\left[(\mathbf{y}-\mathbf{m}_y)(\mathbf{y}-\mathbf{m}_y)^T\right] = \mathbf{R}_y \quad (10.19)$$

$$E\left[(\mathbf{x}-\mathbf{m}_x)(\mathbf{y}-\mathbf{m}_y)^T\right] = \mathbf{R}_{xy}. \quad (10.20)$$

Then, the correlation of the linear combinations between η and ψ is

$$\rho(\mathbf{a},\mathbf{b}) = \frac{\mathbf{a}^T\mathbf{R}_{xy}\mathbf{b}}{(\mathbf{a}^T\mathbf{R}_x\mathbf{a}\mathbf{b}^T\mathbf{R}_y\mathbf{b})^{1/2}}. \quad (10.21)$$

As can be seen, this correlation value $\rho(\mathbf{a},\mathbf{b})$ in the above (10.21) varies with different values of $\mathbf{a}$ and $\mathbf{b}$. Note that the correlation value $\rho(\mathbf{a},\mathbf{b})$ in (10.21) does not depend on the scaling of $\mathbf{a}$ and $\mathbf{b}$.

If we want to maximize the correlation value $\rho(\mathbf{a},\mathbf{b})$, this is equivalent to

$$max_{\mathbf{a},\mathbf{b}}\left(\mathbf{a}^T\mathbf{R}_{xy}\mathbf{b}\right) \quad \text{subject to } \mathbf{a}^T\mathbf{R}_x\mathbf{a} = \mathbf{b}^T\mathbf{R}_y\mathbf{b} = 1. \quad (10.22)$$

Suppose that the covariance matrices $\mathbf{R}_x$ and $\mathbf{R}_y$ are nonsingular and

$$\mathbf{K} = \mathbf{R}_x^{-1/2}\mathbf{R}_{xy}\mathbf{R}_y^{-1/2}, \quad (10.23)$$

where the rank of matrix $\mathbf{K}$ is equal to the rank of matrix $\mathbf{R}_{xy}$. For mathematical simplification, we assume that the eigenvalues of the matrix $\mathbf{K}$ are all distinct, $\lambda_1 >, ..., > \lambda_k > 0$. Then, we define two matrices $\mathbf{N}_1$ and $\mathbf{N}_2$ as

$$\mathbf{N}_1 = \mathbf{K}\mathbf{K}^T \quad (10.24)$$

and

$$\mathbf{N}_2 = \mathbf{K}^T\mathbf{K}, \quad (10.25)$$

where the eigenvalues of $\mathbf{N}_1$ and $\mathbf{N}_2$ are also all distinct, $\lambda_1 >, ..., > \lambda_k > 0$. By the singular value decomposition theorem (see Section

A.8 in the Appendix), the matrix $\mathbf{K}$ can be written in the following form:

$$\begin{aligned}\mathbf{K} &= \mathbf{G}\mathbf{D}\mathbf{Q}^T \\ &= (\mathbf{g}_1, ..., \mathbf{g}_k)\mathbf{D}(\mathbf{q}_1, ..., \mathbf{q}_k)^T,\end{aligned} \tag{10.26}$$

where $\mathbf{G}$ and $\mathbf{Q}$ are orthogonal matrices whose columns are the standardized eigenvectors of $\mathbf{N}_1$ and $\mathbf{N}_2$, respectively, and $\mathbf{D} = diag(\lambda_1^{1/2}, ..., \lambda_k^{1/2})$. Since the matrices $\mathbf{N}_1$ and $\mathbf{N}_2$ are symmetric, the eigenvectors are orthogonal. Thus, we obtain

$$\mathbf{g}_i^T \mathbf{g}_j = \begin{cases} 1, & \text{if } i = j, \\ 0, & \text{otherwise,} \end{cases} \tag{10.27}$$

and

$$\mathbf{q}_i^T \mathbf{q}_j = \begin{cases} 1, & \text{if } i = j, \\ 0, & \text{otherwise.} \end{cases} \tag{10.28}$$

Using the above notation in (10.27) and (10.28), we let

$$\mathbf{a}_i = \mathbf{R}_x^{-1/2}\mathbf{g}_i \tag{10.29}$$

and

$$\mathbf{b}_i = \mathbf{R}_y^{-1/2}\mathbf{q}_i \tag{10.30}$$

where $i = 1, 2, ..., k$. Then the vectors $\mathbf{a}_i$ and $\mathbf{b}_i$ are called the ith canonical correlation vectors for the random vectors $\mathbf{x}$ and $\mathbf{y}$, respectively. The random variables $\eta = \mathbf{a}^T\mathbf{x}$ and $\psi = \mathbf{b}^T\mathbf{y}$ are simply called the ith canonical correlation variables. The $\rho = \lambda_i^{1/2}$ is called the ith canonical correlation coefficient. Note that the covariance of ith and jth canonical correlation variables η_i and η_j is

$$\begin{aligned}C(\eta_i, \eta_j) &= \mathbf{a}_i^T \mathbf{R}_x \mathbf{a}_j \\ &= \mathbf{g}_i^T \mathbf{R}_x^{-1/2} \mathbf{R}_x \mathbf{R}_x^{-1/2} \mathbf{g}_j \\ &= \mathbf{g}_i^T \mathbf{g}_j \\ &= \begin{cases} 1, & \text{if } i = j, \\ 0, & \text{otherwise,} \end{cases}\end{aligned} \tag{10.31}$$

and the covariance of ith and jth canonical correlation variables ψ_i and ψ_j is

$$\begin{aligned} C(\psi_i, \psi_j) &= \mathbf{b}_i^T \mathbf{R}_y \mathbf{b}_j \\ &= \mathbf{q}_i^T \mathbf{R}_y^{-1/2} \mathbf{R}_y \mathbf{R}_y^{-1/2} \mathbf{q}_j \\ &= \mathbf{q}_i^T \mathbf{q}_j \\ &= \begin{cases} 1, & \text{if } i = j, \\ 0, & \text{otherwise.} \end{cases} \end{aligned} \tag{10.32}$$

Thus the ith canonical correlation variables for the random vector $\mathbf{x}$ are uncorrelated and are standardized to have unit variance. This holds similarly for the ith canonical correlation variables for the random vector $\mathbf{y}$.

10.5 Canonical Discriminant Transform

The *canonical discriminant transform*, which has been developed based on the ideas of canonical correlation transform, is a dimensionality reduction technique related to the principal components transform. This technique has certain maximal properties similar to the principal components transform (see Mardia, Kent, and Bibby [1], Miao and Clements [14], and Miao, Benkeser, and Nichols [15]). However, whereas principal components analysis considers the interrelationship within a set of variables, the focus of the canonical discriminant transform is on the relationship between two groups of variables.

If the means on the p-dimensional random vector $\mathbf{x}$ differ over the g groups, it is valuable to examine the variation over the g groups for linear combinations of $\mathbf{x}$ given $y = \mathbf{b}^T\mathbf{x}$. Since the new variable y is a random variable, we may compare the means of y over the g groups. This suggestion is to look for the linear function $\mathbf{b}^T\mathbf{x}$ that *maximizes* the ratio of the between-groups sum of squares to the within-groups sum of squares. That is, let $\mathbf{X}$ be a data matrix and

$$\mathbf{y} = \mathbf{X}\mathbf{b} = \begin{bmatrix} \mathbf{x}_1\mathbf{b} \\ \cdot \\ \cdot \\ \cdot \\ \mathbf{x}_g\mathbf{b} \end{bmatrix}, \tag{10.33}$$

be a linear combination of the columns of matrix $\mathbf{X}$.

To compare the means of the random vector $\mathbf{y}$ over the g groups, we require the F_y ratio test

$$F_y = \frac{\mathbf{b}^T\mathbf{G}\mathbf{b}}{\mathbf{b}^T\mathbf{W}\mathbf{b}}, \tag{10.34}$$

where $\mathbf{G}$ is the between-class scatter matrix and $\mathbf{W}$ is the pooled covariance matrix.

The between-class scatter matrix $\mathbf{G}$ is defined as

$$\mathbf{G} = \frac{1}{g-1}\sum_{k=1}^{g}(\mathbf{m}_k - \overline{\mathbf{m}})(\mathbf{m}_k - \overline{\mathbf{m}})^T, \tag{10.35}$$

where $\overline{\mathbf{m}} = \frac{1}{g}\sum_{i=1}^{g}\mathbf{m}_i$. The pooled covariance matrix $\mathbf{W}$ is defined by

$$\mathbf{W} = \sum_{k=1}^{g}\sum_{i=1}^{N_k}(\mathbf{x}_{ik} - \overline{\mathbf{m}}_{.k})(\mathbf{x}_{ik} - \overline{\mathbf{m}}_{.k})^T, \tag{10.36}$$

where N_k is the number of observations in kth group, $k = 1, 2, ..., g$, and

$$\overline{\mathbf{m}}_{.k} = \frac{1}{N_k}\sum_{i=1}^{N_k}\mathbf{x}_{ik}, \quad k = 1, 2, ...g. \tag{10.37}$$

Note that the F_y ratio in (10.34) is developed based on the canonical correlation function in (10.21). This F_y criterion is called the *canonical discriminant function.* This F_y criterion in (10.34) is intuitively attractive because it is easier to tell the groups apart if the between-groups sum of squares for $\mathbf{y}$ is large relative to the within-groups sum of squares.

To determine a variable y that should characterize differences among groups, we determine the values of the elements of the vector $\mathbf{b}$ that maximize the F_y ratio. That is, we need to determine the vector

$\mathbf{b}$ to maximize the ratio criterion $\frac{\mathbf{b}^T\mathbf{G}\mathbf{b}}{\mathbf{b}^T\mathbf{W}\mathbf{b}}$. Determination of the vector $\mathbf{b}$ that maximizes the F_y criterion involves solving the eigenvalue and eigenvector equations given in the two methods explained below.

10.5.1 Discriminant Function Differentiation Without Constraints

We want to maximize the F_y ratio with respect to the vector $\mathbf{b}$. As a usual practice, we take the partial derivative of F_y with respect to vector $\mathbf{b}$ and set it equal to zero

$$\frac{\partial F_y}{\partial \mathbf{b}} = \frac{2\mathbf{G}\mathbf{b}}{\mathbf{b}^T\mathbf{W}\mathbf{b}} - \frac{2\mathbf{W}\mathbf{b}}{\mathbf{b}^T\mathbf{W}\mathbf{b}}\frac{\mathbf{b}^T\mathbf{G}\mathbf{b}}{\mathbf{b}^T\mathbf{W}\mathbf{b}} = 0. \tag{10.38}$$

After mathematical simplification, we obtain

$$\frac{\mathbf{b}^T\mathbf{G}\mathbf{b}}{\mathbf{b}^T\mathbf{W}\mathbf{b}} = \frac{\mathbf{G}\mathbf{b}}{\mathbf{W}\mathbf{b}} = \lambda. \tag{10.39}$$

Therefore, it can be seen that the maximum value of λ is the largest eigenvalue of the matrix $\mathbf{W}^{-1}\mathbf{G}$, and $\mathbf{b}$ is its corresponding eigenvector:

$$(\mathbf{W}^{-1}\mathbf{G} - \lambda\mathbf{I})\mathbf{b} = 0. \tag{10.40}$$

The vector $\mathbf{b}$ in the canonical discriminant function is the eigenvector of $\mathbf{W}^{-1}\mathbf{G}$ corresponding to the largest eigenvalue λ_1.

As can be seen, when there are two or more groups of observations with measurements on several quantitative variables provided, the canonical discriminant transform can obtain a linear combination of the variables that summarizes between-class variation in much the same way that the principal components transform summarizes total variation. The highest possible multiple correlation with the groups is called the *first canonical correlation.* The coefficients of the linear combination are the canonical coefficients. The first canonical component is defined by the variable of the linear combination.

The second linear combination uncorrelated with the first canonical component that has the highest possible multiple correlation within the groups is considered the *second canonical correlation.* Until the number of canonical components equals the number of cluster groups minus one, the process of extracting canonical components can be repeated.

10.5.2 Discriminant Function Differentiation with Constraints

From (10.39), we can see that the largest eigenvalue λ_1 maximizes the F_y ratio

$$\lambda_1 = \frac{\mathbf{b}^T\mathbf{G}\mathbf{b}}{\mathbf{b}^T\mathbf{W}\mathbf{b}}. \tag{10.41}$$

It is customary to normalize vector $\mathbf{b}$ such that $\mathbf{b}^T\mathbf{W}\mathbf{b} = 1$, which implies that the pooled covariance matrix of $\mathbf{y}$ has unit variance. We want to maximize the F_y ratio

$$F_y = max_{\mathbf{b}}[\mathbf{b}^T\mathbf{G}\mathbf{b}], \quad \text{subject to } \mathbf{b}^T\mathbf{W}\mathbf{b} = 1. \tag{10.42}$$

This condition is convenient so that $\lambda_1 = \mathbf{b}^T\mathbf{G}\mathbf{b}$, and hence the eigenvalue measures among the group sum of squares for the variable $y_1 = \mathbf{b}_1^T\mathbf{x}$. Then we let $\mathbf{b} = \mathbf{W}^{-1/2}\mathbf{a}$, and write an equation as

$$\begin{aligned} F_y &= \mathbf{b}^T\mathbf{G}\mathbf{b} - \lambda(\mathbf{b}^T\mathbf{W}\mathbf{b} - 1) \\ &= \mathbf{a}^T\mathbf{W}^{-1/2}\mathbf{G}\mathbf{W}^{-1/2}\mathbf{a} - \lambda(\mathbf{a}^T\mathbf{a} - 1), \end{aligned} \tag{10.43}$$

where λ is a nonzero *Lagrange* multiplier. We take the partial derivative with respect to vector $\mathbf{a}$ and set it equal to zero

$$\frac{\partial F_y}{\partial \mathbf{a}} = 2\mathbf{W}^{-1/2}\mathbf{G}\mathbf{W}^{-1/2}\mathbf{a} - 2\lambda\mathbf{a} = 0. \tag{10.44}$$

Therefore, the eigenvalue and eigenvector problem is as follows:

$$[\mathbf{W}^{-1/2}\mathbf{G}\mathbf{W}^{-1/2} - \lambda\mathbf{I}]\mathbf{a} = 0. \tag{10.45}$$

Solving (10.45) for λ gives

$$\lambda = \frac{1}{\mathbf{W}^{1/2}\mathbf{G}^{-1}\mathbf{W}^{1/2}}. \tag{10.46}$$

Finally, the ith canonical discriminant coefficients are given by

$$\mathbf{b}_i = \mathbf{W}^{-1/2}\mathbf{a}_i. \tag{10.47}$$

It can be shown that $\mathbf{W}^{-1}\mathbf{G}$ in (10.40) and $\mathbf{W}^{-1/2}\mathbf{G}\mathbf{W}^{-1/2}$ in (10.45) have the same eigenvalue by using the maximization and minimization theorem (see Section A.10 in the Appendix). In general,

$\mathbf{W}^{-1}\mathbf{G}$ and $\mathbf{W}^{-1/2}\mathbf{G}\mathbf{W}^{-1/2}$ all have $r = min\{(g-1), p\}$ nonzero eigenvalues, which are denoted by λ_i $(i = 1, 2, ..., r)$ in decreasing order. The corresponding eigenvectors, denoted by $\mathbf{b}_i$ $(i = 1, 2, ..., r)$, which are the canonical discriminant coefficients, define the first, second, and subsequent canonical discriminant features. The first k canonical discriminant features, $k \leq min\{(g-1), p\}$, are useful for summarizing the difference between the groups in k dimensional components spaces.

The relative magnitudes of the respective eigenvalues, λ_i, give a descriptive index important for each of the discriminant axes. By expressing each eigenvalue as a percentage of the "total variance" accounted for

$$L_k = \frac{\lambda_k}{\sum_{j=1}^{r} \lambda_j}, \quad k = 1, 2, ..., r, \tag{10.48}$$

we can determine which discriminant axes capture the major sources of variation separating the cluster groups.

10.5.3 Properties of the Canonical Discriminant Transform

The properties of the canonical discriminant transform derive from the fact that it has certain maximal properties similar to the principal components transform. In addition, the canonical discriminant transform can take the discrimination of cluster groups into consideration. Indeed, this procedure preserves the separation between the cluster groups. On the other hand, due to normalizing the canonical discriminant coefficients, the pooled within-group variance matrix of the canonical discriminant features is one, that is, $\mathbf{b}^T\mathbf{W}\mathbf{b} = 1$ in (10.42). Recall that the principal components transform does not have this convenient invariance property.

Another property that the canonical discriminant features possess is that they are uncorrelated whether the correlation is calculated from the total sample or from the pooled within-class correlation. Note that the canonical discriminant coefficients are not orthogonal, and do not represent perpendicular directions through the space of the original measurement features, because the $\mathbf{W}^{-1}\mathbf{G}$ is usually no longer a symmetric matrix. However, this problem can be overcome by using the additional constraint of orthogonality. This leads to the optimal discriminant transform, which we will discuss in the next section.

A problem of the canonical discriminant transform is that it needs the pattern samples with labeled information before this approach can be employed on the data set to reduce dimensionality. In other words, in unsupervised statistical signal classification, this approach can be applied only after clustering analysis has been employed on the training data set.

10.6 Optimal Discriminant Transform

The 2-D optimal discriminant transform was developed based on the same criterion as that of the canonical discriminant transform (see Sammon [16] and Miao and Clements [17]). The first eigenvector $\mathbf{b}_1$ in the optimal discriminant feature can be obtained in such a way that we generate it in the canonical discriminant transform. However, the second eigenvector $\mathbf{b}_2$ has to maximize the F_y criterion with the additional condition of orthogonality, that is, $\mathbf{b}_1^T\mathbf{b}_2 = 0$. This is equivalent to maximizing the following function,

$$F_y(\mathbf{b}_2) = \frac{\mathbf{b}_2^T\mathbf{G}\mathbf{b}_2}{\mathbf{b}_2^T\mathbf{W}\mathbf{b}_2} + \phi\mathbf{b}_1^T\mathbf{b}_2, \tag{10.49}$$

where ϕ is a nonzero Lagrange multiplier. A solution to the second eigenvector $\mathbf{b}_2$ in (10.49) was contributed by Sammon [16], based on the direct observation

$$\mathbf{b}_2 = \mathbf{b}_1 - \frac{\mathbf{b}_1^T\mathbf{b}_1\mathbf{W}^{-1}\mathbf{b}_1}{\mathbf{b}_1^T\mathbf{W}^{-1}\mathbf{b}_1}, \tag{10.50}$$

where $\mathbf{b}_1$ is the first eigenvector in the canonical discriminant transform.

Note that (10.49) can be solved by using the partial derivative techniques. We will provide a generalized solution to this problem in the following section. In addition, we will show that the optimal discriminant transform is a special case in the generalized optimal declustering transform.

10.7 Generalized Optimal Declustering Transform

The generalized optimal declustering transform uses a modified criterion of the canonical discriminant transform but with the condition of orthogonality of eigenvectors imposed. This transform also introduces the scalar factors that can be used to control the spread effects of between-class and within-class covariance matrix for cluster groups (see Miao [9]). It can be viewed as an extension of the optimal discriminant transformation (see Sammon [6], Duchene and Leclercq [16], Liu, Cheng, and Yang [17], and Miao and Clements [18]). In other words, an optimal discriminant transform can be considered as a special case in the generalized optimal declustering transform.

To establish a solution of the generalized optimal declustering transform, we require the F_y ratio in the following form:

$$F_y(\mathbf{b}) = \frac{\mathbf{b}^T(\mathbf{B} + \sum_{i=1}^{g} \alpha_i \mathbf{S}_i)\mathbf{b}}{\mathbf{b}^T[\sum_{i=1}^{g}(1-\beta_i)\mathbf{S}_i]\mathbf{b}}, \quad 0 \leq \alpha_i;\ 0 \leq \beta_i \leq 1;\ i = 1, ..., g, \tag{10.51}$$

where $\mathbf{B}$ is the between-class covariance matrix $\mathbf{B} = (\mathbf{m}_i - \mathbf{m}_j)(\mathbf{m}_i - \mathbf{m}_j)^T$ and $\mathbf{S}_i$ is the within-class covariance matrix for cluster group i, and α_i and β_i are scalar spread effects. We refer to α_i and β_i as *dynamic weighting factors.*

The first eigenvector $\mathbf{b}_1$ that maximizes the F_y ratio comes from the solution of the following eigenproblem:

$$\left\{\left(\mathbf{B} + \sum_{i=1}^{g} \alpha_i \mathbf{S}_i\right) - \lambda_1 \left[\sum_{i=1}^{g}(1-\beta_i)\mathbf{S}_i\right]\right\}\mathbf{b}_1 = 0. \tag{10.52}$$

However, the second eigenvector $\mathbf{b}_2$ has to maximize the F_y criterion with the additional condition of orthogonality. That is, $\mathbf{b}_1^T\mathbf{b}_2 = 0$ (or $\mathbf{b}_2^T\mathbf{b}_1 = 0$). This is equivalent to maximizing the following function

$$F_y(\mathbf{b}_2) = \frac{\mathbf{b}_2^T(\mathbf{B} + \sum_{i=1}^{g} \alpha_i \mathbf{S}_i)\mathbf{b}_2}{\mathbf{b}_2^T[\sum_{i=1}^{g}(1-\beta_i)\mathbf{S}_i]\mathbf{b}_2} + \phi \mathbf{b}_2^T\mathbf{b}_1, \tag{10.53}$$

where ϕ is a nonzero Lagrange multiplier.

For simplicity, we let $\mathbf{H}_1 = \mathbf{B} + \sum_{i=1}^{g} \alpha_i \mathbf{S}_i$ and $\mathbf{H}_2 = \sum_{i=1}^{g}(1-\beta_i)\mathbf{S}_i$. Then, (10.53) can be rewritten as the following:

$$F_y(\mathbf{b}_2) = \frac{\mathbf{b}_2^T\mathbf{H}_1\mathbf{b}_2}{\mathbf{b}_2^T\mathbf{H}_2\mathbf{b}_2} + \phi \mathbf{b}_2^T\mathbf{b}_1. \tag{10.54}$$

Note that (10.54) is known as the *optimal discriminant transform.*

In order to solve this problem, we take the partial derivative with respect to the vector $\mathbf{b}_2$ and set it to zero:

$$\begin{aligned}\frac{\partial F_y(\mathbf{b}_2)}{\partial \mathbf{b}_2} &= \frac{\mathbf{H}_1\mathbf{b}_2 + \mathbf{b}_2^T\mathbf{H}_1}{\mathbf{b}_2^T\mathbf{H}_2\mathbf{b}_2} - \frac{[\mathbf{H}_2\mathbf{b}_2 + \mathbf{b}_2^T\mathbf{H}_2]\mathbf{b}_2^T\mathbf{H}_1\mathbf{b}_2}{[\mathbf{b}_2^T\mathbf{H}_2\mathbf{b}_2]^2} + \phi\mathbf{b}_1 \\ &= \frac{2\mathbf{H}_1\mathbf{b}_2}{\mathbf{b}_2^T\mathbf{H}_2\mathbf{b}_2} - \frac{2\mathbf{H}_2\mathbf{b}_2}{\mathbf{b}_2^T\mathbf{H}_2\mathbf{b}_2}\frac{\mathbf{b}_2^T\mathbf{H}_1\mathbf{b}_2}{\mathbf{b}_2^T\mathbf{H}_2\mathbf{b}_2} + \phi\mathbf{b}_1 = 0. \qquad (10.55)\end{aligned}$$

When both sides of (10.55) are multiplied by $\mathbf{b}_1^T\mathbf{H}_2^{-T}$, we obtain:

$$2\frac{\mathbf{b}_1^T\mathbf{H}_2^{-T}\mathbf{H}_1\mathbf{b}_2}{\mathbf{b}_2^T\mathbf{H}_2\mathbf{b}_2} + \phi\mathbf{b}_1^T\mathbf{H}_2^{-T}\mathbf{b}_1 = 0, \qquad (10.56)$$

the second term vanishes due to the orthogonality $\mathbf{b}_1^T\mathbf{b}_2 = 0$. After solving (10.56) and substituting ϕ in (10.55), we get the following equation:

$$\left(\mathbf{H}_1\mathbf{b}_2 - \frac{\mathbf{b}_2^T\mathbf{H}_1\mathbf{b}_2}{\mathbf{b}_2^T\mathbf{H}_2\mathbf{b}_2}\mathbf{H}_2\mathbf{b}_2 - \frac{\mathbf{b}_1\mathbf{b}_1^T\mathbf{H}_2^{-T}}{\mathbf{b}_1^T\mathbf{H}_2^{-1}\mathbf{b}_1}\mathbf{H}_1\mathbf{b}_2\right)\left(\frac{1}{\mathbf{b}_2^T\mathbf{H}_2\mathbf{b}_2}\right) = 0. \qquad (10.57)$$

Equation (10.57) can be written in the following form after reducing by the $\frac{1}{\mathbf{b}_2^T\mathbf{H}_2\mathbf{b}_2}$ term

$$\left[\mathbf{I} - \frac{\mathbf{b}_1\mathbf{b}_1^T\mathbf{H}_2^{-T}}{\mathbf{b}_1^T\mathbf{H}_2^{-1}\mathbf{b}_1}\right]\mathbf{H}_1\mathbf{b}_2 = \frac{\mathbf{b}_2^T\mathbf{H}_1\mathbf{b}_2}{\mathbf{b}_2^T\mathbf{H}_2\mathbf{b}_2}\mathbf{H}_2\mathbf{b}_2. \qquad (10.58)$$

Therefore, both sides of (10.58) are divided by $\mathbf{H}_2\mathbf{b}_2$, and we obtain

$$\frac{\mathbf{b}_2^T\mathbf{H}_1\mathbf{b}_2}{\mathbf{b}_2^T\mathbf{H}_2\mathbf{b}_2} = \frac{\left[\mathbf{I} - \frac{\mathbf{b}_1\mathbf{b}_1^T\mathbf{H}_2^{-T}}{\mathbf{b}_1^T\mathbf{H}_2^{-1}\mathbf{b}_1}\right]\mathbf{H}_1\mathbf{b}_2}{\mathbf{H}_2\mathbf{b}_2} = \lambda_2. \qquad (10.59)$$

As can be seen, determination of the vector $\mathbf{b}_2$ that maximizes the F_y criterion involves solving for the eigenvalue and eigenvector of the following eigen-problem:

$$\left[\left(\mathbf{I}-\frac{\mathbf{b}_1\mathbf{b}_1^T\mathbf{H}_2^{-T}}{\mathbf{b}_1^T\mathbf{H}_2^{-1}\mathbf{b}_1}\right)\mathbf{H}_1-\lambda_2\mathbf{H}_2\right]\mathbf{b}_2=0, \qquad (10.60)$$

where λ_2 is again the greatest eigenvalue, and enginvector $\mathbf{b}_1$ is a first optimal declustering feature.

Therefore, the optimum declustering directions, two eigenvectors associated with two greatest eigenvalues, will cause one cluster group to be well concentrated and the other to be maximally spread over the space while keeping them as far from each other as possible. In addition, the dynamic weighting factors α_i and β_i can be used to control the separation between the cluster groups.

10.8 Transform-Based Statistical Signal Classifiers

To apply principal components, canonical correlation, canonical discriminant, optimal discriminant, or optimal declustering transforms to the training data matrix $\mathbf{X}$, the N original p-dimensional measurement feature vectors $\mathbf{x}$ are transformed to the component features $\mathbf{b}^T\mathbf{x}$ (where $\mathbf{b}$ is a $p \times 1$ vector), which can be represented in one-dimensional component feature space. If $\mathbf{B}$ is a $(d \times p)$ linear transformation matrix, $\mathbf{y} = \mathbf{B}\mathbf{x}$ has d-dimensional random vector.

The properties of the new random vector $\mathbf{y}$ can be derived based on properties of the random vector $\mathbf{x}$. If the random vector $\mathbf{x}$ has a Gaussian distribution, $N_p(\mathbf{m}, \Sigma)$, with mean vector $\mathbf{m}$ and variance Σ, then the random vector $\mathbf{y}$ has the following properties:

- If $\mathbf{x} \sim N_p(\mathbf{m}, \Sigma)$, and $\mathbf{y} = \mathbf{B}\mathbf{x} + \mathbf{c}$, where $\mathbf{B}$ is any $(q \times p)$ matrix, then $\mathbf{y}$ has a q-variate Gaussian distribution, $\mathbf{y} \sim N_q(\mathbf{B}\mathbf{m} + \mathbf{c}, \mathbf{B}\Sigma\mathbf{B}^T)$.
- If $\mathbf{x} \sim N_p(\mathbf{m}, \Sigma)$, and $\Sigma > 0$, then $\mathbf{y} = \Sigma^{-1/2}(\mathbf{x} - \mathbf{m}) \sim N_q(0, \mathbf{I})$, and $\mathbf{y} = (\mathbf{x} - \mathbf{m})^T\Sigma^{-1}(\mathbf{x} - \mathbf{m}) = \sum_i y_i^2 \sim N_q\chi_p^2$, a chi-square distribution.
- If $\mathbf{x} \sim N_p(\mathbf{m}, \sigma^2\mathbf{I})$, and $\mathbf{G}$ is any row-orthonormal matrix with $(q \times p)$, satisfying $\mathbf{G}\mathbf{G}^T = \mathbf{I}_q$, then $\mathbf{y} = \mathbf{G}\mathbf{x} \sim N_q(\mathbf{G}\mathbf{m}, \sigma^2\mathbf{I})$.
- If $\mathbf{x} \sim N_p(\mathbf{m}_i, \Sigma_i)$, $i = 1, 2, ..., M$, is a multivariate Gaussian distribution, then $\mathbf{y} = \mathbf{B}\mathbf{x}$ is also a multivariate Gaussian distribution, $\mathbf{y} \sim N_d(\mathbf{m}_{yi}, \Sigma_{yi})$, where the mean vector $\mathbf{m}_{yi} =$

$\mathbf{Bm}_i$ and covariance matrix $\Sigma_{yi} = \mathbf{B}\Sigma_i\mathbf{B}^T$, $i = 1, 2, ..., M$.

Therefore, transform-based statistical signal classifications for Gaussian models are

$$D_i(\mathbf{y}) = \ln[p(\omega_i)] - \frac{1}{2}\ln|\Sigma_{yi}| - \frac{1}{2}[(\mathbf{y} - \mathbf{m}_{yi})^T\Sigma_{yi}^{-1}(\mathbf{y} - \mathbf{m}_{yi})], \quad (10.61)$$

where $i = 1, 2, ..., M$, for cluster groups. In terms of the transform-based features, (10.61) assigns an unclassified pattern vector $\mathbf{x}$ to the cluster group ω_i where the sample mean is the closest in the transform feature space to $\mathbf{y} = \mathbf{Bx}$.

Note that if the multivariate Gaussian density functions with the same covariance matrix (the sample means are collinear) and the method of discriminant function differentiation with constraints (Section 10.5.2) in the canonical discriminant transform is used, the decision rule based on the nonparametric method (Section 9.5.1) or based on the Bayes' Gaussian rule in (10.61) is the same. Thus, we might hope that the optimal decision rule using the canonical discriminant features would be appropriate for populations where the hypothesis of normality is not exactly satisfied. Recall that principal components features do not have this property.

In the next section, we will introduce the application and performance evaluations of the 2-D transform-based statistical signal classification, providing many detailed illustrations.

10.9 Application and Performance Evaluations

In this section, we provide the experimental results of applying 2-D transform-based statistical signal classification systems to digital waveform classification from an alpha-constant of the digital air monitor, and illustrate the effectiveness of the transformed techniques. To show that the employment of the 2-D transform features reduces the structural complexity of the statistical signal classification without degrading the results of classification performance, test studies were also carried out for the multidimensional statistical signal classification system (see Miao and Clements [14]).

10.9.1 Data Description

The training data set had 749,940 samples consisting of 3,656 digital waveforms sampled with a 100-MHz A/D converter at 10-bit resolution. These were obtained from an alpha-constant of the digital air monitor with radon and plutonium sources. A sample plot of digital waveforms acquired from a digital air monitor is shown in Figure 9.5(a). Note that no a priori labeled information was provided in the training data set. However, knowledge of the underlying physics in this case dictates that there should be three distinct populations of events. Therefore, it was assumed that each event was independent of all the others.

10.9.2 Methodology

A preprocessing algorithm in Example 9.2 was developed to extract 15 measurement features from each digital waveform of incoming radon and plutonium events as shown in Figure 9.5(b). The final selection of the subset of quantitative key features from the initial fifteen measurement features was based on the stepwise discriminant technique along with cross-validation. The seven measurement features that best reveal differences among digital waveform classes were finally generated. Then, a modified k-means algorithm was used to group similar aspects together and generate the labeled information of training vectors.

The cluster results of the seven measurement features in the training data are summarized in Table 10.1. As can be seen, the value of the within-cluster scatter in each cluster is much less than the value of the between-cluster distances (centroid distance). This indicates that the clustering had fair separability. The probability of misclassification error should be low if Gaussian models are assumed.

In order to evaluate and compare the performance of the probability of misclassification, statistical signal classifiers were established on seven measurement features, 2-D principal features, 2-D canonical features, and 2-D optimal declustering features, by using the same training data and the same test data. Thus, we refer to the seven-measurement-feature classifier as a multidimensional statistical signal classifier, and the 2-D transform-based classifier as a transform-based statistical signal classifier.

Table 10.1 Cluster Summary of Training Data Using Seven Measurement Features

Cluster	Number of Vectors	Within-Cluster Scatter	Centroid Distance 1	Centroid Distance 2	Centroid Distance 3
1	221	0.7696	0.0000	5.7940	4.0515
2	255	0.8249	5.7940	0.0000	4.7054
3	1353	0.5646	4.0515	4.7054	0.0000

The holdout method for estimating the probability of misclassification error was applied for the applications. One 50% portion (called the training data) of the training data set per waveform class was used to design the statistical signal classifiers, and the remaining 50% portion (called the test data) was used to estimate the probability of misclassification error.

For reasons of clarity, we discuss the application results of each method individually. The evaluation section will then relate the performances, giving both conclusions and recommendations for practical use.

10.9.3 Multidimensional Statistical Signal Classifier

The multidimensional statistical signal classifier was established on the Bayes' optimum decision functions for Gaussian models. We first computed the mean vector $\widehat{\mathbf{m}}_i$, the covariance matrix $\widehat{\mathbf{C}}_i$ for the patterns of class ω_i $(i = 1, 2, 3)$, and the pooled covariance matrix $\widehat{\mathbf{C}}$ using the training data. Second, the pooled covariance matrix $\widehat{\mathbf{C}}$ was selected for the classification criterion based on the homogeneity test for covariance matrices. Thus, the decision boundary in (9.87) was used for the implementation.

The estimated probability of misclassification error of the Bayes' optimal classifier was evaluated by using a posterior probability of membership in each cluster. The test accuracy of the Bayes' Gaussian model to classify three classes of digital waveforms is 97.59% by using the test data. Table 10.2 shows the detailed test results of the number of observations, percent classified, and error count estimates for each cluster in the multidimensional statistical signal classifier.

Table 10.2 Test Results of the Probability of Misclassification Using the Multidimensional Feature-Based Statistical Signal Classifier

Clusters	Class 1	Class 2	Class 3	Total Vectors
Class 1	210	1	10	221
Class 2	2	246	6	254
Class 3	23	2	1327	1352
Errors	4.98%	3.15%	1.85%	**2.41%**
Priors	0.1208	0.1394	0.7397	

10.9.4 2-D Principal Feature-Based Statistical Signal Classifier

The principal components transform was applied to the training data to reduce its dimensionality, not only making visual interpretation of cluster graphs more straightforward, but also generating a set of principal component features. As a result, the 2-D scatter plot with the first two principal component features and its corresponding three-dimensional (3-D) mesh plot of the estimate of the underlying probability density function are shown in Figure 10.2(a) and (b), respectively. Note that the obvious boundaries among classes suggest that the clustering had good separability. In this case, the probability of misclassification error would have been sufficiently low to allow its use.

Seven orthogonal principal component features in principal components transform were determined from the seven measurement features. Each of the principal component features is a linear combination of the original seven measurement features. The eigenvalues of the correlation matrix of the training data are summarized in Table 10.3. As can be seen, the first two principal component features together accounted for 86.50% of the total variation among the seven measurement features. As a result, these seven principal component features define seven uncorrelated measures of digital waveform characteristics.

The Bayes' optimum decision function for Gaussian models was built on the first two principal component features. The test accuracy of the 2-D principal feature-based classifier to discriminate digital waveforms among the three classes was 95.73% by using the same

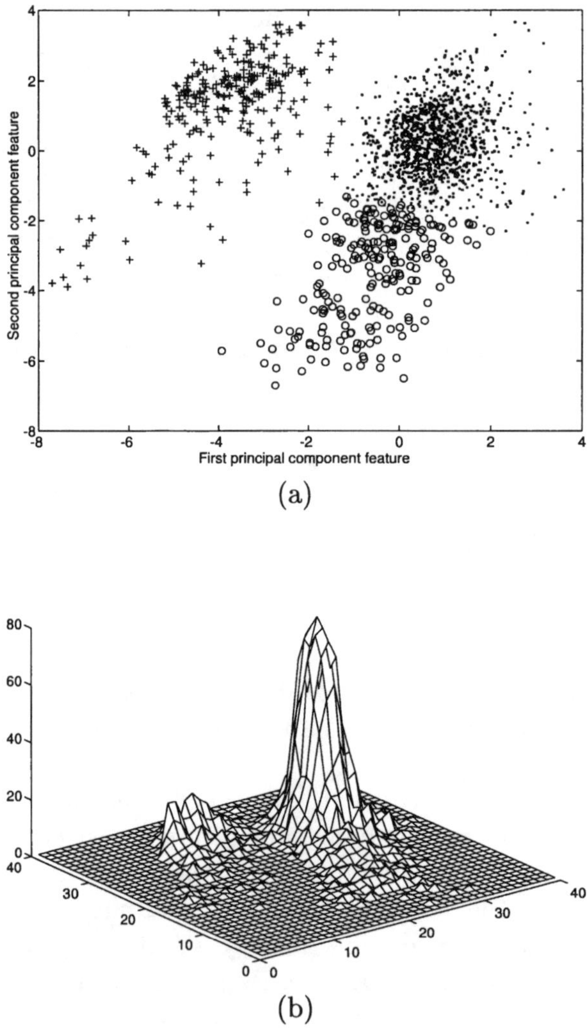

(a)

(b)

Figure 10.2 (a) Clustering graph with the first two principal component features. "o" and "+" represent digital waveforms from radon. "·" indicates digital waveforms from plutonium. (b) 3-D mesh plot of estimation of the underlying probability density function based on the first two principal component features.

Table 10.3 Eigenvalues of the Correlation Matrix of the Training Data Using the Principal Components Transform

	Eigenvalue	Difference	Proportion	Cumulative
Prin1	3.1782	0.3015	45.40%	45.40%
Prin2	2.8767	2.0339	41.10%	86.50%
Prin3	0.8428	0.7674	12.04%	98.54%
Prin4	0.0753	0.0636	1.08%	99.61%
Prin5	0.0117	0.0029	0.17%	99.78%
Prin6	0.0088	0.0023	0.13%	99.91%
Prin7	0.0065	.	0.09%	100%

Table 10.4 Test Results of the Probability of Misclassification Using a 2-D Principal Component Feature-Based Classifier

Clusters	Class 1	Class 2	Class 3	Total Vectors
Class 1	159	1	61	221
Class 2	2	238	14	254
Class 3	0	0	1352	1352
Errors	28.05%	6.30%	0.00%	**4.27%**
Priors	0.1208	0.1394	0.7397	

test data. Table 10.4 shows the detailed test results of the number of observations, percent classified, and error count estimates for each cluster by using the 2-D principal feature-based classifier.

10.9.5 2-D Canonical Feature-Based Statistical Signal Classifier

The canonical discriminant transform, which is a dimensionality-reduction technique related to the principal component transform, was applied to derive two uncorrelated linear combinations of the seven measurement features in the training data. This analysis simplified the structure of the covariance matrix and was applied to the training data to make visual interpretation of cluster graphs more straightforward and to assess the results of cluster performance. Figure 10.3(a) and (b) shows the 2-D scatter plot and its corresponding 3-D mesh plot of the estimated underlying probability density function based on the first two canonical

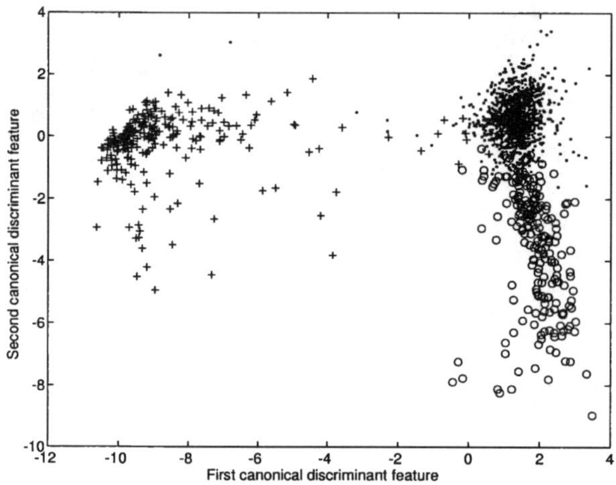

(a)

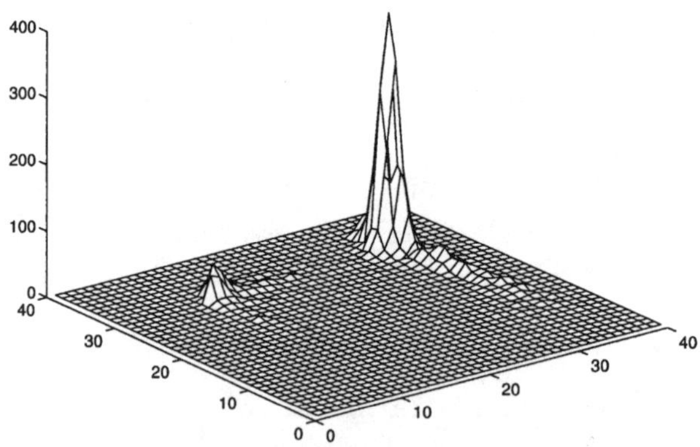

(b)

Figure 10.3 (a) Clustering graph with the first two canonical discriminant features. "o" and "+" denote digital waveforms from radon. "." represents digital waveforms from plutonium. (b) 3-D mesh plot of estimation of the underlying probability density function based on the first two canonical discriminant features.

Table 10.5 Eigenvalue Proportion Among Sum of Squares by the Canonical Discriminant Transform

	Eigenvalue	Difference	Proportion	Cumulative
Can1	13.1087	11.4053	88.49%	88.49%
Can2	1.7034	1.7029	11.50%	99.99%
Can3	0.0005	0.0001	0	100%
Can4	0.0004	0.0001	0	.
Can5	0.0003	.	0	.
Can6	0.0000	.	0	.
Can7	0.0000	.	0	.

discriminant features, respectively. Note that these graphs display the three clusters as fairly separable.

The eigenvalues, which denote the proportion among group sums of squares by the canonical discriminant function, are summarized in Table 10.5. As can be seen, the first and second canonical discriminant features have 88.49% and 11.50% of the total variation, respectively. Those two canonical discriminant features together represent almost 100% of the total variation among the seven measurement features.

An optimum decision function was generated based on the Bayes' Gaussian classifier by using the first two canonical discriminant features. The performance of the 2-D canonical feature-based Bayes' Gaussian classifier to discriminate among the plutonium class and radon subclasses was evaluated by using the same test data.

The estimated probability of misclassification was evaluated based on the posterior probability of membership for each cluster by using the Bayes' optimal classifier. The test results show that the Bayes' Gaussian classifier can classify the digital waveforms among three classes about 96.77% accurary. Table 10.6 shows the detailed test results of the number of observations, percent classified, and error count estimates for each cluster by using the 2-D canonical feature-based classifier. These results of this technique, which demonstrates clearly improved measurement conditions, are therefore very promising.

Table 10.6 Test Results of the Probability of Misclassification Using a 2-D Canonical Feature-Based Classifier

Cluster	Class 1	Class 2	Class 3	Total Vectors
Class 1	195	2	24	221
Class 2	1	243	10	254
Class 3	19	3	1330	1352
Errors	11.76%	4.33%	1.63%	**3.23%**
Priors	0.1208	0.1394	0.7397	

10.9.6 2-D Optimal Declustering Feature-Based Statistical Signal Classifier

The optimal declustering transform was applied to derive two orthogonal linear combinations from the seven measurement features in the training data. This analysis not only can simplify the structure of the covariance to make visual interpretation of cluster graphs more straightforward but it can also enhance the separation among the cluster groups. Figure 10.4(a) and (b) show the 2-D scatter plot and its corresponding 3-D mesh plot of the estimated underlying probability density function based on the first two optimal declustering features. As can be seen, the optimum declustering directions, the two eigenvectors associated with the two largest eigenvalues, will cause one radon class to be well concentrated and the other to be maximally spread over the space, while keeping them as far from each other as possible.

The optimum decision function was then generated by using the Bayes' Gaussian classifier based on the 2-D optimal declustering features. The performance of 2-D optimal declustering feature-based Bayes' Gaussian classifier to discriminate among the plutonium class and radon subclasses was evaluated by using the same test data. The tested accuracy of this Bayes' Gaussian model to classify the digital waveforms among three classes is 98.52%. Table 10.7 shows the detailed test results of the number of observations, percent classified, and error count estimates for each cluster by using the 2-D optimal declustering feature-based classifier. As a result, the 2-D optimal declustering feature-based classifier currently shows the best performance in this application.

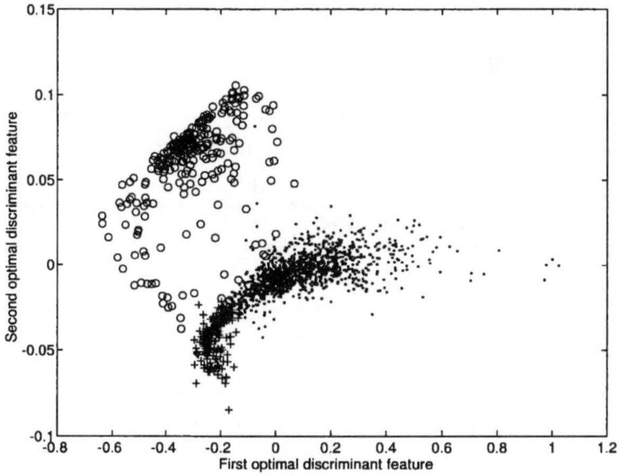

(a)

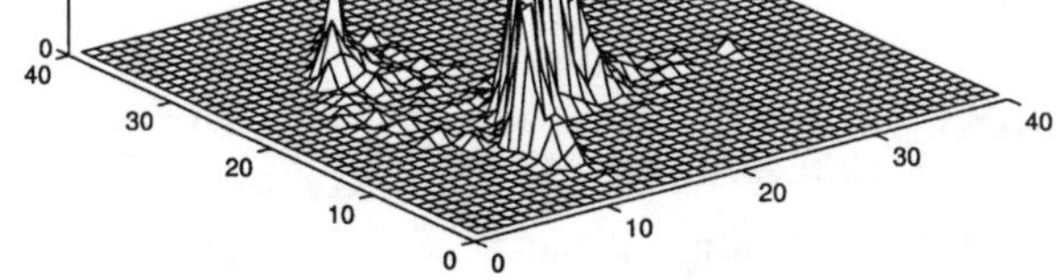

(b)

Figure 10.4 (a) Clustering graph with the first two optimal declustering features. "o" and "+" denote digital waveforms from radon. "." represents digital waveforms from plutonium. (b) 3-D mesh plot of estimation of the underlying probability density function based on the first two optimal declustering features.

Table 10.7 Test Results of the Probability of Misclassification by Using a 2-D Optimal Declustering Feature-Based Classifier

Cluster	Class 1	Class 2	Class 3	Total Vectors
Class 1	216	0	5	221
Class 2	0	240	14	254
Class 3	6	2	1344	1352
Errors	2.26%	5.51%	0.59%	**1.48%**
Priors	0.1208	0.1394	0.7397	

10.9.7 Evaluations of Statistical Signal Classifiers

In conclusion, the test accuracy of the 2-D transform-based statistical signal classifiers in classifying digital waveforms is almost the same as the test accuracy of the multidimensional statistical signal classifier. The comparison results of misclassification error are summarized in Figure 10.5, for the four types of statistical signal classifiers. Note that the covariance matrix of the multidimensional space has a complex structure, not only making visual interpretation of the digital-waveforms sampled data more difficult but also leading to high computational complexity and large storage requirements. Principal components, canonical discriminant, and optimal declustering transforms can simplify the structure of the covariance matrix to make visual interpretation of cluster graphs more straightforward and to assess the results of cluster performance. In fact, the performance of the 2-D transform-based statistical signal classifiers do not degrade compared with the multidimensional statistical signal classifier, while the computational efficiency and the storage of the data memory are significantly improved. Therefore, the 2-D transform-based methods lead to economies in the design of digital hardware and software for statistical signal classification. This property is particularly important for real-time applications in analyzing signal sources with count rates of 1,000 per second, especially when the waveforms are sampled at extremely high rates in the digital system.

Among the three methods of the 2-D transform-based statistical signal classifiers, the misclassification error of the 2-D optimal declustering feature-based statistical signal classifier is the smallest.

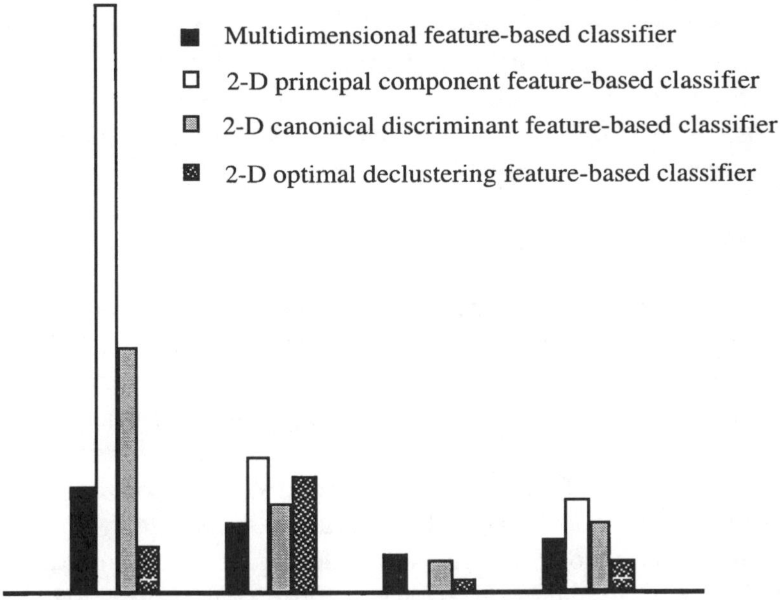

	Class 1	Class 2	Class 3	Overall
■	4.98%	3.15%	1.85%	2.41%
□	28.05%	6.30%	0.00%	4.27%
▒	11.76%	4.33%	1.63%	3.23%
▩	2.26%	5.51%	0.59%	1.48%

Figure 10.5 Summary results of the probability of misclassification error based on four types of statistical signal classifiers.

The reason for this is that the optimal declustering features not only have certain maximal properties similar to the principal component features and the canonical discriminant features but also can take discrimination of classes into consideration. Furthermore, the optimum declustering directions, the two eigenvectors associated with the two largest eigenvalues, will cause one class to be well concentrated and the other to be maximally spread over the space, while keeping them as far from each other as possible. Those properties are useful for reducing the misclassification of error. Therefore, the 2-D optimal declustering feature-based statistical signal classifier has great value in practical applications.

Note that the performance results from the 2-D optimal

declustering feature-based classifier show the best of the systems evaluated. This assumed that all the optimal decision classifiers are built on the assumption of training data distribution with a multivariate Gaussian model. However, if other distributions or if no distributions of training data were used, we would have marginally different results.

10.10 Summary

In this chapter, we presented transform-based statistical signal classification. We discussed technical methods including principal components, canonical correlation, canonical discriminant, optimal discriminant, and optimal declustering transforms. We also provided mathematical treatments for the above-mentioned techniques and gave the properties of the various schemes. We did not attempt to cover all transforms proposed in the literature, particularly nonlinear transforms, but chose instead to emphasize the fundamental principles and practical application value of linear transforms.

The linear transforms in the statistical signal classification system are used for examining relationships among multidimensional measurement features and for summarizing data as well as for detecting linear combinations of the multidimensional measurement features. The operation of the linear transform usually allows for the substitution of very-high-dimensional measurement features by 2-D transform features while minimizing the information lost in the training and test data. The 2-D transform-based statistical signal classifiers can simplify the structure of the covariance matrix to make visual interpretation of cluster graphs more straightforward and to assess the results of classification performance. Usually, the performance of the 2-D transform-based statistical signal classification does not degrade when compared to the multidimensional statistical signal classification, while the efficiency of computational complexity and the storage requirements are significantly improved.

The principal components transform has robust properties and can be applied to any data sets without cluster identity information. However, the principal components transform must be carried out with extreme caution because of the weighting and scaling effects that it can introduce. In addition, the principal components transform does nothing either to conserve or to enhance the separation between the cluster groups. This procedure may easily be thrown off if irrelevant noise features are added.

The canonical discriminant transform is derived based on the ideas of the canonical correlation transform. The canonical discriminant transform has certain maximal properties similar to principal components transform. The focus of the canonical discriminant transform is on the relationship between two groups of random variables, while the principal components transform considers the interrelationship within a set of random variables. However, the canonical discriminant coefficients are not orthogonal, which can be overcome by adding the constraint of orthogonality. This leads to the optimal discriminant transform. It has been shown that the optimal discriminant transform is a special case in the generalized optimal declustering transform. The dynamic weighting factor in the optimal declustering transform can be used for controlling the scatter of the cluster group. This transform tends to have better performance in terms of the accuracy of classification. Furthermore, in contrast to the principal component transform, these linear transforms usually need the labeled information of cluster groups before application to the data set. These linear transforms tend to maintain the separation among the cluster groups, and even enhance the separation between cluster groups for statistical signal classification.

References

[1] Mardia, K. V., J. T. Kent, and J. M. Bibby, *Multivariate Analysis*, 3d printing, Academic Press Inc., New York, 1982.

[2] Miao, J., "Component Feature-Based Digital Waveform Analysis and Classification," Ph.D. Dissertation, Georgia Institute of Technology, Atlanta, Georgia, April 1996.

[3] Chien, Y. T., and K. S. Fu, "On the Generalized Karhunen-Loève Expansion," *IEEE Trans. on Information Theory*, pp. 518-520, July, 1967.

[4] Fukunage, K., and W. L. G. Koontz, "Application of the Karhunen-Loève Expansion to Feature Selection and Ordering," *IEEE Trans. on Computers*, Vol. C-19, pp. 311-318, 1970.

[5] Young, Y., "Reliability of Linear Feature Extractors," *IEEE Trans. on Computers*, Vol. C-20, pp. 967-971, September, 1971.

[6] Duchene, J., and S. Leclercq, "An Optimal Transformation for Discriminant and Principal Component Analysis," *IEEE Trans. on Pattern Analysis and Machine Intellgence*, Vol. 10, No. 6, pp. 978-983, November, 1988.

[7] Oja, E., *Subspace Methods of Pattern Recognition*, Research Studies Press, Letchworth, U.K., 1983.

[8] Young, T. Y., and T. W. Calvert, *Classification, Estimation and Pattern Recognition*, American Elsevier Publishing Company, Inc., New York, 1974.

[9] Miao, J., *Component Feature-Based Digital Waveform Analysis and Classification*, UMI, Ann Arbor, Michigan, 1996b.

[10] Nagy, G., "State of the Arts in Pattern Recognition," *Proc. IEEE*, Vol. 56, No. 5, pp. 836-863, 1968.

[11] Siedlecki, W., K. Siedlecka, and J. Sklansky, "An Overview of Mapping Techniques for Exploratory Pattern Analysis," *Pattern Recognition*, Vol. 21, No. 5, pp. 411-429, 1988.

[12] Liu, D., Y. Yammashita, and H. Ogana, "Pattern Recognition in the Presence of Noise," *Pattern Recognition*, Vol. 28, No. 7, pp. 989-995, 1995.

[13] Shum, H. Y., K. Ikeuchi, and R. Reddy, "Principal Component Analysis with Missing Data and Its Application to Polyhedral Object Modeling," *IEEE trans. on Pattern Analysis and Machine Intelligence*, Vol. 17, No. 9, pp. 854-867, September 1995.

[14] Miao, J., and M. A. Clements, "Unsupervised Pattern Recognition for Digital Waveform Classification from Radiation Detectors," *Proc. IEEE Int. Conf. on Acoustics, Speech and Signal Processing*, Vol. 4, pp. 2751-2754, Detroit, Michigan, May 1995.

[15] Miao, J., P. J. Benkeser, and F. T. Nichols, "A Computer-Based Statistical Pattern Recognition for Doppler Spectral Waveforms of Intracranial Blood Flow," *Computer in Biology and Medicine: An International Journal*, Vol. 26, No. 1, pp. 53-63, February 1996.

[16] Sammon, J. W., "An Optimal Discriminant Plane," *IEEE Trans. on Computers*, C-19, pp. 826-829, 1970.

[17] Liu, K., Y. Q. Cheng, and J. Y. Yang, "A Generalized Optimal Set of Discriminant Vectors," *Pattern Recognition*, Vol. 25, No. 7, pp. 731-739, 1992.

[18] Miao, J., and M. A. Clements, "Optimal Discriminant Feature-Based Waveform Classification with Neural Networks," *Proc. IEEE Int. Conf. on Acoustics, Speech and Signal Processing*, Vol. 4, pp. 3462-3465, Atlanta, Georgia, May, 1996.

Appendix: Matrix Algebra of Linear Transformation

A.1 Introduction

This appendix gives a summary of fundamental definitions and results in matrix algebra for linear transforms. Matrix algebra of linear transforms provides a foundation for developing transform-based statistical signal classification and discrete-time transformation. Therefore, in this appendix, we present the details, necessary definitions, and results along with proofs.

A.2 Vectors

A vector is an array of real-valued numbers or complex-valued numbers and is denoted by the lowercase bold letter $\mathbf{x}$. A vector $\mathbf{x}$ is assumed to be column vector. If the value in a vector $\mathbf{x}$ is real, then the vector $\mathbf{x}$ is said to be a *real vector*. If the value in a vector $\mathbf{x}$ is complex, then the vector $\mathbf{x}$ is said to be a *complex vector*. If a vector $\mathbf{x}$ containing N values is referred to as an N-dimensional vector, the vector $\mathbf{x}$ is as follows:

$$\mathbf{x} = \begin{bmatrix} x_1 \\ x_2 \\ \vdots \\ x_N \end{bmatrix}. \tag{A.1}$$

The transpose of a vector, denoted by $\mathbf{x}^T$, is a row vector

$$\mathbf{x}^T = [x_1, x_2, ..., x_N]. \quad (A.2)$$

The *Hermitian transpose*, denoted by $\mathbf{x}^H$, is the complex conjugate of the transpose of the vector $\mathbf{x}$ as follows:

$$\begin{aligned} \mathbf{x}^H &= \left\{\mathbf{x}^T\right\}^* \\ &= [x_1^*, x_2^*, ..., x_N^*]. \end{aligned} \quad (A.3)$$

A.3 Matrices

A matrix, denoted by $\mathbf{A}$, is a rectangular array of numbers. If a matrix $\mathbf{A}$ has n rows and p columns, then the matrix $\mathbf{A}$ is said to be $(n \times p)$ matrix as follows:

$$\mathbf{A} = \{a_{ij}\} = \begin{bmatrix} a_{11} & a_{12} & \cdots & a_{1p} \\ a_{21} & a_{22} & \cdots & a_{2p} \\ \vdots & \vdots & \vdots & \vdots \\ a_{n1} & a_{n2} & \cdots & a_{np} \end{bmatrix}, \quad (A.4)$$

where a_{ij} is the value in row i and column j of the matrix $\mathbf{A}$, for $i = 1, 2, ..., n$ and $j = 1, 2, ..., p$. The transpose of a matrix $\mathbf{A}$ is generated by interchanging the rows and columns,

$$\mathbf{A} = \begin{bmatrix} a_{11} & a_{21} & \cdots & a_{n1} \\ a_{12} & a_{22} & \cdots & a_{n2} \\ \vdots & \vdots & \vdots & \vdots \\ a_{1p} & a_{2p} & \cdots & a_{np} \end{bmatrix}. \quad (A.5)$$

A square matrix $(n \times n)$ $\mathbf{A}$, which has all of its elements equal to zero except those elements along the main diagonal, is referred to as a *diagonal matrix*. Thus, a diagonal matrix $\mathbf{A}$ has the following form,

$$\mathbf{A} = \begin{bmatrix} a_{11} & 0 & \cdots & 0 \\ 0 & a_{22} & \cdots & 0 \\ \vdots & \vdots & \vdots & \vdots \\ 0 & 0 & \cdots & a_{nn} \end{bmatrix}. \quad (A.6)$$

A diagonal matrix with the diagonal elements equal to one is referred to as the *identity matrix*,

$$\mathbf{I} = \begin{bmatrix} 1 & 0 & \cdots & 0 \\ 0 & 1 & \cdots & 0 \\ \vdots & \vdots & \vdots & \vdots \\ 0 & 0 & \cdots & 1 \end{bmatrix}. \tag{A.7}$$

A square matrix $(n \times n)$ $\mathbf{A}$ is said to be a *Toeplitz matrix* if all the elements along with each of diagonals have the same value. For example, a (4×4) Toeplitz matrix is

$$\mathbf{A} = \begin{bmatrix} 1 & 2 & 6 & 8 \\ 3 & 1 & 2 & 6 \\ 4 & 3 & 1 & 2 \\ 5 & 4 & 3 & 1 \end{bmatrix}. \tag{A.8}$$

If a Toeplitz matrix is symmetric, then all of the elements in the matrix $\mathbf{A}$ can be determined by using either the first column or the first row of the matrix. For example, a (4×4) Toeplitz matrix is symmetric below,

$$\mathbf{A} = \begin{bmatrix} 1 & 2 & 6 & 8 \\ 2 & 1 & 2 & 6 \\ 6 & 2 & 1 & 2 \\ 8 & 6 & 2 & 1 \end{bmatrix}. \tag{A.9}$$

A.4 The Data Matrix

A $(n \times p)$ data matrix, denoted by $\mathbf{X}$, has n observations and p variables as follows,

$$\mathbf{X} = \{x_{ij}\} = \begin{bmatrix} a_{11} & a_{12} & \cdots & a_{1p} \\ a_{21} & a_{22} & \cdots & a_{2p} \\ \vdots & \vdots & \vdots & \vdots \\ a_{n1} & a_{n2} & \cdots & a_{np} \end{bmatrix}, \tag{A.10}$$

where x_{ij} is a data in the observation i and the variable j. The rows of data matrix $\mathbf{X}$ is written as $\mathbf{x}_1^T, \mathbf{x}_2^T, ..., \mathbf{x}_n^T$, where $\mathbf{x}_i$ denotes the

ith row of data matrix $\mathbf{X}$ written as a column. Thus, we may write

$$\mathbf{X} = \begin{bmatrix} \mathbf{x}_1^T \\ \mathbf{x}_2^T \\ \vdots \\ \mathbf{x}_n^T \end{bmatrix}, \tag{A.11}$$

where $\mathbf{x}_i$ is given by

$$\mathbf{x}_i = \begin{bmatrix} x_{i1} \\ x_{i2} \\ \vdots \\ x_{ip} \end{bmatrix}, \quad i = 1, 2, ..., n. \tag{A.12}$$

Note that $\mathbf{x}_i$ is sometimes referred to as the p-dimensional vector denoting the p data on the ith observation.

A.5 Orthogonal Matrices and the Trace

Let $\mathbf{A}$ be a square matrix of dimension $(p \times p)$. If the square matrix $\mathbf{A}\mathbf{A}^T = \mathbf{I}$, then, matrix $\mathbf{A}$ is orthonormal. The orthogonal matrix $\mathbf{A}$ has the following properties:

1. $\mathbf{A}^{-1} = \mathbf{A}^T$.
2. $\mid \mathbf{A} \mid = \pm 1$.
3. $\mathbf{a}_i^T \mathbf{a}_j = 1$ for $i = j$, and $\mathbf{a}_i^T \mathbf{a}_j = 0$ for $i \neq j$, where $\mathbf{a}_i$ is the columns of matrix $\mathbf{A}$.
4. If $\mathbf{A}$ and $\mathbf{B}$ are orthogonal matrices, then, $\mathbf{C} = \mathbf{AB}$ is also orthogonal.

The trace of $\mathbf{A}$, defined by $tr(\mathbf{A}) = \sum_{i=1}^{p} a_{ii}$, satisfies the following properties for square matrices $\mathbf{A}(p \times p)$, and $\mathbf{B}(p \times p)$ as well as scalar β:

$$tr(\beta) = \beta, \tag{A.13}$$

$$tr(\mathbf{A} \pm \mathbf{B}) = tr(\mathbf{A}) \pm tr(\mathbf{B}), \tag{A.14}$$

$$tr(\beta \mathbf{A}) = \beta tr(\mathbf{A}), \tag{A.15}$$

$$tr \sum \mathbf{x}_i^T \mathbf{A} \mathbf{x}_i = tr(\mathbf{A} \sum \mathbf{x}_i \mathbf{x}_i^T), \tag{A.16}$$

and

$$tr(\mathbf{AB}) = tr(\mathbf{BA}) = \sum_i \sum_j a_{ij} b_{ji}. \tag{A.17}$$

As a special case of (A.17) when square matrix $\mathbf{B} = \mathbf{A}^T$, we have

$$tr(\mathbf{A}\mathbf{A}^T) = tr(\mathbf{A}^T\mathbf{A}) = \sum a_{ij}^2. \tag{A.18}$$

If square matrix $\mathbf{A}$ is orthogonal, then (A.18) is given by

$$tr(\mathbf{A}\mathbf{A}^T) = tr(\mathbf{I}) = \sum_{i=1}^{p} I_{ii} = p. \tag{A.19}$$

Also, for $\mathbf{C}$ and $\mathbf{D}$ with $n \times p$ and $p \times n$, respectively, the trace function is given by

$$tr(\mathbf{CD}) = tr(\mathbf{DC}) = \sum_i \sum_j c_{ij} d_{ji} = \sum_j \sum_i d_{ji} c_{ij}. \tag{A.20}$$

A.6 Matrix Differentiation

Let $\mathbf{X}$ be an $n \times p$ matrix and $f(\mathbf{X})$ be a function of matrix $\mathbf{X}$, then the derivative of matrix function $f(\mathbf{X})$ with respect to matrix $\mathbf{X}$ is given by

$$\frac{\partial f(\mathbf{X})}{\partial \mathbf{X}} = \frac{\partial f(\mathbf{X})}{\partial x_{ij}} \tag{A.21}$$

where $i = 1, 2, ..., n$ and $j = 1, 2, ..., p$, and the derivative is also $n \times p$. We develop some results of matrix differentiation as follows:

1. $\frac{\partial \mathbf{b}^T \mathbf{x}}{\partial \mathbf{x}} = \mathbf{b}.$
2. $\frac{\partial \mathbf{x}^T \mathbf{x}}{\partial \mathbf{x}} = 2\mathbf{x}.$
3. $\frac{\partial \mathbf{x}^T \mathbf{B} \mathbf{x}}{\partial \mathbf{x}} = \mathbf{B}\mathbf{x} + \mathbf{x}^T\mathbf{B}.$
4. $\frac{\partial \mathbf{x}^T \mathbf{B} \mathbf{y}}{\partial \mathbf{x}} = \mathbf{B}\mathbf{y}.$

A.7 Eigenvalues and Eigenvectors

Eigenvalues and eigenvectors provide useful and important information about a matrix. They also contribute to an important representation for matrices known as eigenvalue decomposition. This decomposition will be very useful in our discussions of linear transformation.

Let $\mathbf{A}$ be any square matrix of dimension $n \times n$. The function

$$E(\lambda) = \mid \mathbf{A} - \lambda \mathbf{I} \mid \tag{A.22}$$

is an nth order polynomial in λ, where the n roots of $E(\lambda)$, $\lambda_1, \lambda_2, ..., \lambda_n$ are called eigenvalues of the square matrix $\mathbf{A}$.

For each eigenvalue, λ_i, $i = 1, 2, ..., n$, the matrix function $(\mathbf{A} - \lambda\mathbf{I})$ is singular. There is at least one nonzero vector, $\mathbf{v}_i$, such that

$$\mathbf{A}\mathbf{v}_i = \lambda_i \mathbf{v}_i. \tag{A.23}$$

These vectors, $\mathbf{v}_i$, are called the eigenvectors of square matrix $\mathbf{A}$. An eigenvector $\mathbf{v}_i$ is called normalized if it has unit norm, $\| \mathbf{v}_i \| = \mathbf{v}^T\mathbf{v} = 1$.

Furthermore, if $\mathbf{B}$ is a nonsingular square matrix, then matrices $\mathbf{A}$ and $\mathbf{B}\mathbf{A}\mathbf{B}^{-1}$ have the same eigenvalues, λ_i. In addition, matrix $\mathbf{B}\mathbf{A}\mathbf{B}^{-1}$ has the eigenvector $\mathbf{u}_i = \mathbf{B}\mathbf{v}_i$. The proofs of the above definition are in the following:

Proof The function in (A.22) can be rewritten,

$$\begin{aligned} E(\lambda) &= \mid \mathbf{A} - \lambda\mathbf{I} \mid \\ &= \mid \mathbf{A} - \lambda\mathbf{B}^{-1}\mathbf{B} \mid \\ &= \mid \mathbf{B} \mid\mid \mathbf{A} - \lambda\mathbf{B}^{-1}\mathbf{B} \mid\mid \mathbf{B}^{-1} \mid \\ &= \mid \mathbf{B}\mathbf{A}\mathbf{B}^{-1} - \lambda\mathbf{I} \mid . \end{aligned} \tag{A.24}$$

Thus, (A.24) proves that matrices $\mathbf{A}$ and $\mathbf{B}\mathbf{A}\mathbf{B}^{-1}$ have the same eigenvalues. Since $\mathbf{v}_i$ is an eigenvector of matrix $\mathbf{A}$ for eigenvalue λ_i, then there exists a nonzero vector $\mathbf{u}$ satisfying the condition

$$\mathbf{B}\mathbf{A}\mathbf{B}^{-1}(\mathbf{B}\mathbf{v}_i) = \lambda_i(\mathbf{B}\mathbf{v}_i). \tag{A.25}$$

Therefore, the eigenvector $\mathbf{u}_i = \mathbf{B}\mathbf{v}_i$ is an eigenvector of matrix $\mathbf{B}\mathbf{A}\mathbf{B}^{-1}$ for eigenvalue λ_i.

A.8 Theorem of Spectral Decomposition

Given any symmetric matrix $\mathbf{A}$ of dimension $(p \times p)$, then $\mathbf{A}$ can be written as follows

$$\begin{aligned} \mathbf{A} &= \mathbf{R}\Lambda\mathbf{R}^T \\ &= \sum_{i=1}^{p} \lambda_i \mathbf{r}_i \mathbf{r}_i^T, \end{aligned} \tag{A.26}$$

where Λ is a diagonal matrix consisting of the eigenvalues of $\mathbf{A}$; $\mathbf{R}$ is an orthogonal matrix whose columns are normalized eigenvectors; and the $\mathbf{r}_i$ are the corresponding eigenvectors. The theorem is called the *spectral decomposition theorem.* It is sometimes referred to as the *Jordan decomposition theorem* in the literature.

Proof Suppose we can find orthonormal eigenvectors $\mathbf{r}_1, \mathbf{r}_2, ..., \mathbf{r}_p$ such that $\mathbf{A}\mathbf{r}_i = \lambda_i \mathbf{r}_i$ for some numbers λ_i. Then

$$\begin{aligned} \mathbf{r}_i^T \mathbf{A} \mathbf{r}_j &= \lambda_i \mathbf{r}_i^T \mathbf{r}_j \\ &= \begin{cases} \lambda_i, & \text{if } i = j, \\ 0, & \text{otherwise.} \end{cases} \end{aligned} \tag{A.27}$$

In the matrix form, we have

$$\mathbf{R}^T \mathbf{A} \mathbf{R} = \Lambda. \tag{A.28}$$

Multiplying by $\mathbf{R}$ and $\mathbf{R}^T$ on both sides of (A.28) yields

$$\mathbf{R}\mathbf{R}^T \mathbf{A} \mathbf{R} \mathbf{R}^T = \mathbf{R}\Lambda\mathbf{R}^T. \tag{A.29}$$

Because $\mathbf{R}\mathbf{R}^T = \mathbf{I}$, we have

$$\mathbf{A} = \mathbf{R}\Lambda\mathbf{R}^T. \tag{A.30}$$

From (A.24), we can see that $\mathbf{A}$ and $\mathbf{R}\Lambda\mathbf{R}$ have the same eigenvalues. So the elements of Λ are exactly the eigenvalues of symmetric matrix $\mathbf{A}$ with the same multiplicity.

In addition, if $\mathbf{A}$ is a nonsingular symmetric matrix, then for any integer n, we have

$$\Lambda^n = diag(\lambda_i^n), \tag{A.31}$$

and in the form of matrix algebra,

$$\mathbf{A}^n = \mathbf{R}\Lambda^n\mathbf{R}^T. \tag{A.32}$$

Moreover, if all the eigenvalues of symmetric matrix **A** are positive, we can define the rational powers as follows:

$$\Lambda^{p/q} = diag(\lambda_i^{p/q}), \tag{A.33}$$

and

$$\mathbf{A}^{p/q} = \mathbf{R}\Lambda^{p/q}\mathbf{R}^T. \tag{A.34}$$

for integers q and p, where $q > 0$. If some of the eigenvalues of matrix **A** are zero, then (A.32) and (A.34) still hold if the exponents p/q are restricted to be positive.

Proof Based on the spectral decomposition theorem, we have

$$\mathbf{A} = \mathbf{R}\Lambda\mathbf{R}^T. \tag{A.35}$$

Then we can take the power of 2 of both sides of (A.35),

$$\begin{aligned} \mathbf{A}^2 &= (\mathbf{R}\Lambda\mathbf{R}^T)^2 \\ &= \mathbf{R}\Lambda\mathbf{R}^T\mathbf{R}\Lambda\mathbf{R}^T \\ &= \mathbf{R}\Lambda^2\mathbf{R}^T, \end{aligned} \tag{A.36}$$

and since $\mathbf{R}^T\mathbf{R} = \mathbf{I}$, then,

$$\mathbf{A}^{-1} = \mathbf{R}\Lambda^{-1}\mathbf{R}^T, \tag{A.37}$$

where $\Lambda^{-1} = diag(\lambda_i^{-1})$. By using induction, (A.32) can be easily proved.

To prove (A.34), note that

$$\begin{aligned} \mathbf{A}^p &= (\mathbf{A}^{p/q})^q \\ &= \mathbf{R}\Lambda^{p/q}\mathbf{R}^T \dots \mathbf{R}\Lambda^{p/q}\mathbf{R}^T \\ &= \mathbf{R}\Lambda^{p/q}\mathbf{R}^T. \end{aligned} \tag{A.38}$$

Important special cases of (A.34), when $p = 1$ and $q = 2$, are

$$\mathbf{A}^{1/2} = \mathbf{R}\Lambda^{1/2}\mathbf{R}^T, \tag{A.39}$$

where $\Lambda^{1/2} = diag(\lambda_i^{1/2})$ and $\lambda_i \geq 0$ for all i, and

$$\mathbf{A}^{-1/2} = \mathbf{R}\Lambda^{-1/2}\mathbf{R}^T, \tag{A.40}$$

where $\Lambda^{-1/2} = diag(\lambda_i^{-1/2})$ and $\lambda_i > 0$ for all i. The decomposition form of (A.39) or (A.40) is called the *symmetric square root decomposition* of matrix **A**.

Furthermore, for any symmetric matrix **A**, there exists an orthogonal transformation

$$\mathbf{y} = \mathbf{R}^T\mathbf{x}, \tag{A.41}$$

such that

$$\mathbf{x}^T\mathbf{A}\mathbf{x} = \sum_i \lambda_i y_i^2. \tag{A.42}$$

Proof Consider the spectral decomposition theorem where $\mathbf{A} = \mathbf{R}\Lambda\mathbf{R}^T$. Then, we can write

$$\begin{aligned} \mathbf{x}^T\mathbf{A}\mathbf{x} &= \mathbf{x}^T\mathbf{R}\Lambda\mathbf{R}^T\mathbf{x} \\ &= \mathbf{y}^T\mathbf{R}^T\mathbf{R}\Lambda\mathbf{R}^T\mathbf{R}\mathbf{y} \\ &= \mathbf{y}^T\Lambda\mathbf{y} \\ &= \sum_i \lambda_i y_i^2, \end{aligned} \tag{A.43}$$

where **R** has as its columns the eigenvectors of matrix **A** and $\lambda_1, ..., \lambda_p$ are the corresponding eigenvalues.

A.9 Theorem of Singular Value Decomposition

Suppose that **S** is an $(n \times p)$ matrix with rank r. Then **S** can be written as

$$\mathbf{S} = \mathbf{U}\mathbf{D}\mathbf{V}^T, \tag{A.44}$$

where **U** and **V** are orthonormal matrices of dimensions $(n \times r)$ and $(p \times r)$, respectively, and **D** is a diagonal matrix with positive elements. Equation A.44 is referred to as the *theorem of singular value decomposition.*

Proof Since **U** and **V** are orthonormal matrices of dimensions $(n \times r)$ and $(p \times r)$, respectively, then we have

$$\mathbf{U}^T\mathbf{U} = \mathbf{I}_r, \tag{A.45}$$

and

$$\mathbf{V}^T\mathbf{V} = \mathbf{I}_r. \tag{A.46}$$

Let $\mathbf{Y} = \mathbf{S}^T\mathbf{S}$, then $\mathbf{Y}$ is a symmetric matrix, which also has rank r. Using the spectral decomposition theorem, we write

$$\begin{aligned} \mathbf{Y} &= \mathbf{S}^T\mathbf{S} \\ &= \mathbf{V}\Lambda\mathbf{V}^T, \end{aligned} \tag{A.47}$$

where $\mathbf{V}$ is a column orthonormal matrix of eigenvectors of $\mathbf{Y}$ and $\Lambda = diag(\lambda_1, \lambda_2, ..., \lambda_r)$ with $\lambda_i > 0$ since

$$\begin{aligned} \lambda_i &= \mathbf{v}_i^T\mathbf{S}^T\mathbf{S}\mathbf{v}_i \\ &= \parallel \mathbf{S}\mathbf{v}_i \parallel \\ &> 0. \end{aligned} \tag{A.48}$$

Now let $d_i = \lambda_i^{1/2}$, for $i = 1, 2, ..., r$. This implies that $\mathbf{D} = diag(d_1, d_2, ..., d_r)$. In addition, the $\mathbf{U}$ matrix of dimension $(n \times r)$ is defined by

$$\mathbf{u}_i = d_i^{-1}\mathbf{S}\mathbf{v}_i, \qquad \text{i} = 1,\ 2,\ ...,\ \text{r}. \tag{A.49}$$

Then, we have the following results,

$$\begin{aligned} \mathbf{u}_i^T\mathbf{u}_j &= d_i^{-1}d_j^{-1}\mathbf{v}_i^T\mathbf{S}^T\mathbf{S}\mathbf{v}_i \\ &= \lambda_i d_i^{-1}d_j^{-1}\mathbf{v}_i\mathbf{v}_j \\ &= \begin{cases} 1, & \text{if } i = j, \\ 0, & \text{otherwise.} \end{cases} \end{aligned} \tag{A.50}$$

As can be seen, matrix $\mathbf{U}$ is also an orthonormal matrix.

Assuming that $\mathbf{x}$ is any $p \times 1$ vector, then it can be written as

$$\mathbf{x} = \sum_i \alpha_i\mathbf{v}_i. \tag{A.51}$$

Let $\mathbf{e_i}$ be a $r \times 1$ vector that has 1 in the ith position and 0 elsewhere. Based on (A.49), we have the results:

$$\begin{aligned}\mathbf{UDV}^T\mathbf{x} &= \sum_i \alpha_i \mathbf{UDe}_i \\ &= \sum_i \alpha_i d_i \mathbf{u}_i \\ &= \sum_i \alpha_i \mathbf{Sv}_i \\ &= \mathbf{Sx}. \end{aligned} \tag{A.52}$$

We have proved that $\mathbf{UDV}^T = \mathbf{S}$ because this above formula in (A.52) holds for any $\mathbf{x}$.

A.10 Quadratic Forms

The quadratic form of a symmetric matrix $\mathbf{A}$ is defined by

$$\mathcal{Q}(\mathbf{x}) = \mathbf{x}^T\mathbf{A}\mathbf{x} = \sum_{i=1}^{n}\sum_{j=1}^{n} a_{ij}x_i x_j, \tag{A.53}$$

where $\mathbf{x} = [x_1, x_2, ..., x_n]$ is a vector of n variables.

If $\mathcal{Q}(\mathbf{x}) > 0$ for all $\mathbf{x} \neq \mathbf{0}$, then $\mathcal{Q}(\mathbf{x})$ is said to be a positive definite quadratic form. If $\mathcal{Q}(\mathbf{x}) \geq 0$ for all $\mathbf{x} \neq \mathbf{0}$, then $\mathcal{Q}(\mathbf{x})$ is called a positive semidefinite quadratic form. A symmetric matrix $\mathbf{A}$ is said to be positive definite or positive semidefinite if $\mathcal{Q}(\mathbf{x})$ is a positive definite quadratic form or positive semidefinite quadratic form, respectively.

A.11 Maximization and Minimization Theorem

Let $\mathbf{A}$ and $\mathbf{B}$ be two symmetric matrices and $\mathbf{B} > 0$. Then, the maximum or minimum problem of

$$\mathbf{x}^T\mathbf{A}\mathbf{x} \qquad \text{subject to } \mathbf{x}^T\mathbf{B}\mathbf{x} = 1 \tag{A.54}$$

can be solved when $\mathbf{x}$ is the eigenvector of $\mathbf{B}^{-1}\mathbf{A}$ corresponding to the largest or smallest eigenvalue of $\mathbf{B}^{-1}\mathbf{A}$. In addition, Let λ_1 and λ_p be the largest and smallest eigenvalues of $\mathbf{B}^{-1}\mathbf{A}$, respectively.

Optimization subject to the constraint (A.54) gives the following:

$$max_x(\mathbf{x}^T\mathbf{A}\mathbf{x}) = \lambda_1, \tag{A.55}$$

and

$$min_x(\mathbf{x}^T\mathbf{A}\mathbf{x}) = \lambda_p. \tag{A.56}$$

Proof Let $\mathbf{B}^{1/2}$ denote the symmetric square root of symmetric matrix $\mathbf{B}$, and let linear transform $\mathbf{y} = \mathbf{B}^{1/2}\mathbf{x}$. Then, (A.54) can be rewritten as

$$\mathbf{y}^T\mathbf{B}^{-1/2}\mathbf{A}\mathbf{B}^{-1/2}\mathbf{y} \qquad \text{subject to } \mathbf{y}^T\mathbf{y} = 1. \tag{A.57}$$

However, by using a spectral decomposition of the symmetric matrix, the matrix $\mathbf{B}^{-1/2}\mathbf{A}\mathbf{B}^{-1/2}$ can be written in the following form:

$$\mathbf{B}^{-1/2}\mathbf{A}\mathbf{B}^{-1/2} = \mathbf{G}\Lambda\mathbf{G}, \tag{A.58}$$

where $\mathbf{G}$ is an orthogonal matrix and Λ is a diagonal matrix of the eigenvalues of $\mathbf{B}^{-1/2}\mathbf{A}\mathbf{B}^{-1/2}$ with $\lambda_1 \geq \lambda_2 \geq, ..., \geq \lambda_p \geq 0$.

Let $\mathbf{z} = \mathbf{G}^T\mathbf{y}$. Then

$$\begin{aligned} \mathbf{z}^T\mathbf{z} &= \mathbf{y}^T\mathbf{G}\mathbf{G}^T\mathbf{y} \\ &= \mathbf{y}^T\mathbf{y}. \end{aligned} \tag{A.59}$$

So, using (A.58) and (A.59), (A.57) can be rewritten in the following form:

$$max_z(\mathbf{z}^T\Lambda\mathbf{z}) = max_z(\sum \lambda_i z_i^2) \qquad \text{subject to } \mathbf{z}^T\mathbf{z} = 1. \tag{A.60}$$

If the eigenvalues are in descending order, then (A.60) satisfies the following,

$$max_z(\sum \lambda_i z_i^2) \leq \lambda_1 max_z(\sum z_i^2) = \lambda_1. \tag{A.61}$$

Therefore, this theorem is proved for the maximization result.

For the minimization problem, (A.57) can be rewritten in the following form after using (A.58) and (A.59):

$$min_z(\mathbf{z}^T\Lambda\mathbf{z}) = min_z(\sum \lambda_i z_i^2) \qquad \text{subject to } \mathbf{z}^T\mathbf{z} = 1. \tag{A.62}$$

If the eigenvalues are in descending order, then (A.62) satisfies the following:

$$min_z(\sum \lambda_i z_i^2) \geq \lambda_p min_z(\sum z_i^2) = \lambda_p. \qquad \text{(A.63)}$$

Again, this theorem is proved for the minimization result.

We have collected together all of the linear algebra and matrix theories that are useful for understanding the theory development of digital signal processing and statistical classification in the textbook. The reader is encouraged to look up the relevant topics further. Of course, there are many textbooks on linear algebra and matrix theory. We recommend [1-5] in the references below.

References

[1] Kailath, T., *Linear Systems*, Prentice-Hall, Inc., Englewood Cliffs, New Jersey, 1980.

[2] Mardia, K. V., J. T. Kent, and J. M. Bibby, *Multivariate Analysis*, 3d printing, Academic Press Inc., New York, 1982.

[3] Miao, J., *Component Feature-Based Digital Waveform Analysis and Classification*, UMI, Ann Arbor, Michigan, 1996.

[4] Neter, J., W. Wasserman, and M. H. Kutner, *Applied Linear Regression Models*, Richard D. Irwin, Inc., Homewood, Illinois, 1983.

[5] Strang, G., *Linear Algebra and Its Applications*, Academic Press Inc., New York, 1980.

About the Authors

George J. Miao holds a B.Eng. from Shanghai University of Science and Technology, Shanghai, China, an M.S. from Columbia University, New York, and a Ph.D. in electrical engineering from the Georgia Institute of Technology, Atlanta, Georgia. He has 14 years of experience and specializes in digital signal processing with applications in multimedia, wireless, and wireline digital communications. He has worked on DSP-based communication systems and architecture design for many companies, including Motorola, Fujitsu Network Communications, Hewlett-Packard, and Intel. He has published over 40 technical papers, and holds several U.S. patents pending. Dr. Miao has been listed in the *International Who's Who of Information Technology* since 1999.

Mark A. Clements is a professor in the School of Electrical and Computer Engineering at the Georgia Institute of Technology, and director of the Interactive Media Technology Center. He holds B.S., S.M., E.E., and Sc.D. degrees, all in electrical engineering and computer science, from the Massachusetts Institute of Technology, Cambridge, Massachusetts. He has published over 100 technical papers and holds several U.S. patents. He also coauthored *Objective Measures of Speech Quality* (Prentice Hall, 1988). He has served as a member of the IEEE Speech Technical Committee and has been an associate editor for speech in *IEEE Transactions on Speech and Audio.* Dr. Clements was the technical program cochairman of Speech Tech in 1986, and the technical program chair of the *IEEE International Conference on Acoustics, Speech, and Signal Processing* in 1996.

Index

Recent Titles in the Artech House Signal Processing Library

Computer Speech Technology, Robert D. Rodman

Digital Signal Processing and Statistical Classification, George J. Miao and Mark A. Clements

Handbook of Neural Networks for Speech Processing, Shigeru Katagiri, editor

Hilbert Transforms in Signal Processing, Stefan L. Hahn

Phase and Phase-Difference Modulation in Digital Communications, Yuri Okunev

Signals, Oscillations, and Waves: A Modern Approach, David Vakman

Statistical Signal Characterization, Herbert L. Hirsch

Statistical Signal Characterization Algorithms and Analysis Programs, Herbert L. Hirsch

Voice Recognition, Richard L. Klevans and Robert D. Rodman

For further information on these and other Artech House titles, including previously considered out-of-print books now available through our In-Print-Forever® (IPF®) program, contact:

Artech House
685 Canton Street
Norwood, MA 02062
Phone: 781-769-9750
Fax: 781-769-6334
e-mail: artech@artechhouse.com

Artech House
46 Gillingham Street
London SW1V 1AH UK
Phone: +44 (0)171-973-8077
Fax: +44 (0)171-630-0166
e-mail: artech-uk@artechhouse.com

Find us on the World Wide Web at:
www.artechhouse.com